eTELEMED 2019

The Eleventh International Conference on eHealth, Telemedicine, and Social Medicine

February 24 – 28, 2019

Athens, Greece

eTELEMED 2019 Editors

Yoshitoshi Murata, Iwate Prefectural University, Japan
Oliver Heinze, University Hospital Heidelberg, Germany

Printed from e-media with permission by:

Curran Associates, Inc.
57 Morehouse Lane
Red Hook, NY 12571

proceedings.com

Some format issues inherent in the e-media version may also appear in this print version.

Printed by Curran Associates, Inc. (2019)

International Academy, Research, and Industry Association (IARIA)
412 Derby Way
Wilmington, DE 19810

Phone: (408) 893-6407
Fax: (408) 527-6351

petre@iaria.org

Additional copies of this publication are available from:

Curran Associates, Inc.
57 Morehouse Lane
Red Hook, NY 12571 USA
Phone: 845-758-0400
Fax: 845-758-2633
Email: curran@proceedings.com
Web: www.proceedings.com

Table of Contents

eTELEMED 2019

Forward

The eleventh edition of The International Conference on eHealth, Telemedicine, and Social Medicine (eTELEMED 2019), held in Athens, Greece, February 24 - 28, 2019, considered advances in techniques, services, and applications dedicated to a global approach of eHealth.

Development of wireless homecare, of special types of communications with patient data, of videoconferencing and telepresence, and the progress in image processing and date protection increased the eHealth applications and services, and extended Internet-based patient coverage areas. Social and economic aspects as well as the integration of classical systems with the telemedicine systems are still challenging issues.

eTELEMED 2019 provided a forum where researchers were able to present recent research results and new research problems and directions related to them. The topics covered aspects from classical medicine and eHealth integration, systems and communication, devices, and applications.

We take this opportunity to thank all the members of the eTELEMED 2019 Technical Program Committee as well as the numerous reviewers. The creation of such a broad and high-quality conference program would not have been possible without their involvement. We also kindly thank all the authors who dedicated much of their time and efforts to contribute to the eTELEMED 2019. We truly believe that, thanks to all these efforts, the final conference program consists of top quality contributions.

This event could also not have been a reality without the support of many individuals, organizations, and sponsors. We are grateful to the members of the eTELEMED 2019 organizing committee for their help in handling the logistics and for their work to make this professional meeting a success.

We hope that eTELEMED 2019 was a successful international forum for the exchange of ideas and results between academia and industry and for the promotion of progress in eHealth and Telemedicine research.

We also hope that Athens provided a pleasant environment during the conference and everyone saved some time for exploring this beautiful city.

eTELEMED 2019 Chairs

eTELEMED Steering Committee

Lisette Van Gemert-Pijnen, University of Twente - Enschede, the Netherlands
Marike Hettinga, Windesheim University of Applied Sciences, the Netherlands
Joseph Finkelstein, Center for Bioinformatics and Health Analytics - Columbia University, USA
Åsa Smedberg, Stockholm University, Sweden
Yoshitoshi Murata, Iwate Prefectural University, Japan
Anthony Glascock, Drexel University, USA
Britt Östlund, Royal Institute of Technology (KTH), Sweden

eTELEMED Industry/Research Advisory Committee

Salah H. Mandil, eStrategies & eHealth for WHO and ITU - Geneva, Switzerland
Bobby Gheorghiu, Canada Health Infoway/Inforoute Santé du Canada, Canada

eTELEMED 2019

COMMITTEE

eTELEMED Steering Committee

Lisette Van Gemert-Pijnen, University of Twente - Enschede, the Netherlands
Marike Hettinga, Windesheim University of Applied Sciences, the Netherlands
Joseph Finkelstein, Center for Bioinformatics and Health Analytics - Columbia University, USA
Åsa Smedberg, Stockholm University, Sweden
Yoshitoshi Murata, Iwate Prefectural University, Japan
Anthony Glascock, Drexel University, USA
Britt Östlund, Royal Institute of Technology (KTH), Sweden

eTELEMED Industry/Research Advisory Committee

Salah H. Mandil, eStrategies & eHealth for WHO and ITU - Geneva, Switzerland
Bobby Gheorghiu, Canada Health Infoway/Inforoute Santé du Canada, Canada

eTELEMED 2019 Technical Program Committee

Don Adjeroh, West Virginia University, USA
Nitin Agarwal, University of Arkansas at Little Rock, USA
Najeeb Al-Shorbaji, eHealth Development Association of Jordan | FIAHSI, Amman, Jordan
Davide Aloini, University of Pisa, Italy
Hessah A. Alsalamah, King Saud University, Saudi Arabia
James G. Anderson, Purdue University, USA
Adina Astilean, Technical University of Cluj-Napoca, Romania
Luigi Benedicenti, University of Regina, Canada
Javad Birjandtalab, The University of Texas at Dallas, USA
Andrzej Cacko, Medical University of Warsaw, Poland
Manuel Campos Martínez, University of Murcia, Spain
Elena Cardillo, Institute of Informatics and Telematics - National Research Council, Italy
Keith Chan, The Hong Kong Polytechnic University, Hong Kong
Maria Teresa Chiaravalloti, National Research Council | Institute of Informatics and Telematics, Rende, Italy
Marcos C. d'Ornellas, Universidade Federal de Santa Maria, Brazil
Isabel de la Torre Díez, University of Valladolid, Spain
Gert-Jan de Vreede, University of South Florida, USA
Kerstin Denecke, Bern University of Applied Sciences, Switzerland
Cecilia Dias Flores, Universidade Federal de Ciências da Saúde de Porto Alegre, Brazil
Talko Dijkhuis, Hanze University of Applied Sciences, Netherlands
Aranka Do, Hanze University of Applied Sciences, Netherlands
Jianqiang Du, Jiangxi University of Traditional Chinese Medicine, Nanchang, China
Dmitriy Dubovitskiy, De Montfort University, UK
Georg Duftschmid, Medical University Vienna, Austria
Leila Eadie, University of Aberdeen, UK
Claudio Eccher, FBK-Fondazione Bruno Kessler, Italy
Christiane Eichenberg, Sigmund Freud PrivatUniversität Wien, Austria

Copyright Information

Quantitative Analysis of Facial-Expression Training Application for Medical Doctors

Akemi Oe, Yoshitoshi Murata
Graduate School of Software and Information Science
Iwate Prefectural University
Iwate, Japan
e-mails: akemioe@gmail.com, y-murata@iwate-pu.ac.jp

Abstract— **Establishing trust between a patient and a doctor depends as much on their relationship as on the doctor's medical abilities. One of important factors in building and maintaining a relationship is whether the doctor produces facial expressions appropriate for the patient's condition. We developed a facial-expression training application for medical students and tested its effectiveness as a training tool. In this application, Microsoft Cognitive Services Emotion API is used to analyze the facial expression. Prior to developing this application, we studied what kind of facial expressions were appropriate in a doctor-patient greeting situation. We focused on four conditions for practicing facial expressions. Three of them are greetings given by doctors at the beginning of medical interviews of adult patients in a general ward of a hospital. The fourth is a medical scenario in which a doctor puts a stethoscope to a patient's chest, i.e., auscultation. To verify the training application's effectiveness, the facial expressions before and after learning with the application were evaluated by potential patients. The results confirmed the application's utility.**

Keywords-doctor-patient interaction; facial expression; nonverbal communication; facial-expression training application.

I. INTRODUCTION

Patient satisfaction is an important component of medical care [1]. Improving patient satisfaction enhances trust and the relationship between patient and doctor, which leads to stronger adherence to prescribed protocol, such as taking medicine, and to enhanced therapeutic effects [2][3]. Many studies and reviews have shown that the main determinant of patient satisfaction is the patient-doctor relationship [4]–[8] and that patient satisfaction is higher when the patient communicates with a doctor who has strong nonverbal-communication skills [9][10]. However, young inexperienced doctors and medical students often have trouble producing appropriate facial expressions when greeting a patient. The first author of this paper, a lecturer on medical communication, often hears young doctors complaining that, though they intend to smile, patients say that they seem angry.

In response to these troubles, we developed a facial-expression training application that physicians can use for independent-study. In this application, Microsoft Cognitive Services Emotion API is used to analyze the facial expression. Prior to developing this application, we quantitatively clarified acceptable expressions required when doctors treat patients by comparing human evaluation results with those of a computer system [11]. The results were incorporated into the application as model expressions. These model expressions are standards to compare with those of learners using the application. Then, to verify the effectiveness of the application, learners' facial expressions before and after learning with the application were evaluated by potential patients. The evaluation results confirmed the utility of the application.

After introducing related work in Section II and previous work in Section III, we describe the system of the facial-expression training application in Section IV. We discuss the results of our experiments in Section V. Section VI gives our conclusion, and Section VII considers future work.

II. RELATED WORK

A. *Nonverbal Communication*

Medical interviews have traditionally focused on gathering relevant information from patients [12]. In contemporary medicine, the focus has expanded to building a trusting relationship, sharing decision-making, responding to the patient's emotional state, and supporting actions related to the patient's condition and treatment; this requires the doctor to have a wide range of communication skills [13]. These skills include "looking at a patient not as a case but as a human being" [14] and "building and maintaining a good relationship between doctor and patient" [15]. It has been shown that such skills have a greater effect on patient satisfaction than the doctor's medical skills, the medicine prescribed, the information provided, the questions asked, and the advice and instructions given. In particular, a patient's satisfaction is positively related to the doctor being warm [14]–[16], empathic [14][16]–[18], friendly [16], and giving the impression of being human [17].

Nonverbal communication is a means of communicating these emotional aspects of oneself. Patient satisfaction is higher when the doctor has a strong ability to express his or her emotions and to read the emotions of others through nonverbal cues such as facial expressions, gaze, posture, and tone of voice [9][19][20]. In short, a doctor's nonverbal communication is an important aspect of patient care.

B. *Learning Facial Expressions*

Natural, unconscious facial expressions can be seen in humans from infancy [21], but eventually conscious expressions appear. Conscious expressions might also be thought of as "false" expressions. These expressions are skills that form the basis of more complicated expressive behaviors, such as emotional expressions performed in accordance with rules in communication situations [22].

Facial expressions as expressions of emotion are thought to be founded in an understanding of emotions, and there are many developmental studies about understanding emotions. However, studies about learning facial expressions are mostly focused on social-skills training for children with developmental disabilities, and there are not many studies for adults.

In one study, participants were required to express other people's facial expressions by recalling emotions and showing photos of faces with emotions, the showing photos effect was shown about "anger" [23].

In research to develop a smile-training system with the goal of training participants to produce celebrity-like smiles (a facial expression recognized as attractive) [24], researchers created a smile-fitted deformation considering the characteristics of a celebrity-like smile in participants' expressionless faces. As a result of training, participants were able to express a smile highly likely to be recognized as attractive by others.

From the above, it is clear that presenting appropriate facial-expression images is effective for facial-expression training as both a method of expressing emotions and a learning method.

III. Previous Work

A previous study [11] has revealed what are considered acceptable facial expressions for a doctor in accordance with the patient's condition. One way to analyze appropriate facial expressions for doctors is to define the facial expressions of experienced doctors. However, their facial expressions are not always deemed appropriate. Also, because many young physicians have trouble presenting appropriate facial expressions when greeting a patient, we chose to find facial expressions that would be acceptable for most patients, including potential patients, from facial expressions that medical students think are suitable. We chose greeting patients in the general ward of a hospital as the target situation.

The procedure was as follows. The participant roleplaying the patient portrayed three conditions: a patient who feels physically healthy (a "bright patient"), one whose physical condition is unknown (an "expressionless patient"), and one who feels badly and is suffering pain (a "patient in pain"). We photographed these roleplayed conditions. Then, a plurality of medical students greeted the three photographs with the facial expressions that each student deemed appropriate, and we recorded video of them doing so. Comparing an evaluation by adults and analytical results from a computer of the medical student's greeting movies revealed what were considered the acceptable facial expressions in accordance with the patient's condition. The acceptable facial expressions when a young doctor greets an adult patient who is hospitalized in a general ward are as follows. For patients who feel physically healthy, the most acceptable facial expression is Figure 1; "continuous happiness" (expressed more as a laugh rather than simply a smile). For patients without a facial expression, the most acceptable facial expression is Figure 2; initially "happiness" (expressed as a smile) and then "neutrality" (expressionlessness).

Time	Happiness	Anger	Contempt	Disgust	Fear	Sadness	Surprise	Neutral
00:00.1	1.00	0.00	0.00	0.00	0.00	0.00	0.00	0.00
00:00.7	1.00	0.00	0.00	0.00	0.00	0.00	0.00	0.00
00:01.1	1.00	0.00	0.00	0.00	0.00	0.00	0.00	0.00
00:02.2	1.00	0.00	0.00	0.00	0.00	0.00	0.00	0.00
00:01.7	1.00	0.00	0.00	0.00	0.00	0.00	0.00	0.00
00:02.8	1.00	0.00	0.00	0.00	0.00	0.00	0.00	0.00
00:03.3	1.00	0.00	0.00	0.00	0.00	0.00	0.00	0.00
00:03.9	1.00	0.00	0.00	0.00	0.00	0.00	0.00	0.00

Figure 1. Acceptable facial expression for patients who feel physically healthy.

Time	Happiness	Anger	Contempt	Disgust	Fear	Sadness	Surprise	Neutral
00:00.1	0.66	0.00	0.00	0.00	0.00	0.00	0.00	0.34
00:00.7	0.71	0.00	0.00	0.00	0.00	0.00	0.00	0.29
00:01.2	0.79	0.00	0.01	0.00	0.00	0.00	0.00	0.20
00:01.6	0.94	0.00	0.00	0.00	0.00	0.00	0.00	0.06
00:02.2	0.35	0.00	0.01	0.00	0.00	0.00	0.00	0.63
00:02.7	0.93	0.00	0.00	0.00	0.00	0.00	0.00	0.07
00:03.2	0.24	0.00	0.00	0.00	0.00	0.00	0.00	0.76
00:03.8	0.29	0.00	0.00	0.00	0.00	0.00	0.00	0.70
00:04.0	0.40	0.00	0.00	0.00	0.00	0.00	0.00	0.59

Figure 2. Acceptable facial expression for patients without a facial expression.

Time	Happiness	Anger	Contempt	Disgust	Fear	Sadness	Surprise	Neutral
00:00.7	0.00	0.00	0.00	0.00	0.00	0.07	0.00	0.93
00:01.1	0.00	0.00	0.00	0.00	0.00	0.06	0.02	0.92
00:01.6	0.00	0.00	0.00	0.00	0.00	0.04	0.00	0.96
00:02.2	0.00	0.00	0.00	0.00	0.00	0.05	0.01	0.92
00:02.6	0.00	0.00	0.00	0.00	0.00	0.33	0.00	0.66
00:00.1	0.00	0.00	0.00	0.00	0.00	0.00	0.00	1.00
00:03.2	0.00	0.01	0.02	0.00	0.00	0.13	0.00	0.84
00:03.6	0.00	0.00	0.01	0.00	0.00	0.05	0.00	0.94
00:03.0	0.00	0.00	0.01	0.00	0.00	0.06	0.00	0.92

Figure 3. Acceptable facial expression for patients in bad physical condition suffering pain.

For patients in bad physical condition suffering pain, the most acceptable facial expression is Figure 3; "neutrality" with a little "sadness" or "surprise". The cells in the Figure1-3 corresponding to 0 or more and less than 0.2 are shown in blue, 0.2 or more and less than 0.4 in green, 0.4 or more and less than 0.6 in yellow, 0.6 or more and less than 0.8 in orange, and 0.8 or more in red. The total for all emotions is 1, and the value for neutrality is obtained by subtracting the total value for the seven emotions from 1.

IV. FACIAL-EXPRESSION TRAINING APPLICATION SYSTEM

A. *Requirements and Design Concept*

Learners study in the following order so that the appropriate facial expressions can be learned efficiently and repeatedly. When starting, to enable learners to notice what their facial expressions are before training, the system has them greet with expressions that they think are suitable without any specific instructions.

Step 1: A learner chooses one of the model patients.

Step 2: The learner greets the model patient with a facial expression that the learner thinks appropriate, and this greeting is recorded.

Step 3: A video of the model doctor greeting the patient with the appropriate facial expression and analysis data are displayed, and important points to notice about the appropriate facial expressions are introduced.

Step 4: The video recorded in Step 2 and its analysis data are displayed.

Step 5: The learner repeats from Steps 2 to 4 until satisfied.

Step 6: If the learner is satisfied, the learner selects a patient with a different condition and returns to Step 1.

In the developed system, in addition to being able to check the facial expressions the learner performed by recording them as a video, the learner can also check the quality of his or her expression on a frame-by-frame basis. Each learner can save images as learning data and facial-expression analysis results by applying security processing because these data and analysis results are important as research data.

B. *System Configuration*

We developed and used a system that quantitatively analyzes changes in facial expression. It is based on the Cognitive Services Emotion API [25] provided by Microsoft's Azure cloud service, and a facial-expression-emotion detection system for video images. Our facial-expression-emotion analysis system calculates the ratio for seven emotions (happiness, anger, contempt, disgust, fear, sadness, and surprise) reflected in the input video image and for neutrality. The total for all emotions is 1, and the value for neutrality is obtained by subtracting the total value for the seven emotions from 1. The configuration of this application is as shown in Figure 4. A video camera is controlled by the Open CV [26]. Recoded video data are converted the Motion-JPEG, and send to the Cognitive Services Emotion API.

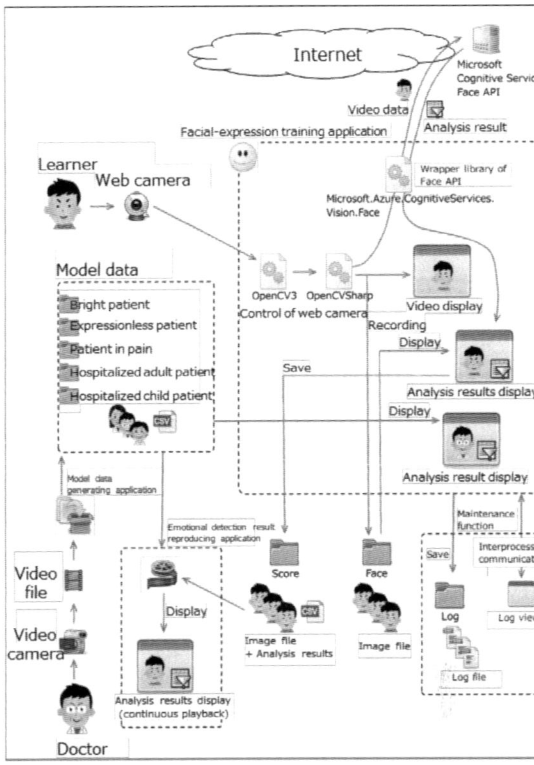

Figure 4. Configuration of facial-expression training application.

At first, the learner selects a patient to be a training partner in Figure 5. After clicking the recording button, the learner talks to the model patient. When the greeting and consultation is over, the learner clicks the stop button followed by the next button in Figure 6.

The emotional values of the model doctor's facial expression are displayed. By selecting a timeline, the corresponding facial expression is displayed in Figure 7.

The emotional values of the learner's facial expression are displayed. By selecting a timeline, the corresponding facial expression is displayed in Figure 8.

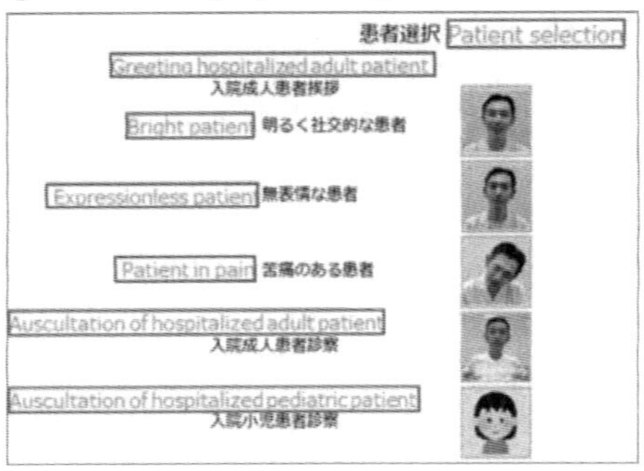

Figure 5. Step 1: Patient-selection screen.

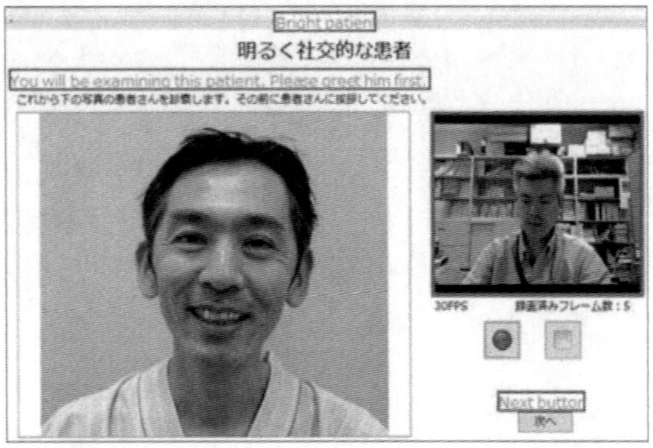

Figure 6. Step 2: Learner greets and examines patient while screen is recorded.

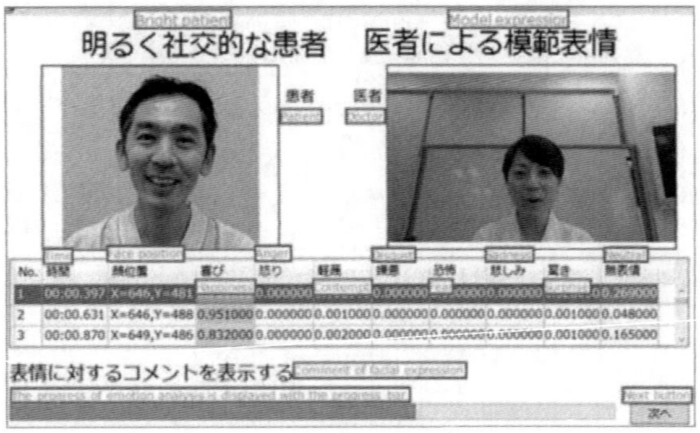

Figure 7. Step 3: Model doctor's appropriate facial expression and analysis results.

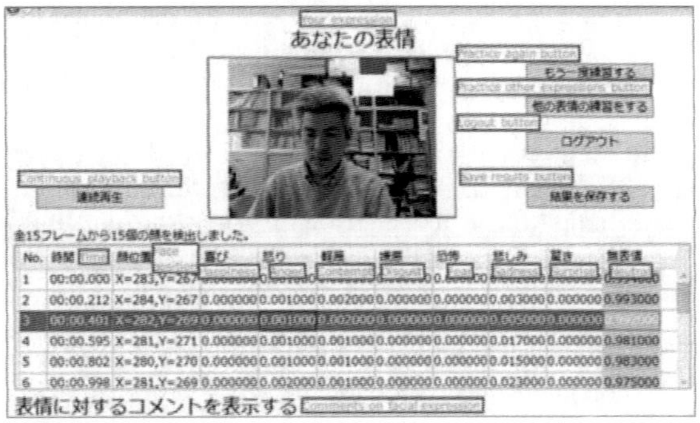

Figure 8. Step 4: Analysis of image taken in Step 2.

C. *Model Doctors*

Videos of model doctors were prepared with the following procedure. In a previous study [11], the acceptable doctors' expressions for the patients were measured and selected. The model doctors (one male and one female) reproduced the acceptable facial expressions, and their reproductions were recorded as videos. Seventy-nine people evaluated these videos of Figure 9–11 and confirmed that they were appropriate, so we adopted them as the model doctor videos.

(1) Woman (2) Man

Figure 9. Model doctor's appropriate facial expression for bright patiets.

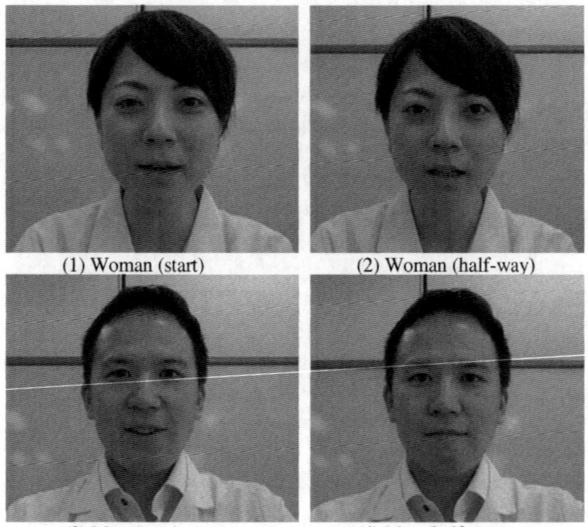

Figure 10. Model doctor's appropriate facial expression for expressionless patiets.

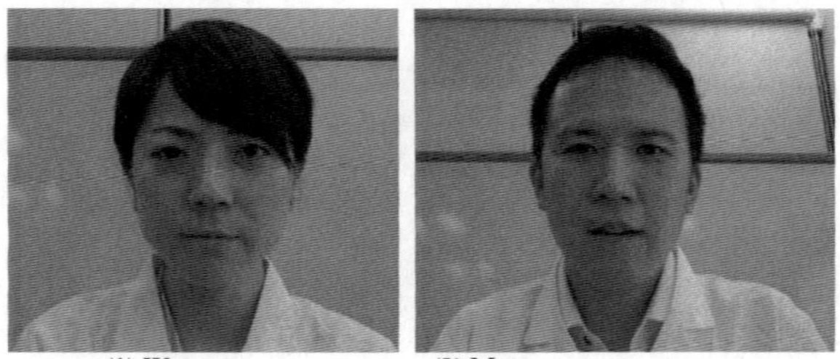

(1) Woman (2) Man

Figure 11. Model doctor's appropriate facial expression for patiets in pain.

V. EVALUATION

A. *Experiment*

Six participants including three medical students played doctors and greeted patients with facial expressions that they thought appropriate for each condition of Figure 5. The application recorded them (before training)., the participants repeated the facial-expressions training so that they could approximate the analysis results of the model doctors. After practice, we recorded their facial expressions again (after training). Although evaluation by actual patients is best, it would have been difficult to request actual patients' participation. Hence, we asked 30 general healthy adults who had been hospitalized in the past or would be in the future. We showed the video recordings (before and after training) to 10 men and 20 women (average age 41.0 years) without sound. We asked them to judge whether the doctor's facial expression was appropriate for the condition on a 3-point scale (1: appropriate, 2: neutral, 3: not appropriate). We asked them to also comment on anything they felt or noticed. We showed the recordings without sound because we wanted them to focus on the appropriate facial expressions in medical communication conditions, and emotion is easier to read from speech than from facial expressions.

B. *Effects of Training*

The graph of Figure 12–15 showing the result of the rating before and after the training is as follows.

In order to examine the effects of training with the application, we conducted 2 (using the application: before and after) × 6 (doctors) two-way analysis of variance with the evaluation results as the dependent under four conditions. The results showed a significant difference at the 1% level under all conditions (in order from condition 1, $F(59, 295) = 3.19$, $P<.01$; $F(59, 295) = 4.51$, $P<.01$; $F(59, 295) = 4.52$, $P<.01$; $F(59, 177) = 5.26$, $P<.01$).

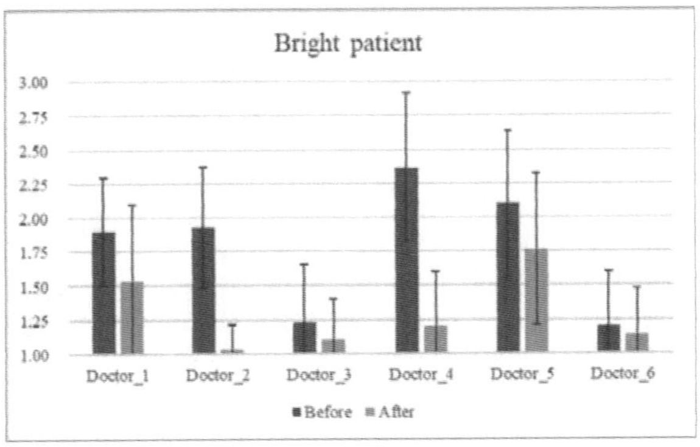

Figure 12. Average and standard deviation of each doctor's evaluation score for bright patients.

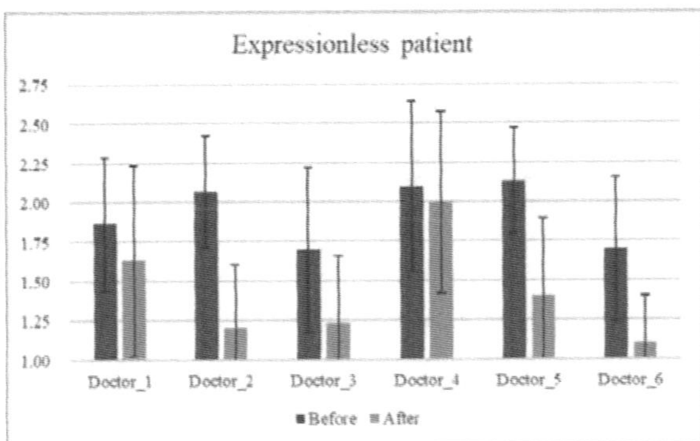

Figure 13. Average and standard deviation of each doctor's evaluation score for expressionless patients.

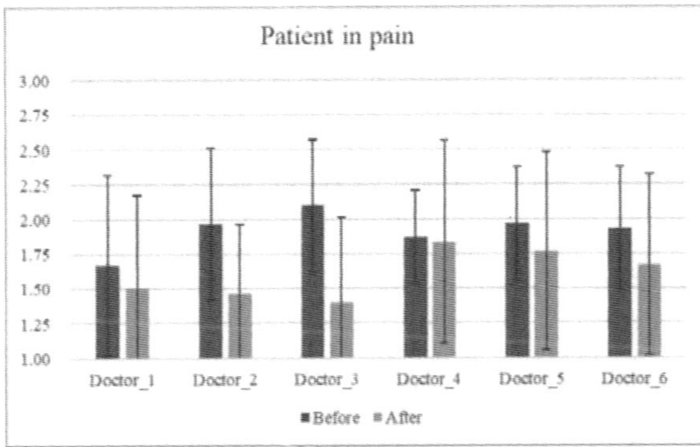

Figure 14. Average and standard deviation of each doctor's evaluation score for patients in pain.

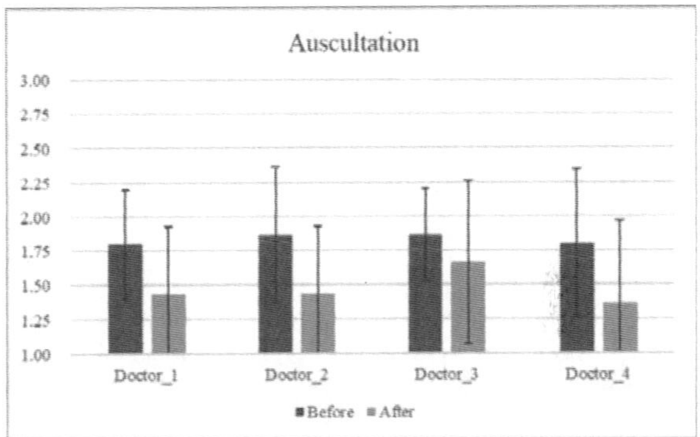

Figure 15. Average and standard deviation of each doctor's evaluation score for auscultation.

We consider the effectiveness of this application to have been confirmed because the evaluation scores improved with use of the application under all conditions. We also noticed the following two points during the experiment. The first point is that the evaluation for "patients in pain" was different for each person. Most persons regarded the doctor's serious expression as empathy and evaluated that as adequate. However, a few persons felt that the doctor's expression caused unease and worry, and that person evaluated the expression as not adequate. Therefore, the model expressions for patients in pain may not be limited to one type. The second point is that the facial expressions practiced in the application are unnatural for a few persons. By training in the application, people playing the role of doctor created expressions close to those of the model doctor's videos, and accordingly the evaluation score rose. However, when a facial expression that the participants learned was far from facial expressions that they always do, a few evaluators felt that the expression looked like "a pretended expression" or "artificial expression."

VI. Conclusion

We developed a facial-expression training application that physicians can use for independent-study. Six participants practiced using this application to express appropriate facial expressions. To verify the effectiveness of this application, the facial expressions that participants made before and after training with it were evaluated by 30 people. On the basis of the results, we consider the usefulness of the facial-expression training application to be verified. We think most learners could express their facial expression naturally by practicing repeatedly.

In the future, we will improve the application so that the on-screen instructions will guide the learner's expressions more properly. In addition, because requirements for a doctor who cares for children are very different from those of one who cares for adults, we plan to add learning content for pediatricians.

REFERENCES

[1] A. Donabedian, The Definition of Quality and Approaches to Its Assessment. Ann Arbor, MI: Health Administration Press, 1980.

[2] S. H. Kaplan, S. Greenfield, and W. E. Ware, Jr, "Assessing the effects of physician-patient interactions on the outcomes of chronic disease," Med. Care, vol. 27, pp. S110–S127, March 1989.

[3] L. M. Noble, A. Kubacki, J. Martin, and M. Lloyd, "The effect of professional skills training on patient-centredness and confidence in communicating with patients," Med. Educ., vol. 41, pp. 432–440, May 2007.

[4] J. A. Hall, D. L. Roter, and N. R. Katz, "Meta-analysis of correlates of provider behavior in medical encounters," Med. Care, vol. 26, pp. 657–675, July 1988.

[5] J. A. Hall, J. A. Harrigan, and R. Rosenthal, "Nonverbal behavior in clinician-patient interaction," Appl. Prev. Psychol., vol. 4, pp. 21–37, Winter 1995.

[6] L. M. Ong, J. C. de Haes, A. M. Hoos, and F. B. Lammes, "Doctor-patient communication: A review of the literature," Soc. Sci. Med., vol. 40, pp. 903–918, April 1995.

[7] T. L. Thompson, "Interpersonal communication and health care," in Handbook of Interpersonal Communication, 2nd ed., M. L. Knapp and G. R. Miller, Eds. Thousand Oaks, CA: Sage, pp. 696–725, 1994.

[8] S. Williams, J., Weinman, J, and J. Dale, "Doctor-patient communication and patient satisfaction: A review," Family Practice, vol. 15, pp. 480–492, October 1998.

[9] M. R. DiMatteo, A. Taranta, H. S. Friedman, and L. M. Prince, "Predicting patient satisfaction from physicians' nonverbal communication skills," Med. Care, vol. 18, pp. 376–387, April 1980.

[10] M. R. DiMatteo, R. D. Hays, and L. M. Prince, "Relationship of physicians' nonverbal communication skill to patient satisfaction, appointment noncompliance, and physician workload," Health Psychol., vol. 5, pp. 581–594, 1986.

[11] A. Oe and Y. Murata, "Quantitative Analysis of Facial Expression for Medical Doctors," IARIA, proceeding of Global Health 2018, pp.8–13, 2018.

[12] J. Bird, S. A. and Cohen-Cole, "The three-function model of the medical interview. An educational device," Advances in Psychosomatic Medicine., vol. 20, pp. 65–88, 1990.

[13] H. de Haes and J. Bensing, "Endpoints in medical communication research, proposing a framework of functions and outcomes," Patient Educ. Couns., vol. 74, pp. 287–294, March 2009.

[14] Z. Ben-Sira, "Affective and instrumental components in the physician-patient relationship: An additional dimension of interaction theory," J. Health Soc. Behav., vol. 21, pp. 170–180, June 1980.

[15] M. K. Buller and D. B. Buller, "Physicians' communication style and patient satisfaction," J. Health Soc. Behav., vol. 28, pp. 375–388, December 1987.

[16] B. M. Korsch and V. F. Negrete, "Doctor-patient communication," In Physician-patient Communication: Readings and Recommendations, G. Henderson, Ed. Springfield, IL: Charles C. Thomas, pp. 29–40, 1981.

[17] S. J. Williams and M. Calnan, "Convergence and divergence: Assessing criteria of consumer satisfaction across general practice, dental and hospital care settings," Soc. Sci. Med., vol. 33, pp. 707–716, 1991.

[18] R. C. Wasserman, T. S. Inui, R. D. Barriatua, W. B. Carter, and P. Lippincott, "Pediatric clinicians' support for parents makes a difference: An outcome-based analysis of clinician-parent interaction," Pediatrics, vol. 74, pp. 1047–1053, December 1984.

[19] R. C. Wasserman, T. S. Inui, R. D. Barriatua, W. B. Carter, and P. Lippincott, "Pediatric clinicians' support for parents makes a difference: An outcome-based analysis of clinician-parent interaction," Pediatrics, vol. 74, pp. 1047–1053, December 1984.

[20] H. S. Friedman, M. R. DiMatteo, and A. Taranta, "A study of the relationship between individual differences in nonverbal expressiveness and factors of personality and social interaction," J. Res. Pers., vol. 14, pp. 351–364, September 1980.

[21] L. A. Camras, and M. M, Shuster, "Current emotion research in developmental psychology," Emotion review, 5, pp. 321-329, 2013.

[22] M. Lewis, M. W. Sullivan, and A. Vasen, "Making faces: Age and emotion differences in the posing of emotional expression," Developmental Psychology, 23, pp. 690-697, 1987.

[23] N. Takahashi and I. Daibo, "The facial expressions by emotion-instruction and photograph-instruction, and the effect of presence of another person in the facial expressions of anger and sadness," Japanese Journal of Interpersonal and Social Psychology, 3, pp. 65-72, 2003.

[24] F. Kameyama and Y. Muraoka, "An Interactive Design System of an attractive smile," J IEICE technical report. Human information processing, 95(554), pp. 15-22, 1996.

[25] Microsoft Emotion API, https://azure microsoft.com/en-us/services/cognitive-services/emotion/, [retrieved: January 2019].

[26] Open CV, https://opencv.org/, [retrieved: January 2019].

A System for Collecting Motion Data on Patients' Activities of Daily Living

Rintaro Takahashi, Yoshitoshi Murata, Oky Dicky Ardiansyah Prima
Faculty of Software and Information Science
Iwate Prefectural University
Takizawa, Japan
e-mail: g031n100@s.iwate-pu.ac.jp, {y-murata, prima}@iwate-pu.ac.jp,

Abstract— **Functionary recovery is administered to hemiplegic patients as rehabilitation. Some patients who recover their functions better in a rehabilitation facility cannot do any activities in their home. Therefore, recovering Activities of Daily Living (ADL) has become more significant than functionary recovery recently. Since existing ADL recovery level indices are based on questionnaires, recovery level judgments are easily affected by an evaluator's subject. In this paper, we describe a system we propose that collects and stores motion data on daily life activities to quantitatively evaluate ADL recovery levels.**

Keywords-rehabilitation; functionary recovery; activities in daily living; ADL; BLE beacon; Google Firebase.

I. INTRODUCTION

Most patients suffering from cerebrovascular disease have paralysis on one side of the body, and their bodies lean and twist to the paralyzed side. Also, because of unusual muscle strain, their hands and feet become stiff. In some cases, muscles of the upper body go into convulsions. Functionary recovery is administered to hemiplegic patients as rehabilitation. However, some patients are not always to live less inconveniently in their home. Some patients who recover hand and arm functionality better in a rehabilitation facility cannot eat meals better in their home. Therefore, recovering Activities of Daily Living (ADL) has recently become more significant than recovering functionaries.

The Barthel Index, which is based on questionnaires, is popularly used to quantitatively evaluate ADL recovery levels [1][2]. With questionnaires, however, recovery level judgments easily change in accordance with the evaluator's subject. Each recovery level is digitized to a few levels. For example, answers for feeding include "unable," "needs help cutting, spreading butter, etc., or requires modified diet" and "independent." Each answer is scored 0, 5, or 10. However, the recovery level for feeding with help ranges from "a patient eating food directly from dishes without using a spoon or fork" to "a patient eating a meal with a knife and fork in almost the same way as a healthy person." Also, it takes too much time to ask and observe whether a patient can do an activity independently without needing help.

Functional Independence Measure (FIM) [3] and Katz Index [4]-[6] scores are also used to evaluate ADL. FIM scores cover not only functional disease but also mental disease. Scores are broken down into seven levels for each activity, including feeding. Katz Index scores are usually applied to cure elder patients or those suffering from chronic disease.

The question formats for these evaluation methods are basically the same, and an evaluator needs much time to ask questions and observe a patient. We think that a quantitative evaluation system with a computer is needed to evaluate patients objectively without needing to ask them any questions and/or observe them.

Judging ADL recovery levels is based on whether patients can do tasks, such as eating, getting dressed, bathing, washing, and discharging bodily waste by themselves. Therefore, a system that collects motion data of patients in daily life needs to not only measure and collect the motions of body parts but also detect which activities are performed. However, it is very difficult to estimate these merely from changes in acceleration and/or gyro sensor data obtained from devices attached to body parts. Therefore, we estimate activities by using information about places, such as a dining table, bathroom, dressing room, or bedroom. We used the BLE beacon [7] to detect places in this system.

Most surgeons also think that postoperative patient functions assessed by ADL and quality of life have become especially important ways to measure surgical treatment outcomes for the elderly [8].

In this study, we developed a system to collect and store patients' motion data to quantitatively judge **the recovery level of activities in daily living**. A patient's name, measured location, sensor-attached body parts and time-stamps are described as the file names for each measured data file in this system. Since we wish to use this system not only to evaluate patients' ADL but also to develop algorithms for detecting whether patients can do designated activities, we designed the system so that it can store video data of recorded patients' motions.

The system only requires that recognized medical doctors or physiotherapists can access measured data to maintain security. To ensure this, we developed a data collecting system based on Google Firebase [9]. Since the Firebase application can be independently implemented for any organization, high level security can be maintained.

Evaluation algorithms are needed to judge recovery level from measured data. However, in this paper we describe a system to collect measured data rather than evaluation algorithms. We plan to introduce the latter in another paper.

After introducing related work in Section II, we describe the system's design concept and its implementation in Sections III and IV. Examples of measured data for a participant played a patient who has paralysis on one side are introduced in Section V. Section VI concludes with a summary of key points.

II. RELATED WORK

To develop a quantitative evaluation system for the recovery level of activities in daily living of hemiplegic patients, we have to know how to evaluate ADL quantitatively, existing life log systems and healthcare information cloud service.

A. Evaluation index for function level in daily living

Three indexes to evaluate function level in daily living are widely used: the Barthel Index, the Functional Independence Measure (FIM) and the Katz Index. They are basically questionnaires for daily life activities, such as feeding. The Barthel Index and FIM are popularly applied to evaluate function levels for rehabilitation patients, such as those afflicted with cerebrovascular disease. There are ten question items in the Barthel Index: Feeding, Moving from wheelchair to bed and return, Personal toilet (washing face, combing hair, shaving, cleaning teeth), Getting on and off the toilet (handling clothes, wiping, flushing), Bathing self, Walking on level surfaces, Ascending and descending stairs, Dressing (includes tying shoes, fastening fasteners), Controlling bowels and Controlling bladder [1] [2]. A score of independently doing an activity is usually 10 points, doing it with help is usually 5 points, and not doing it is 0 points.

FIM evaluates not only physical functions but also social abilities, such as communication or social recognition [3]. The number of questions covers 18 issues; 13 for physical functions and five for social abilities. Questions about physical functions are more segmentized. For example, the dressing function is divided into dressing the upper body and the lower body, moving activities are divided into the moving between a wheelchair and a bed/chair, and sitting on a toilet seat and moving to a bathtub. Scores are given on a seven-point system. Independently doing an activity gets seven points, doing it with full help gets one point, and doing it with partial help gets scores ranging from two to six points.

The Katz Index is usually applied to elder patients or those suffering from chronic disease in a variety of care settings [4 - 6]. The index ranks adequacy of performance in six activities: bathing, dressing, toileting, transferring, continence, and feeding. Clients are scored yes/no for independence in each of the six functions.

Every three indexes evaluate whether a patient can do activities in daily living. Therefore, our proposed system must know what kinds of activities a patient tries to do.

B. Life log system

Over the years, many researchers have tried to estimate daily life human activities, such as walking and sitting up and down from acceleration and/or gyro sensor data obtained from wearable devices and/or smartphones. In this paper, we refer to the research done respectively by Khan et al. and Wang et al. [10] [11]. Only a few motions were given in this research; distinctions among activities were not recognized. In contrast, Debraj et al. tried to recognize 19 daily living activities [12]. They collected environment information, such as that for temperature and location in addition to activity information. They used GPS and BLE beacons to identify places. However, they did not consider the Barthel Index or other indices and consequently their target activities did not correspond to activities in the index of function recovery levels.

C. Healthcare cloud serveice

Zhang al. developed a cyber-physical system for patient-centric healthcare applications and services. They called it Health-CPS. It was built on cloud and big data analytics technologies [13]. It consisted of a data collection layer, a data management layer and an application service layer to collect and follow up on many kinds of big data. It used a security tag to maintain security.

Doukas et al. proposed a mobile system that enables electronic healthcare data storage, update and retrieval using cloud computing [14]. A mobile application was developed using Google's Android OS and Amazon's S3 to provide management of patient health records and medical images.

We developed a cloud service whose collecting function for medical data is basically the same as that for the above systems. However, our system is specialized so that it can collect activity and place information to functionally evaluate recovery levels that correspond to existing evaluation methods, such as the Barthel Index. In this paper, we show how we implemented the system with SONY Smart Watch 3 [15] as the sensor node, as well as Android smartphone, BLE beacon, and Google Firebase.

III. SYSTEM DESIGN CONCEPT

We designed the proposed system so that it could not only evaluate ADL for a patient, but also develop algorithms for detecting whether a patient can do a designated activity. The system collects and stores sensor data and video data synchronously and allows appropriate persons to access stored data. We designed the system while taking the following issues into consideration:

1) Suppressing battery consumption for wearable sensor devices and smartphones
2) Suppressing recorded data and collecting necessary data
3) Maintaining security.

Google Firebase service provides many functions, including authentication and real-time database functions, to enable systems to be managed effectively, such as through the means of allowing access to authorized persons. Since any organization can independently implement Firebase applications, it becomes possible to maintain high level security. This is why we implemented our data collecting system on Google Firebase.

The image of a data collecting system that collects data about the motions that a patient performs daily is shown in Figure 1. The system we propose consists of sensor devices, a sensor relay unit (smartphone), BLE beacons and Google Firebase. A smartphone is used as the sensor relay unit that controls sensor devices and temporally stores and forwards measured data to the Firebase.

BLE beacons are placed in various locations: under a dining table, on top of a toilet, in a bathroom, in a bedroom, in a closet. When the smartphone receives a BLE beacon signal level that exceeds the threshold level, it sends a message to sensor devices to start measuring data. And when the smartphone receives a receiving signal level lower than the threshold level, it sends a message to sensor devices telling them to stop measuring data. Sensor devices and smartphones are managed by the Realtime Database. Security is maintained by enabling only authorized persons using the system, including patient, readers, such as medical doctor and installation personnel, such as nurse are also managed by the Realtime Database is used to maintain security. In this system, measured data are downloaded for pre-registered persons from the web server.

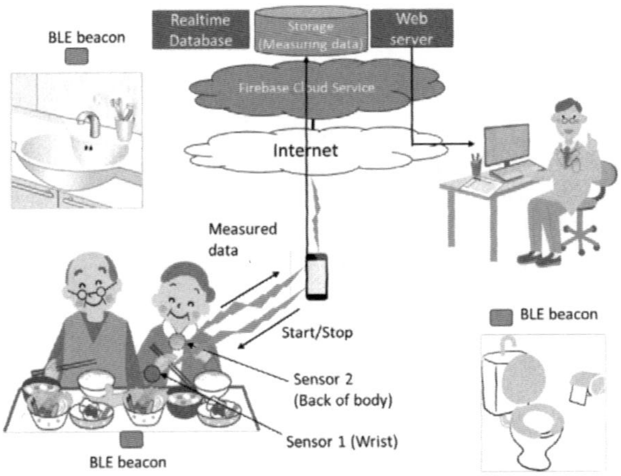

Figure 1. Image of the data collecting system for patient's daily life motions.

IV. System Implementation

We developed a PatientApp program that works on the sensor relay unit and a DataCollectionServer program that works on Firebase. The PatientApp manages sensor devices, gets measured data from sensor devices and uploads the data file to the DataCollectionServer.

A. *PatientApp*

This time, we developed a PatientApp program based on the Android Framework. With this program, a developer must first access the Firebase and download a configuration file. An Android application package file (Apk File) is then made as a building application and is connected to Firebase. This makes it possible to securely download the Apk File for each organization.

Before starting to measure sensor data and/or video data, it is necessary to enter a patient's name, bind a sensor with a body part, bind a BLE beacon with a place of activity and select a video recording on/off function. Therefore, we designed a transition diagram of UI pages as shown in Figure 2. There were three alternatives for a user name at the login; the patient's name, the medical worker's name with measuring devices set up, and the medical professional's name with measured data analyzed. For the latter two cases, a patient's name must be entered after the login. Therefore, we decided on the first one, login with a patient's name.

After login, a "List of setting up" page is presented. An example of this page is shown in Figure 3. With it, a user can confirm a state of setting. When the "Change" button is clicked, the page will change to the "Sensor" page to bind a sensor with a body part. When the "Next" button is clicked, the page will change to the "Beacon" page to bind a BLE beacon with a place in activity. When the "Next" button is clicked, the page will change to the "Video" page to select video ON/OFF. When the "Next" button is clicked, the page will change to the "List of setting up" page. When the "Next" button is clicked in the "List of setting up" page, the page will change to the "Measuring" page. When the "Start" button on this page is clicked, the PatientApp sends a message to the sensor messages to start measuring, and the "Start" button changes to the "Stop" button. When the "Stop" button is clicked, the PatientApp sends a message to the sensor messages to stop measuring, and the "Stop" button changes to the "Start" button. When the "End" button is clicked, the PatientApp finishes.

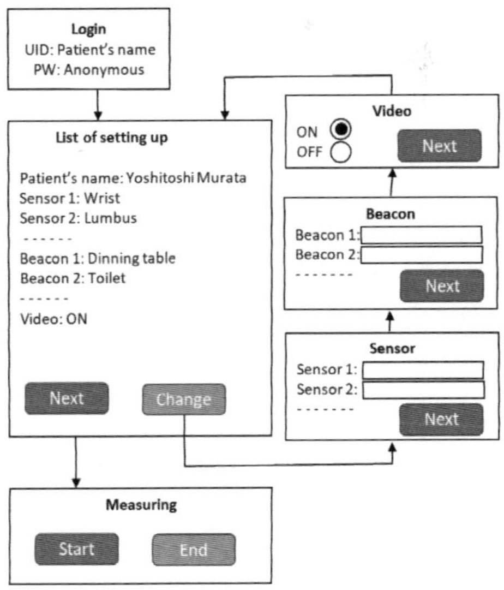

Figure 2. Transitions of UI pages in the PatientApp.

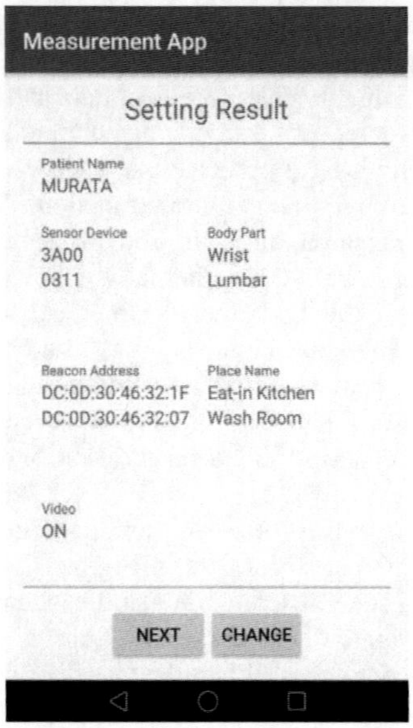

Figure 3. Example of setting up page list.

When a sensor receives a BLE beacon signal, it starts measuring, and, when a sensor loses a BLE beacon signal, it stops measuring. After clicking the "Stop" button, measured data are changed to a measured data file. Its file name is "Patient name_place_body part_timestamp" to recognize its properties. The file is uploaded to the storage in Firebase.

We developed the following six packages of classes to achieve the above proceedings:

- Beacon: receiving beacon signals and handing their information to other classes.
- Mobile2wear: controlling a sensor device and receiving measured data.
- Camera: managing a video camera.
- Firebase: converting measured data and transferring the data to the Firebase storage.
- View: managing transition of pages
- Viewmodel: listening events on buttons or input boxes and handing, such information to other classes.

B. DataCollectionServer

The DataCollectionServer has the following functions:

- Data upload function: The sensor relay unit temporally stores measured data and forwards them to the server.
- Data download function: Authorized persons, such as medical doctors can access the DataCollectionServer and download measured data files securely.

It consists of the Storage and WebSite. The WebSite collaborates with the Storage and provides a file download function to a medical professional through the Web browser.

In this subsection, we mainly introduce how to upload and download measured data file.

1) Data upload function (Figure 4)

After a measured file has been made, the PatientApp uploads the file to the storage server in Firebase as shown in Figure 4. The storage server generates the file download URL, which is managed in the Realtime Database.

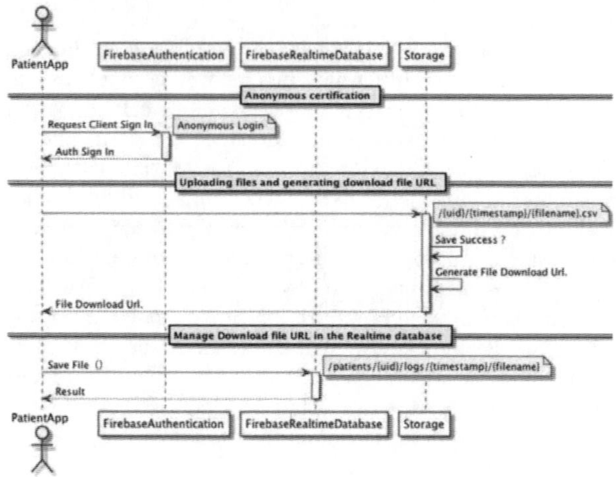

Figure 4. Sequence flow to upload measured files.

2) Measured data download function (Figure 5)

Supervisors input the access account of medical professionals from the management page in Firebase. The sequence flow with which medical professionals download their patients' files is shown in Figure 5. When medical professionals access the Website, they log in with their assigned ID and password on the page of Figure 6 (a).

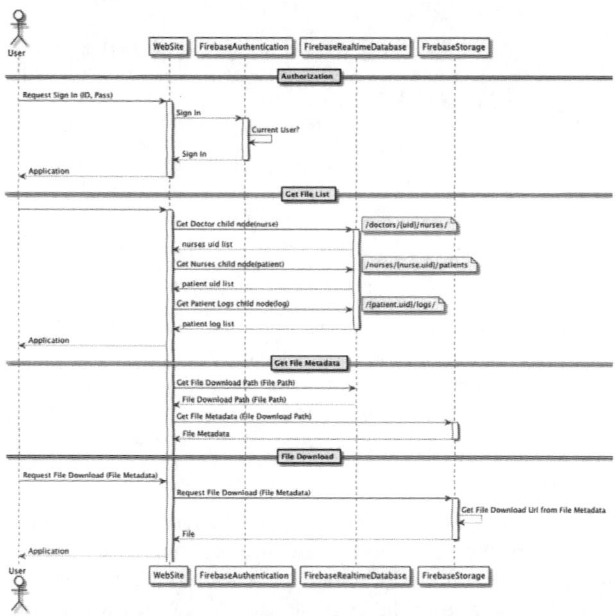

Figure 5. Sequence flow to download measured files.

After login, the Website application accesses the Realtime Database to get information related to nurses and patients. The Website application also gets meta-data such as an access path to a stored file. When a medical professional clicks a file on the page of Figure 6 (b), the Website application accesses the indicated file on the Storage through the access path. Finally, the indicated file is downloaded.

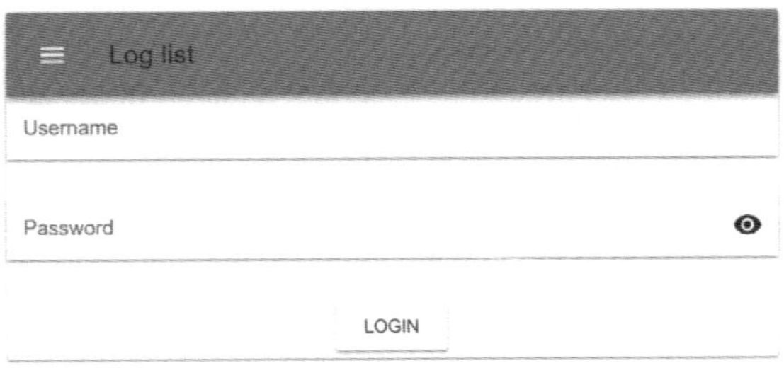

(a) Login page

(2) File list page

Figure 6. WebSite user interface.

V. EXAMPLES OF MEASURED DATA IN DAILY LIVING

We measured motions of eating lunch and brushing teeth to text whether the developed system can measure well. A "normal" participant usually performed these actions first, followed by a person who was paralyzed in the right hand. In this experiment, wrist and lumbar region motions were measured with a 3D-gyro sensor. We used two SONY Smart Watch 3 units as the 3-D gyro sensor [15]. One was attached to the wrist with a wrist band and the other was attached to the lumbar region with a lumbago band as shown in Figure 7.

Figure 8 shows a picture and measured data for eating a curry rice dish with a spoon. In this figure, (a) is a picture of a participant eating freely and (b) shows a participant fixing and twisting his arm. The (c) and (d) depictions are graphs of the measured angle data for (a) and (b). When the participant played the role of a patient, his upper body always leaned forward and while eating he moved his arm less than when he usually did while eating. Such differences are presented in red square frames in (c) and (d). These data show that a participant eats food by himself. However, from these data it is impossible to judge whether a participant is able to eat food by himself without help. On the other hand, the data shows how their joints move.

Figure 9 shows angle data measured during the tooth-brushing activity. The right arm was fixed and twisted, and a wearable device was fixed on the right wrist. Since the participant brushed his teeth with his left hand, his right arm did not move except when he was sticking toothpaste on the brush and putting the toothpaste on a wash stand. These data show that a participant brushes his teeth by himself, but do not show whether the participant brushes his teeth without help from others.

A medical doctor and physiotherapists plan to measure motions for real patients, and introduce their results.

We believe that the obtained results enable us to determine whether with the proposed system we can judge whether a patient can do activities on the basis of Barthel Index or Katz Index judgments. However, it is impossible to judge whether a patient can do activities with or without help. On the other hand, they make it possible to ascertain how affected joints move remotely.

We believe that the results we have obtained will make it possible for us to make a new method to evaluate ADL on the basis of measured angle data rather than on Barthel Index FIM and Katz Index results.

Figure 7. Participant wearing two SONY Smart Watch 3 sets.

(a) Eating with free arm.

(b) Eating with fixed and twisted arm.

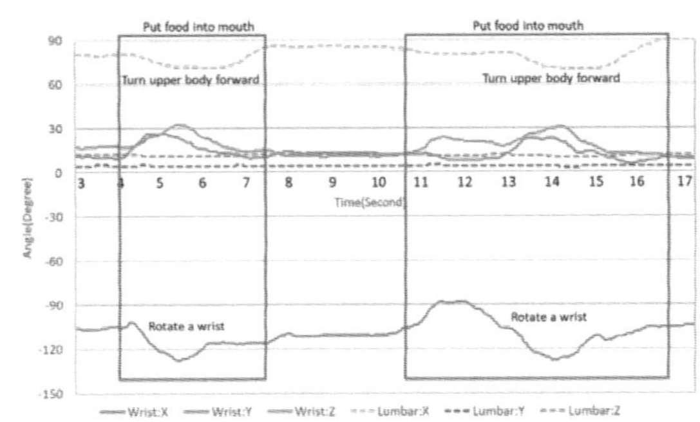

(c) Angle of wrist and lumbar for free arm.

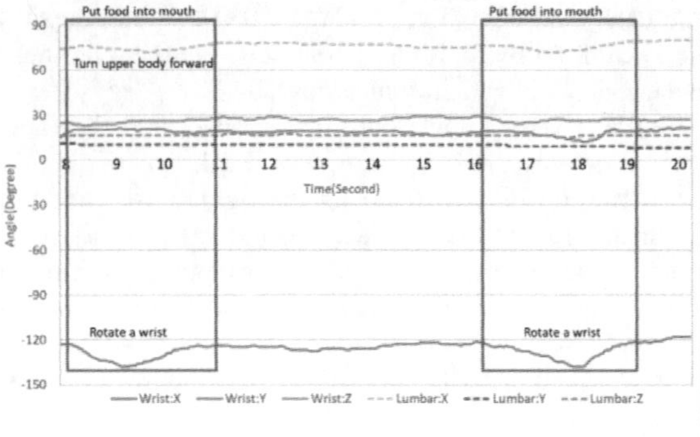

(d) Angle of wrist and lumbar for fixed and twisted arm.

Figure 8. Measured data for the wrist and lumbar region in eating a curry rice dish with a spoon.

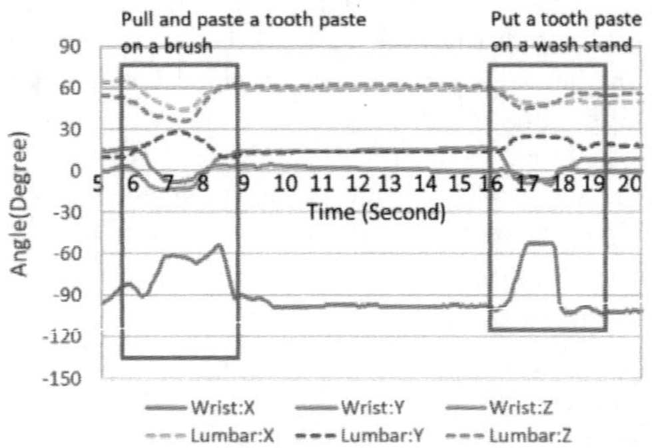

Figure 9. Measured data for the wrist and lumbar region in tooth-brushing.

VI. CONCLUSION AND FUTURE WORK

Existing evaluation indexes for activities in daily living (ADL) recovery levels such as the Barthel index are based on questionnaires. Therefore, the judging of recovery levels can be easily affected by an evaluator's subject. We proposed a system that is able to collect and store motion data about daily life activities to quantitatively evaluate ADL recovery levels. The system was developed on the basis of the Google Firebase. We used information about places such as a dining table, a bathroom to estimate activities. The BLE beacon was adopted for detecting places in this system. Measurement results obtained for a simulated patient demonstrated that it is possible to judge which ADL a patient can or cannot do. However, it is difficult to estimate whether a patient can do them with or without help.

A medical doctor and physiotherapists plan to measure motions of real patients using this system, and introduce their results.

The system we have developed does not correspond to GPS. The next version we plan to develop will correspond to it to measure motion during walking or running.

Our goal is to develop a new index for evaluating ADL recovery levels on the basis of motion data measured for persons performing ADL.

ACKNOWLEDGEMNT

The authors extend thanks to Mr. Kazuhiro Yoshida for the help he provided in performing this research. The research and development was supported by the MIC/SCOPE #181602007.

REFERENCES

[1] The Barthel Index; http://www.strokecenter.org/wp-content/uploads/2011/08/barthel.pdf [retrieved: January, 2019].

[2] F. I. Mahoney and D. Barthel, "Functional evaluation: The Barthel Index.", Maryland State Medical Journal 1965, vol.14, pp. 56-61, 1965.

[3] D. Chumney, et al., "Ability of Functional Independence Measure to accurately predict functional outcome of stroke-specific population: Systematic review," Departmentof Veterans Affairs, Journal of Rehabilitation Research & Development, Volume 47, Number 1, pp. 17-29, 2010.

[4] M. Shelkey and M. Wallace, "Katz index of independence in activity of daily living (ADL)," https://www.semanticscholar.org/paper/Katz-Index-of-Independence-in-Activities-of-Daily-Shelkey-Wallace/fb433f328b79f56d82a52551960c93da74469baa [retrieved: January, 2019].

[5] S. Katz, et al., "Progress in development of the index of ADL," Gerontologist 1970, vol. 10, pp. 20 –30, 1970.

[6] M. Shelkey, and M. Wallace, "Katz Index of Independence in Activities of Daily Living (ADL)," The Hartford Institute for Geriatric Nursing, New York University, College of Nursing, Issue Number 2, 2012.

[7] Bluetooth Technology; https://www.bluetooth.com/bluetooth-technology/radio-versions, [retrieved: January, 2019].

[8] T. Amemiya, et al., "Activities of Daily Living and Quality of Life of Elderly Patients After Elective Surgery for Gastric and Colorectal Cancers," Lippincott Williams & Wilkins, Annals of Surgery, Volume 246, Number 2, pp. 222-228, 2007.

[9] Google Firebase Cloud Messaging, https://firebase.google.com/docs/cloud-messaging/?hl=en, [retrieved: January, 2019].

[10] K. Zhan, S. Faux, and F. Ramos, "Multi-scale Conditional Random Fields for First-Person Activity Recognition on Elders and Disabled patients," ELSEVIER, Pervasive and Mobile Computing, Volume 16, Part B, pp. 251-267, 2015..

[11] L. Wang, T. Gu, X. Tao, H. Chen, and J. Lu, "Recognizing Multi-User Activities Using Wearable Sensors in a Smart Home," ELSEVIER, Pervasive and Mobile Computing, Volume 7, Number 3, pp.287–298, 2011.

[12] D. Debraj, B. Pratool, K. D. Sajal, and C. Sriram, "Multimodal Wearable Sensing for Fine-Grained Activity Recognition in Healthcare," IEEE, Internet Computing, pp. 26-35, 2015.

[13] Y. Zhang, et al., "Health-CPS: Healthcare Cyber-Physical System Assisted by Cloud and Big Data," IEEE, SystemsJournal, Volume 11, Number 1, pp. 88-95, 2017.

[14] C. Doukas, T. Pliakas, and I. Maglogiannis, "Mobile Healthcare Information Management utilizing Cloud Computing and Android OS," IEEE, 2010 Annual International Conference of the IEEE Engineering in Medicine and Biology, pp. 1037-1040, 2010.

[15] SONY Smart Watch 3, https://www.sonymobile.com/global-en/products/smart-products/smartwatch-3-swr50/#gref, [retrieved: January, 2019].

Single Camera 3D Human Pose Estimation for Tele-rehabilitation

Oky Dicky Ardiansyah Prima, Takashi Imabuchi,
Yuta Ono, Yoshitoshi Murata, Hisayoshi Ito
Graduate School of Software and Information Science
Iwate Prefectural University
Takizawa, Japan
e-mail: prima@iwate-pu.ac.jp, {g236o001, g231q005}@s.iwate-pu.ac.jp,
{y-murata, hito}@iwate-pu.ac.jp

Yukihide Nishimura
Department of Rehabilitation Medicine
Iwate Medical University
Morioka, Japan
e-mail: ynishi@iwate-med.ac.jp

Abstract—**The need of using advanced remote devices to promote effective self-management of rehabilitation has rapidly grown in developed countries. The widely spread camera-equipped mobile devices and Internet of Things (IoT) have been expected to deliver professional services by connecting clinician to client for assessment and consultation. This study proposes an IoT-based Tele-Rehabilitation (TR) framework using a single camera to observe the body joints of the client in three-dimensional (3D) space on performing Activities of Daily Living (ADL). Our experiments show that the proposed framework is capable to measure joint and orientation angles of elbow and knee comparable with the measurements using the Kinect. A waterproof camera was used to show that the proposed system can be extended to do the joint measurements during aquatic therapy and fitness pools.**

Keywords-rehabilitation; 3D human pose; ADL; range of motion; IoT.

I. INTRODUCTION

The growing number of elderly people and the decreasing number of healthcare professionals led to the implementation of Tele-Rehabilitation (TR) initiatives in healthcare regular practice [1]. TR is the provision of rehabilitation services from a distance using communication technologies. It is a developing field of telehealth to increase accessibility and enhancing continuity of care to individuals who are in geographically remote regions. TR enables clinicians to optimize the timing, intensity and duration of therapy effectively. The use of videoconferencing and virtual reality systems allows clinicians to interact with patients in real-time. Therefore, TR is expected to reduce the potential time and cost of rehabilitation services, especially for individuals who have economically disadvantaged. The future of TR is promising as a wide range of services to suit the needs of the individual.

Efforts have been made to conduct TR for physical therapy using low-cost depth sensors, such as Microsoft Kinect. There are two versions of Kinect: Kinect V1 and V2. Hereinafter, we simply referred both versions as Kinect. The Software Development Kit (SDK) enables developers to access body joints positions and orientations. Previous studies have shown that these devices show performance adequate for a range of healthcare imaging applications [2][3]. However, there are some concerns with occlusions and noises during tracking the joints [4]. Data smoothing techniques, such as Kalman filtering are required to minimize the problems. Moreover, a study of validity and reliability of the Kinect on measuring joint angles shows that 95% Limits of Agreement (LOA) between the Kinect and the goniometer exceeded ±5° which suggests a concern of using the Kinect in rehabilitation [5].

Despite the issues of the Kinect, many attempts have been made to use this device to support rehabilitation virtually. The Kinect can be used as a natural user input interface to the Virtual Reality (VR) system on carrying out physical rehabilitation therapies [6]. This system will enable the users to get an audiovisual feedback in real-time whether they are properly doing the specific therapy or not. The users can record their body joint movements during exercise and sent them to clinicians to get more advices on maintaining or improving the rehabilitation stages.

The Kinect is known to have some limitations on measuring body joints if some parts of the body are occluded. The Kinect needs to capture the full body of the user to measure the joint locations properly. Therefore, measuring a target user performing Activities of Daily Living (ADL) tasks where the lower limbs are hidden, such as dining and sleeping would be difficult to achieve.

The needs of TR are not limited to in-room rehabilitation programs. The basic ADL includes a functional mobility to move from one place to another while performing tasks, such as the ability to walk, get in and out of bed, and get into and out of a chair. Since the newest Kinect (Kinect V2) is only capable to measure targets up to 6m and inside 70° by 60° Field of View (FOV) from the its sensor, area measurements of these tasks are limited.

With the recent enhanced techniques utilizing Artificial Intelligent (AI), the state-of-the-art computer vision has enabled to measure body joints in three-dimensional (3D) using a single camera. This measurement takes two steps: joint localization in two-dimensional (2D) image coordinate and 3D coordinate estimation for each 2D joint. Depends on the method to localize joints, method based on bottom-up human pose estimation, such as "OpenPose" enables joint localization without capturing the full body beforehand [7]. Once this method localizes some joints, part affinity fields between joints are calculated to connect corresponding joints to estimate 2D human pose. Many studies have investigated the problem of inferring 3D joints from their 2D projections. These studies involve traditional 2D to 3D methods, which define the bone lengths and use a binary decision tree to estimate the 3D joints [8] or deep-net-based 2D to 3D methods

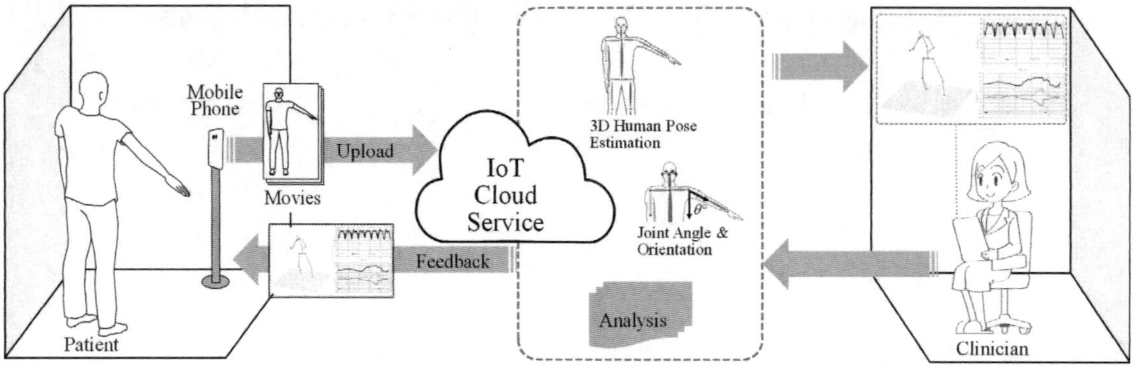

Figure 1. The proposed TR framework in this study.

which estimate 3D joints with Deep Neural Networks (DNN) [9]. The use of DNN may produce a high calculation costs to be performed on dedicated Graphics Processing Units (GPUs). However, this calculation can be performed in a server where the client can send the recorded movie to and get the resulted 3D joints using the Internet of Things (IoT) infrastructures.

This study proposes a framework to enable the use of a single camera 3D human pose estimation to estimate 3D joints and orientation of limbs which can be used in a broad range of TR services. Preliminary evaluations are made to investigate the accuracy of the estimated joint and orientation angles compared to the Kinect and to reveal its ability to handle joint occlusions on estimating the 3D joints. An attempt to estimate 3D joints for a user performing fitness pool is provided to open the further development in the future.

This paper is organized as follows. Section II describes related works on TR. Section III describes methods to measure joint and orientation angles of elbow and knee, and to perform our experiments. Section IV shows the accuracy of joint and orientation angles of elbow and knee measured in this study for a subject performing several tasks. Finally, Section V concludes the achievements and discusses the future prospective of this study.

II. Related Work

A single camera has a prospect to be a useful tool to assess a certain Range of Motion (ROM). DrGoniometer, an iPhone app, has enabled the manual measurement of patient articulation angles and storage of all related information to build up historical data for each articulation and movement [10]. For shoulder proprioception assessment, Mitchell et al. (2014) assessed the validity of DrGoniometer by measuring ROM of participants on performing active shoulder external rotation [11], where DrGoniometer was found to be comparable to the standard goniometer. For elbow proprioception assessment, Ferriero et al. (2011) found that DrGoniometer is reliable for elbow joint goniometry on measuring passive ROM of elbow [12].

Enhanced computer vision techniques enable the automated measurement of head and body pose. For cervical spine proprioception assessment, the Perspective-n-Point camera pose determination (PnP problem) [13] from a single camera can be used to measure head repositioning accuracy to diagnose people with neck disorders. The PnP problem estimates the relative position between camera and head posture from predefined 3D facial feature points (3D landmarks) and their corresponding points in camera coordinates. Head pose estimation based on the PnP problem has been implemented in some open source software libraries, such as OpenCV [14] and Dlib [15]. There are many popular photographic apps on Android and iPhone utilizing this technique to modify facial shapes. However, to our knowledge, there are no reports on the validity of the head pose measured by the PnP problem for a rehabilitation purpose. For ROM assessment, limb joints in 3D can be estimated from a targeted body captured by a single camera. The DNN have boosted the accuracy of the detection of joint locations from an image. OpenPose has enabled to localize multiple human body joint in real time by implementing the Part Affinity Fields (PAFs) to encode the location and orientation of limbs without capturing the full body [7]. Martinez et al. (2017) proposed a relatively simple deep feedforward network over the Human3.6M [16], the largest publicly available 3D human pose dataset containing 3.6 million human poses and corresponding images to estimate 3D joints from their 2D projections. This work has been made available as open source software, namely "3D-pose-baseline [17]."

The 3D joint estimation from a single camera is hard to be run on a mobile device because of its extensive calculation that has to be performed on GPUs. However, with the spread of IoT infrastructures, this calculation can be done in the cloud server. Using services over the Internet, clinicians can conduct physical TR with patients at their homes. This new service will provide not only a support for assistance with exercise but also smart data to maintain or improve the rehabilitation stages.

III. Methods

The proposed TR framework was developed based-on OpenPose and 3D-pose-baseline. The framework processes user's videos and estimates the 3D human pose from the

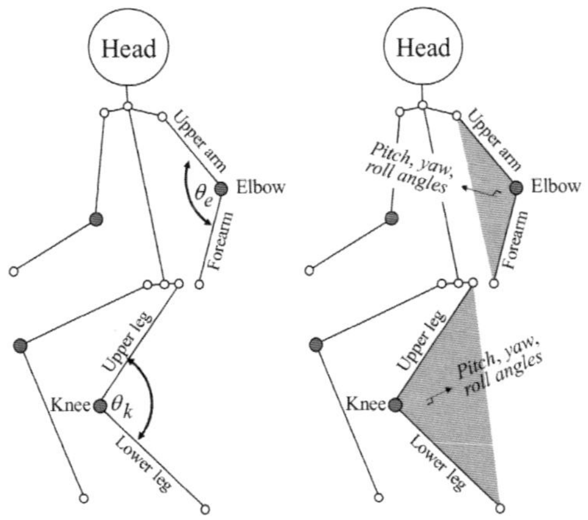

(a) Joint angles (b) Joint orientation angles

Figure 2. Measurements of joint and orientation angles (θ_e, θ_k, *pitch*, *roll*, *yaw*) of elbow and knee in this study.

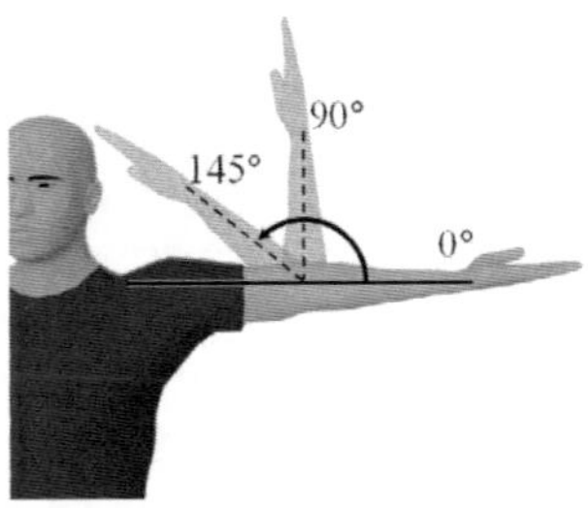

Figure 3. Elbow joint angle during side flexion.

videos. Body joint and orientation angles are measured, and these data are presented as time series plots concurrently with the 3D plot of the body posture. Here, the joint orientation angles consist of pitch, yaw, and roll angles. At this moment, only joint and orientation angles of elbow and knee are measured and visualized by the proposed framework. However, other joints can be measured with the same manners. Python programming language was used to process the calculations.

Our proposed framework, as shown in Figure 1, is based on the client-server model. Using mobile phones, patients can record their movements on performing ADL tasks and send the recorded videos to the IoT cloud service. The IoT will measure the body joint and orientation shortly after it detects a new incoming video. The measurement results are sent to clinicians where they can monitor the rehabilitation stages and give necessary advices. All results are stored in the database which can be accessed by the patients from their mobile devices.

A. Joint Angle Measurement

Joint angle was measured as a relative angle between the longitudinal axis of two adjacent segments. For the elbow joint angle, the adjacent segments are the upper arm and the forearm, respectively. Whereas, for the knee joint angle, the adjacent segments are the upper and the lower legs, respectively. Note that although the lower leg consists of a tibia and a fibula, only a single segment can be estimated from the 3D human pose estimation, as well as the Kinect's result.

Here, Figure 2(a) shows elbow and knee joint angles measured in this study. Let *u* and *v* be vectors representing two adjacent segments, the angle between *u* and *v* is equal to

$$\theta_{joints} = 180° - \frac{\overline{u} \cdot \overline{v}}{|\overline{u}||\overline{v}|}, \qquad (1)$$

where *joints* represent elbow (*e*) and knee (*k*), respectively.

B. Joint Orientation

Joint orientation was measured as an orientation of a triangle plane constructed by two adjacent segments. Let

$$ax + by + cz + d = 0 \qquad (2)$$

is a triangle-plane constructed by three 3D points as shown in Figure 2(b), the orientation (*R*) and translation (*T*) of this plane against pre-defined a reference triangle plane can be calculated using Singular Value Decomposition (SVD) [19]. Euler angles were calculated from *R* to derive the plane's orientation angles: pitch, yaw, roll angles.

C. Data Extraction and Analysis

Firstly, we measured right-hand's elbow joint and orientation angles of a subject while performing side flexion as shown in Figure 3, using a single camera and the Kinect concurrently. Occlusion is not expected to occur on this posture. The measurement results were compared to assess how the resulted measurements from the camera correspond to those from the Kinect. Secondly, we measured elbow and knee joint angles of a subject while performing the following tasks: swinging a tennis racket, running and running in place, and swimming. For swinging a tennis racket, the resulted joint angles from the camera were compared to those from the Kinect. Occlusion was expected to occur during this task. For the rest of the tasks, measurements were conducted using only a single camera which moves in accordance with the movement of the subject.

IV. Results

We measured right-hand's elbow joint and orientation angles of a subject, as shown in Figure 4, while performing elbow side flexion. The subject was asked to stand upright and slowly perform this task until maximum ROM and move back to the initial pose. The minimum and maximum ROMs for the subject measured by a goniometer were 0° and 146°, respectively. During flexion, the camera measured the elbow joint angles which started from 48° and gradually increased to 122° as the peak. After reaching the peak, the angles gradually decreased to the initial angle. On the other hand, the Kinect measures with the same trend as the camera's but shows 10° and 135° for the minimum and maximum ROMs, respectively. This result indicates that the Kinect yields better absolute accuracies in joint angle measurements. For orientation, a

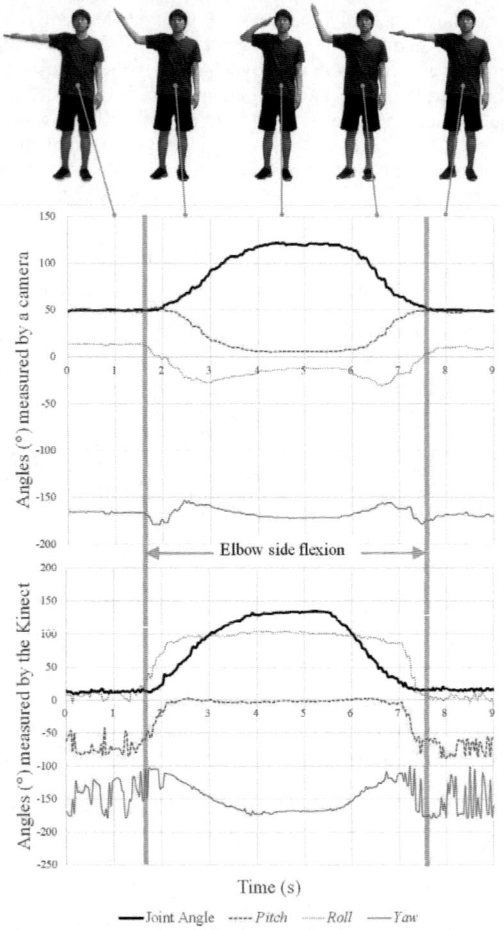

Figure 4. Right-hand's elbow joint and orientation angles of a subject while performing elbow side flexion.

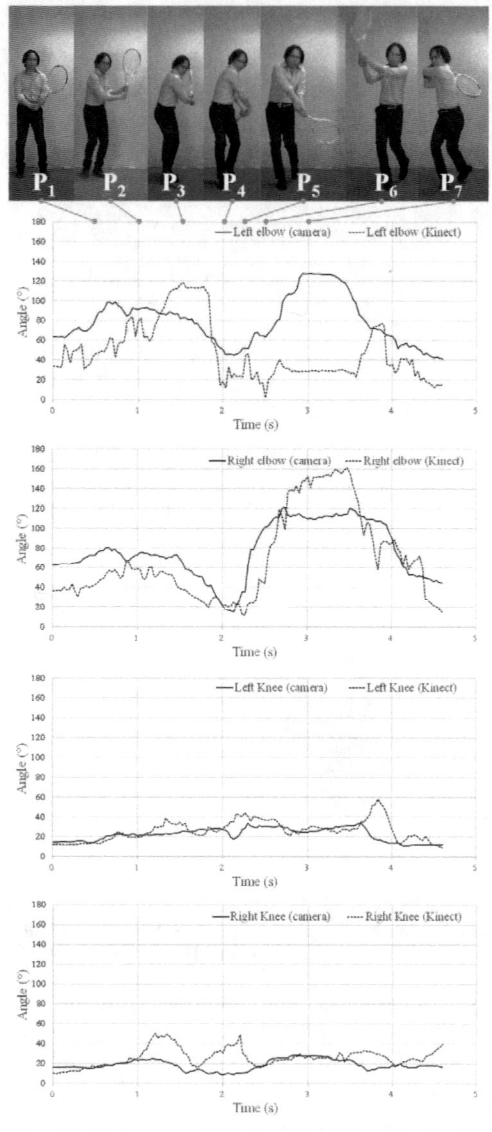

Figure 5. Measurements of elbow and knee joint angles of a subject while swinging a tennis racket.

great agreement was achieved for yaw angles. However, pitch and roll angles measured from the camera were inversely proportional to those from the Kinect. We considered that this problem was caused by the difference in the coordinate system representation that need to be adjusted for body joints between the proposed framework and the Kinect. As a known issue [4], noises were observed from the resulted measurements using the Kinect.

Elbow and knee joint angles of a subject while swinging a tennis racket, as shown in Figure 5, were measured by a single camera and the Kinect. Measurement scenes (P_1 - P_7) indicate occlusions of the subject's arms. Overall, the trend of the resulted joint angles from the camera agrees with those from the Kinect. However, there are gaps as occlusions occurred, such as in scenes from P_3 to P_7. The Kinect failed to measure the left-hand's elbow joint angles, which is highly occluded in these scenes. Table I shows 3D human pose estimated from the camera and the Kinect in P_4 and P_6 scenes. It is obvious that the camera estimated the 3D human pose better than the Kinect because the OpenPose used in this study works better on handling occlusion to determine the joint location. Hence, the joint angles measured from the camera is more reliable the those from the Kinect.

We measured elbow and knee joint angles of a subject on performing two tasks: running and running in place, measured by a single camera. During the measurements, the right arm and the right leg of the subject, as shown in Figure 6, were occluded from time to time. Ideally, fluctuations of left elbow

TABLE I. COMPARISON OF 3D HUMAN POSE ESTIMATION BY A SINGLE CAMERA AND BY THE KINECT

Scene	Posture	Camera	Kinect
P_4		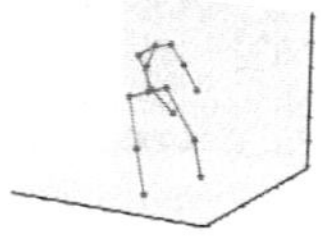	
P_6		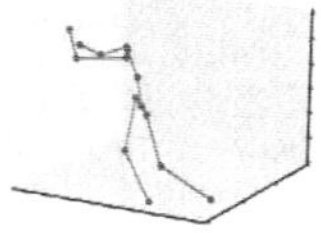	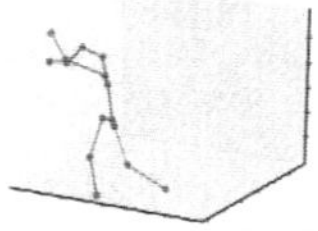

and left knee joint angles are identical with that of right elbow and right knee but half-cycle shifted in phase. On scenes where the entire right arm was mostly occluded by the body, the proposed framework estimated the posture of the right arm to be greatly extended behind the body. Thus, high values of the elbow joint angles were measured during these scenes. On the other hand, despite the right leg was sometimes partly occluded by the left leg, good results of knee joint angles were derived.

Finally, Figure 7 shows joint and orientation angles of elbow and knee of a subject entering a swimming pool before starting to swim. A waterproof camera was used to record the scene. The estimated 3D human pose was presented to show that pose estimation was not affected by the underwater light environment. Measurement of joint and orientation angles can be done in the same manner as on land. At this moment, detailed validation for the measurement results has not been carried out but we will report the analysis results in our next paper.

V. CONCLUSION AND FUTURE WORK

In this study, we have proposed a tele-rehabilitation framework where patients can send recorded videos during performing ADL tasks and get feedbacks to visualize their body joint and orientation angles to maintain or improve the rehabilitation stages. The proposed framework is more robust than the Kinect on handling occlusion; thus, it opens the possibility to measure body joint and orientation angles while doing ADL tasks where a part of limbs is hidden, such as dining and sleeping. The resulted measurements from the proposed framework are less sensitive to noises than those from the Kinect. Although the Kinect shows better results in term of absolute accuracies, the relative accuracy of the proposed framework against the Kinect is acceptable. The proposed framework can be extended to do the same measurements during aquatic therapy and fitness pools.

Future works include conducting experiments to measure body joint and orientation angles of a number of participants while doing ADL tasks where a part of limbs is hidden, and performing details analyses to determine validity of the proposed framework.

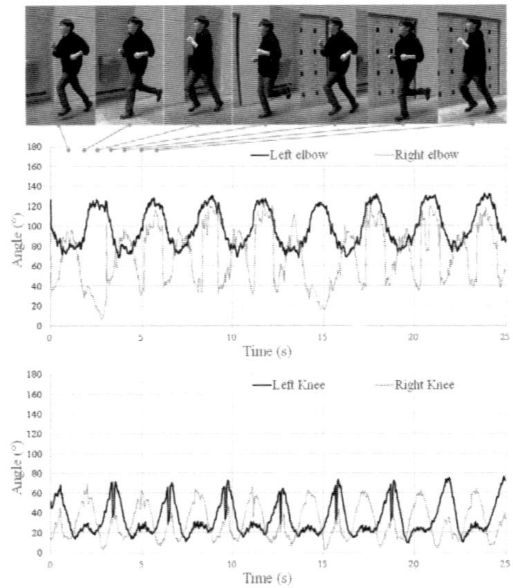

(a) Running

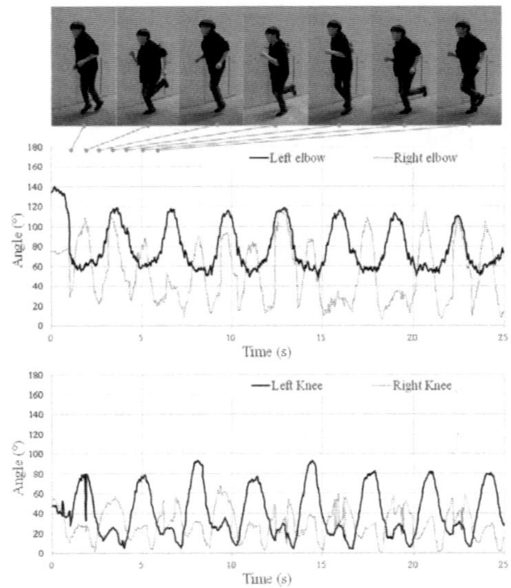

(b) Running in place

Figure 6. Measurements of elbow and knee joint angles of a subject on performing two tasks: running and running in place.

ACKNOWLEDGMENT

We acknowledge the effort from the authors of OpenPose and 3d-pose-baseline to make 3D joint measurements using a single camera possible. This work was supported by the MIC/SCOPE #181602007.

REFERENCES

[1] M. H. A Huis In't Veld, H. van Dijk, H. J. Hermens, and M. M. R. Vollenbroek-Hutten, "A systematic review of the

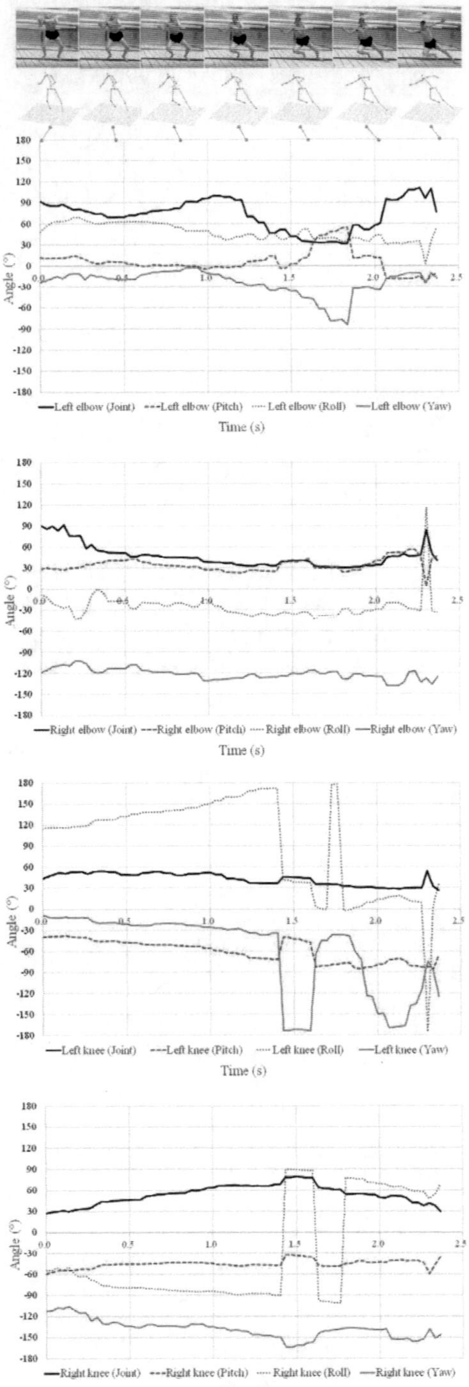

Figure 7. Measurements of elbow and knee joint and orientation angles of a subject entering a swimming pool before starting to swim.

methodology of telemedicine evaluation in patients with postural and movement disorders," Journal of Telemedicine and Telecare, vol. 12, no. 6, pp. 289-97, 2006.

[2] S. Aleesandro, C. Andrea, M. Matteo, and M. T. Lorenzo, "Kinect V2 Performance Assessment in Daily-Life Gestures: Cohort Study on Healthy Subjects for a Reference Database for Automated Instrumental Evaluations on Neurological Patients," Applied Bionics and Biometrics, pp. 1-16, 2017, https://doi.org/10.1155/2017/8567084.

[3] S. H. Lee et al. , "Measurement of Shoulder Range of Motion in Patients with Adhesive Capsulitis Using a Kinect," PLOS ONE vol. 10, no. 6, pp. 1-12, 2015, doi:10.1371/journal.pone.0129398.

[4] H. M. Hondori and M. Khademi, "A Review on Technical and Clinical Impact of Microsoft Kinect on Physical Therapy and Rehabilitation," Journal of Medical Engineering, vol. 2014, pp. 1-16, 2014, http://dx.doi.org/10.1155/2014/846514.

[5] M. E. Huber, A. L. Seitz, M. Leeser, and D. Sternad, "Validity and reliability of Kinect skeleton for measuring shoulder joint angles: a feasibility study," Physiotherapy, vol. 101, no. 4, pp. 389–393, 2015.

[6] M. Pedraza-Hueso, S. Martín-Calzóna, F. J. Díaz-Pernasa, and M. Martínez-Zarzuela, "Rehabilitation Using Kinect-based Games and Virtual Reality," Procedia Computer Science, vol. 75, pp. 161 – 168, 2015.

[7] Z. Cao, T. Simon, S.E. Wei, Y. and Sheikh, "Realtime Multi-person 2D Pose Estimation Using Part Affinity Fields," Computer Vision andPattern Recognition 2017, pp. 1302-1310, 2017.

[8] H. J. Lee and Z. Chen, "Determination of 3D Human Body Postures from a Single View," Computer Vision, Graphics and Image Processing, vol. 30, pp. 148–168, 1985.

[9] J. Martinez, R. Hossain, J. Romero, and J. J. Little, "A Simple Yet Effective Baseline for 3d Human Pose Estimation," arXiv preprint arXiv:1705.03098, pp. 1-10, 2017.

[10] DrGoniometer. http://www.drgoniometer.com/ [retrieved: January, 2019]

[11] K. Mitchell, S. B. Gutierrez, S. Sutton, S. Morton, and A. Morgenthaler, "Reliability and Validity of Goniometric iPhone Applications for the Assessment of Active Shoulder External Rotation," Physiotherapy Theory and Practice, vol. 30, no. 7, pp. 521–525, 2014.

[12] G. Ferriero, F. Sartorio, C. Foti, D. Primavera, E. Brigatti, and S. Vercelli, "Reliability of a New Application for Smartphones (DrGoniometer) for Elbow Angle Measurement.," PM&R, vol. 3, no. 12, pp. 1153–1154, 2011.

[13] L. Kneip, D. Scaramuzza, and R. Siegwart, "A Novel Parametrization of the Perspective-three-point Problem for a Direct Computation of Absolute Camera Position and Orientation," Proc. CVPR, pp. 2969–2976, 2011.

[14] OpenCV. https://opencv.org/ [retrieved: January, 2019]

[15] Dlib C++ Library. http://dlib.net/ [retrieved: January, 2019]

[16] C. Ionescu, D. Papava, V. Olaru, and C. Sminchisescu, "Human3.6M: Large Scale Datasets and Predictive Methods for 3D Human Sensing in Natural Environments," IEEE Transactions on Pattern Analysis and Machine Intelligence, vol. 36, no. 7, pp. 1325-1339, 2014.

[17] 3D-pose-baseline. https://github.com/una-dinosauria/3d-pose-baseline. [retrieved: January, 2019]

[18] S. Obdržálek, G. Kurillo, J. Han, T. Abresch, and R. Bajcsy, "Real-time human pose detection and tracking for tele-rehabilitation in virtual reality," Studies in Health Technology and Informatics, vol. 173, pp. 320-324, 2012.

[19] K. S. Arun, T. S. Huang, and S. D. Blostein, "Least-squares Fitting of Two 3-D Point Sets," IEEE Transactions on Pattern Analysis and Machine Intelligence, vol. 9, pp. 698–700, 1987.

Measurement of Shoulder and Trunk Movements in Hemiplegic Participants Using a System for Collecting Motion Data

Yukihide Nishimura, Hiroyuki Tsuboi,
Department of Rehabilitation Medicine,
Iwate Medical University
Morioka, Japan
e-mail: ynishi@iwate-med.ac.jp, hiroyuki.tsuboi@j.iwate-med.ac.jp,

Yoshitoshi Murata, Oky Dicky Ardiansyah Prima
Faculty of Software and Information Science,
Iwate Prefectural University
Takizawa, Japan
e-mail: {y-murata, prima} @iwate-pu.ac.jp

Abstract—**Upper limb and trunk functionalities are important for performing activities of daily living, such as eating, drinking, clothing and personal care. However, hemiplegic patients impair upper extremity functionalities, the activities are performed by using compensatory movement strategies. Recently, Takahashi and Murata et al. developed a system for collecting motion data with a cloud service. The purpose of this study was to investigated whether clinical application of this system for collecting motion data is available as assessment for hemiplegic patients. Two hemiplegic participants and one healthy participant performed drinking task, and the angles of shoulder abduction and trunk flexion were calculated. Angles of shoulder abduction and trunk flexion during drinking between non-paretic and healthy sides were similar, however, the larger angles of shoulder abduction and trunk flexion during drinking by paretic side showed compared with them by non-paretic and healthy sides. We believe that this system will be available as a quantitative assessment of activities of daily living in patients with disabilities.**

Keywords-rehabilitation treatment; activities of daily living; disability; functional recovery.

I. INTRODUCTION

Upper limb functionality is important for performing Activities of Daily Living (ADL). However, patients who have suffered neurological damage owing to stroke may have impaired upper extremity functionality (hemiplegia), then the activity must be performed by using compensatory movement strategies. Therefore, hemiplegic patients needed the rehabilitation treatments to improve upper limb and compensatory functionalities. Thus, to assess the movements of upper limb and trunk when performing ADL is important to evaluate impairment severity and help in the design of physical and occupational therapy interventions tailored to hemiplegic patients.

Recently, many researchers have tried to estimate physical activities from acceleration and/or gyro sensor data obtained from wearable devices and/or smart phones. In addition, Zhang et al. developed a cyber-physical system for patient-centric healthcare applications and services that was built on cloud and big data analytics technologies [1]. The system consisted of a data collection layer, a data management layer and an application service layer to collect and follow up on many kinds of big data. It used a security tag to maintain security. Doukas et al. also proposed a mobile system that enables electronic healthcare data storage, update and retrieval using cloud computing [2]. A mobile application was developed using Google's Android OS and Amazon's S3 to provide management of patient health records and medical images. Based on the above background, Takahashi et al. developed a system for collecting motion data using Google Firebase service [3]. The system collects and stores sensor and video data synchronously, and allows appropriate persons to access stored motion data. The number of subjects in previous studies for a quantitative assessment of movement, such as three-dimensional (3-D) motion analysis is relatively small [4][5]. However, we believe that the system developed by Takahashi et al. can be expected collecting big data. The collecting big data may allow a better understanding of limbs and trunk functionality for performing ADL and help in the design of physical and occupational therapy interventions tailored to hemiplegic patients.

The purpose of this study was to investigate whether clinical application of a system for collecting motion data is available as assessment on rehabilitation treatment for hemiplegic patients.

In Section II, the method in this study is explained. The results are outlined and discussed in Section III. Conclusion and future work are described in Section IV.

II. METHODS

Two hemiplegic adult participants and one healthy adult participant were participated in this study. The upper limb functionalities in hemiplegic participants were assessed with Fugl–Meyer Assessment of Upper Extremity (FMA-UE). The arm impairment can be classified based on FMA-UE scores between 39 and 57 (moderate impairment) and 58 to 64 (mild impairment) in accordance with previous report [4]. Hemiplegic participant A had moderate arm impairment (FMA-UE scores: 43) and hemiplegic participant B had mild impairment (FMA-UE scores: 59). All participants could drink a cup of water without assistance. Participants performed drinking task three times on both sides, respectively. The angles of shoulder abduction and trunk flexion during second task were calculated. The task was performed in the following steps: 1) reaching out for the cup, 2) grasping and transporting the cup forward to the mouth, 3) drinking, 4) transporting the cup backward to the pickup point, and 5) returning the hand to the initial position [5]. Shoulder and Trunk movements were measured with a 3-D-gyro sensor. We used two SONY Smart Watch 3 units as the 3-D gyro sensor [6]. 3-D gyro sensors were placed on the L1 spinous process and upper arm (see Figure 1).

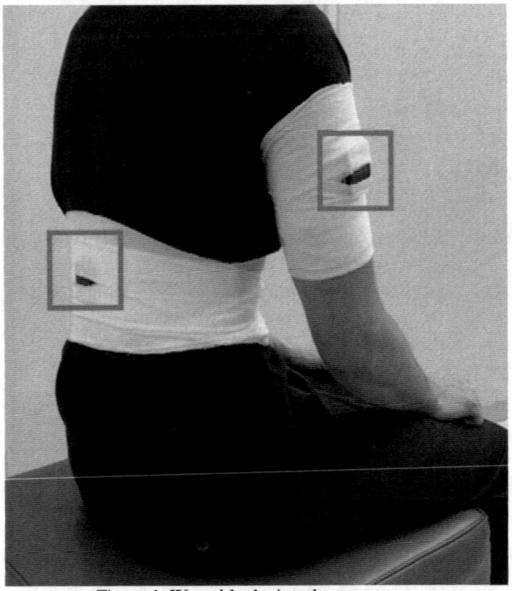

Figure 1. Wearable device placements.

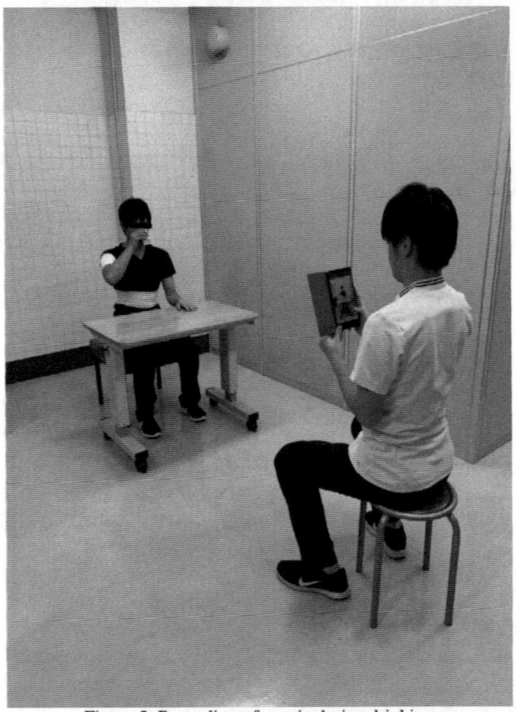

Figure 2. Recording of movie during drinking

The movies during drinking were recorded by the tablet-type device that accepted to 3-D gyro sensors (Figure 2). The data were expressed in relation to the percentage of the drinking task cycle that had lapsed (0–100% of the drinking task cycle).

III. RESULTS AND DISCUSSION

Figure 3 showed the data of changes of angle on shoulder abduction (Figure 3A) and trunk flexion (Figure 3B) during drinking. Time of drinking by paretic side was longer than by non-paretic side in each hemiplegic patient.

Figure 4 showed angle of shoulder abduction (Figure 4A) and trunk flexion (Figure 4B) during drinking on normalized times. Angles of shoulder abduction and trunk flexion during drinking between non-paretic and healthy sides were similar, however, the larger angles of shoulder abduction and trunk flexion during drinking by paretic side showed compared with them by non-paretic and healthy sides.

In this study, we measured shoulder and trunk movements in two hemiplegic participants using a system for collecting motion data with a cloud service, and found that the larger angles of shoulder abduction and trunk flexion during drinking by paretic side compared with non-paretic and healthy sides. In general, 3-D imaging measurement technique is powerful tool for a quantitative assessment of movement in all degrees of freedom, previous studies have reported that movement when drinking for hemiplegic patients using 3-D motion analysis. In fact, some studies revealed that larger shoulder abduction in hemiplegic patients compared with healthy individuals [4][5]. In addition, we insist that angle of shoulder on normalized time during drinking (Figure 4A) in this study was similar to that using 3-D motion analysis [5]. Also, trunk movement has been reported to have a significant correlation with stroke impairment severity [4][7]. This was confirmed in this study, i.e., the difference of trunk flexion between paretic and nonparetic sides in hemiplegic participant with moderate impairment level was larger at 20% during drinking than that in hemiplegic participant mild impairment level (Figure 4B). Alt Murphy et al. [4] purposed that compensatory trunk movements should be considered as an essential part of movement patterns after stroke and included in evaluation models.

Finally, we suggest that using of a system for collecting motion data with a cloud service reflect more real physical activities and/or functionalities.

IV. CONCLUSION AND FUTURE WORK

We believe that a system for collecting motion data with a cloud service will be available as a quantitative assessment of ADL in patients with disabilities. The future work must assess of the reliability of data obtained from this system. In addition, to be more developed this system need to investigate many patients who have suffered neurological damage owing to stroke and cervical spinal cord injuries.

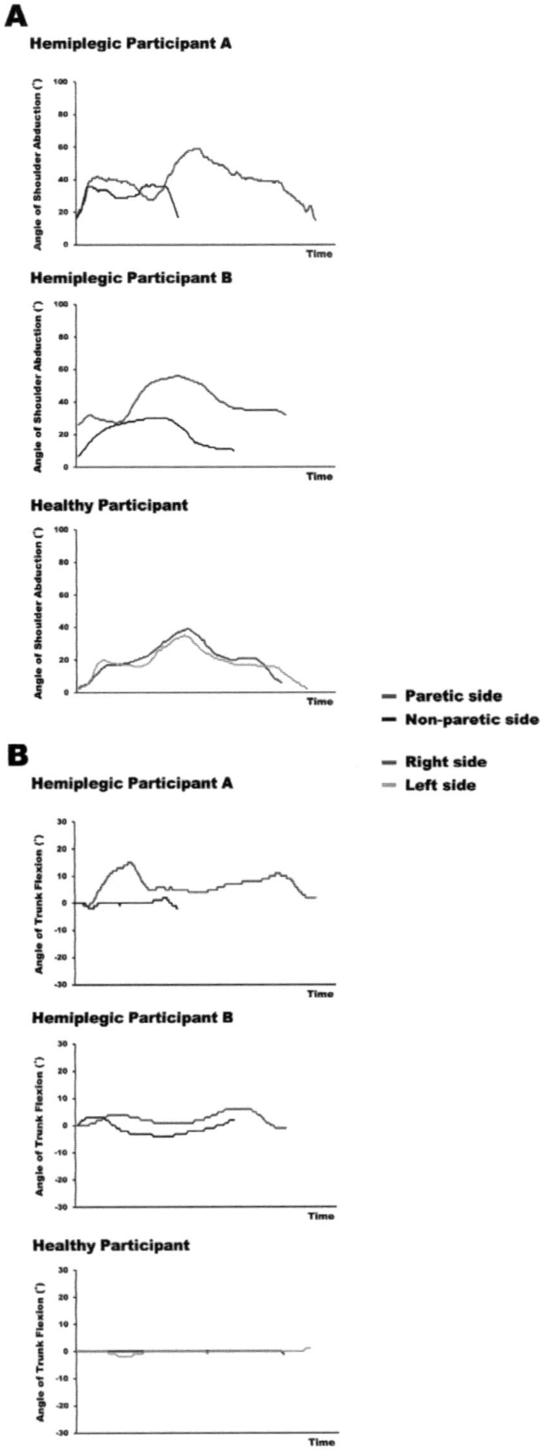

Figure 3. Raw data of shoulder (A) and trunk (B) in two hemiplegic participants and one healthy participant obtained from wearable devices.

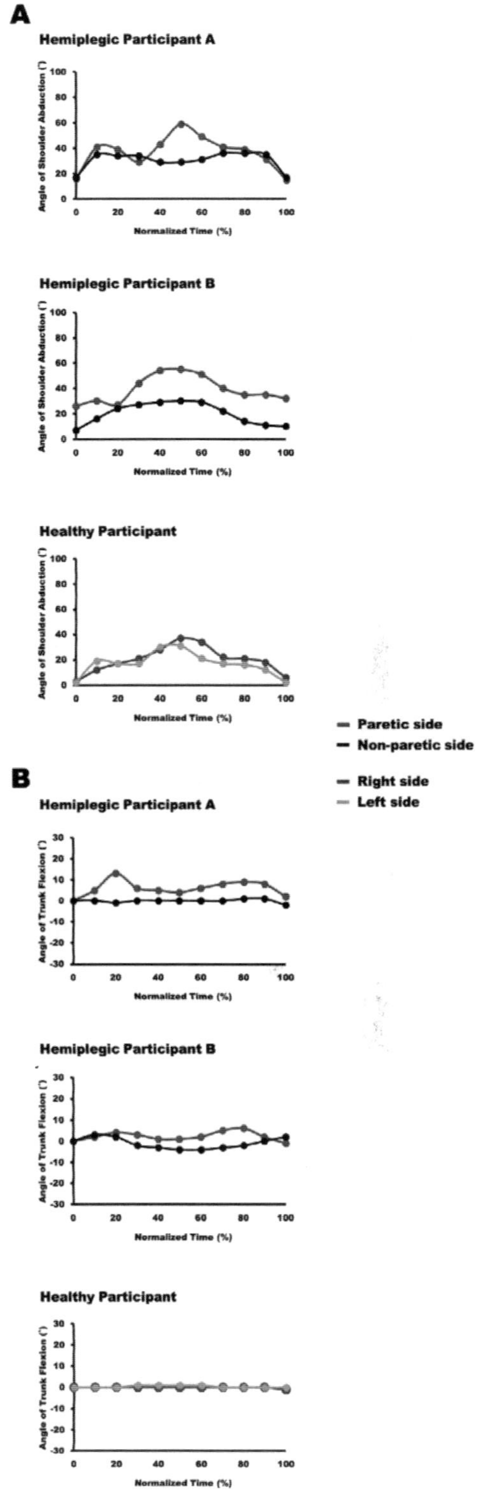

Figure 4. angles of shoulder (A) and trunk (B) on normalized time.

ACKNOWLEDGMENT

We thank Shunpei Okimura, Fumiaki Yamaguchi, Ryo Sato, Katsunori Takahashi, Syuusuke Takahashi, Shota Idobata and Minami Suzuki for the technical assistance. The research and development were supported by the MIC/SCOPE #181602007.

REFERENCES

[1] Y. Zhang, M. Qiu, C. W. Tsai, M. M. Hassan, and A. Alamri, "Health-CPS: Healthcare cyber-physical system assisted by cloud and big data", IEEE, SystemsJournal, vol. 11, no. 1, pp. 88-95, 2017.

[2] C. Doukas, T. Pliakas, and I. Maglogiannis, "Mobile healthcare information management utilizing cloud computing and android OS", IEEE, 2010 Annual International Conference of the IEEE Engineering in Medicine and Biology, pp. 1037-1040, 2010.

[3] R. Takahashi, Y. Murata, and O. D. A. Primal, "A system for collecting motion data on patients' activities of daily living", unpublished.

[4] M. Alt Murphy, C. Willén, and K. S. Sunnerhagen, "Kinematic variables quantifying upper-extremity performance after stroke during reaching and drinking from a glass", Neurorehabilitation and Neural Repair, vol. 25, no. 1, pp. 71-80, 2011.

[5] K. Kim, et al., "Kinematic analysis of upper extremity movement during drinking in hemiplegic subjects", Clinical Biomechanics, vol. 29, no. 3, pp. 248-256, 2014.

[6] SONY Smart Watch 3, https://www.sonymobile.com/global-en/products/smart-products/smartwatch-3-swr50/#gref, [retrieved: January, 2019].

[7] G. Thielman, T. Kaminski, and A. M. Gentile, "Rehabilitation of reaching after stroke: comparing 2 training protocols utilizing trunk restraint", Neurorehabilitation and Neural Repair, vol. 22, no. 6,pp. 697-705, 2008.

Assessment of Joint Range of Motion Measured by a Stereo Camera

Yuta Ono, Oky Dicky Ardiansyah Prima, Takashi Imabuchi,
Yoshitoshi Murata, Hisayoshi Ito
Graduate School of Software and Information Science
Iwate Prefectural University
Takizawa, Japan
e-mail: g231q005@s.iwate-pu.ac.jp, prima@iwate-pu.ac.jp,
g236o001@s.iwate-pu.ac.jp, {y-murata, hito}@iwate-pu.ac.jp

Yukihide Nishimura
Department of Rehabilitation Medicine
Iwate Medical University
Morioka, Japan
e-mail: ynishi@iwate-med.ac.jp

***Abstract*—Many studies have been conducted to measure joint Range of Motion (ROM) to approach Activities of Daily Living (ADL) assessments using nonintrusive three-dimensional (3D) sensors, such as motion capture devices and the Kinect, as alternatives to goniometer. While these sensors have been widely used to measure joint angles, the measuring range is restrictive and the complexity to setup has prevented these devices to be used as self-measurements during home-based training. With the recent progress on human pose estimation using computer vision approach, locations of joints can be detected from a vision camera in real time. This achievement opens a possibility to measure ROMs in a wide area and in any locations as long as the subject is located inside the camera's Field of View (FOV). This study extends the current human pose estimation to capture joints in 3D using a stereo camera and compares the ROMs of shoulder and elbow derived from the proposed method with those derived from the Kinect. Measurements of ROM using both devices show good agreement indicating that the proposed method is valid to measure ROMs, as a replacement for the Kinect.**

Keywords-rehabilitation; computer vision; range of motion; activities of daily living; 3D human pose estimation.

I. INTRODUCTION

Many studies have reported the close relationship between joint Range of Motion (ROM) and Activities of Daily Living (ADL) [1][2]. The loss of ROM may occur at all ages due to injuries, diseases, surgery and normal aging, giving a direct effect on posture and movement. For those who have impaired ROM, the activity needs to be performed by using compensatory strategies [3]. ROM is commonly evaluated as the degree of maximum range of motion. However, the individual difference is big, such as a patient with impaired shoulder flexion motion may not be able to raise his upper limb but may still be able to conduct most ADL tasks.

Traditional methods to measure ROM use apparatus, such as goniometers, inclinometers, and video in several specified directions. To obtain reliable measurements, clinicians are suggested to take repeated ROM measurements. Since the universal goniometer has scale in 5° increments, the measurement fluctuation is usually expected up to ±5°. The use of motion capture devices to measure angles has increased the reliabilities of ROM measurements. With these devices, ROM can be measured while the ADL tasks were performed by a subject.

Some portable and low-cost devices equipped with depth sensors, such Microsoft Kinect and Intel RealSense can be used to track human motions and capture human postures in 3D. Previous studies have shown that these devices show performance adequate for a range of healthcare imaging applications [4][5]. Unlike motion capture systems, ROM measurements using these depth sensors suffer from occlusion problems. Therefore, these sensors have to be positioned in an appropriate location and direction to avoid these problems. However, the tracking accuracy depends on having a perfect 3D model of the subject which requires pose estimation using multiple cameras [6].

The state-of-the-art computer vision techniques have enabled the detection of 2D human limb joints using a single camera. These techniques utilize fine-tuned convolutional network architectures. The "Stacked Hourglass Network (SHN)" was known to have a robust performance to a variety of challenges on joint detections for multiple people [7]. The most recent technique, "OpenPose" uses part affinity fields for the fast detection of multiple people [8]. SHN and OpenPose are available online as open source software for research purposes. Using dual cameras mounted side-by-side (a stereo camera), 3D human limb joints can be calculated by triangulating the corresponding 2D joints detected in each side camera [9]. The fast growing in Virtual Reality (VR) has made stereo cameras available on the markets. Some smartphones have been equipped by stereo cameras to produce perspective effects to the photos.

This study attempts to expand the application of stereo vision as an easy and low-cost tool to measure ROM on performing ADL and to promote basic self-care. Two kinds of ROM are measured using 3D joints detected from a stereo camera and the Kinect, respectively. The measurements of ROM were performed according to the methods and guidelines for the measurement of joint range of motion by the Japanese Association of Rehabilitation Medicine and the Japanese Orthopaedic Association (JARM & JOA) [10]. ROM from each source is analyzed to reveal whether stereo vision is adequate for practical use on quantifying ROM. This study provides a basic framework to build a ROM measurement system. The framework minimalizes the hardware requirement since the measurement of 3D joint is performed at a server.

This paper is organized as follows. Section II describes the related work on human joint detections using the Kinect and

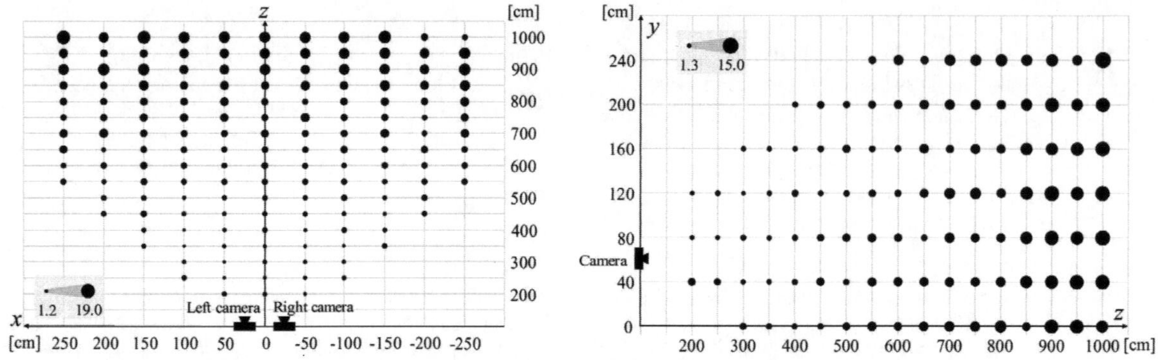

(a) Horizontal errors (X-Z plane)

(b) Vertical errors (Y-Z plane)

Figure 1. Distribution of errors measured at 1,004 control points used in this study.

cameras. Section III describes methods to measure ROM using a stereo camera and the Kinect based on guidelines of JARM and JOA. Section IV shows measurement results of the accuracy of ROM derived from a stereo camera approach. Finally, Section V concludes the achievements and discusses the future prospective of this study.

II. Related Work

The task of estimating the posture of the human body without using markers has been attracting attention in the research area of computer vision. The estimation methods can be broadly classified as either methods using depth or visible cameras.

A low-cost depth camera, the second version of Microsoft Kinect, the Kinect V2, has been used to measure ROM. Since its predecessor, the Kinect has been applied to the rehabilitation field. The Kinect V2 captures depth images between the range of 0.5~8.0m in 30 frame per second of speed. The Software Development Kit (SDK) enables developers to detect 25 joints of up to 6 persons concurrently. Hereinafter, we simply referred both versions as the Kinect. The skeletal joints of the Kinect can be considered as an adequate tool for supporting rehabilitation. However, using these data in clinical applications where precise angle measurements are required needs a significant concern [11].

Studies have been conducted to enable the detection of various joints in complex postures using a camera. Toshev et al. (2014) [12] detects 2D joints using a regression model with a cascade Deep Neural Network (DNN) and associates corresponding joints across the body posture. Newell et al. (2016) [7] proposed SHN to improve the detection by handling a diverse and challenging set of poses with a simple mechanism for reevaluation and assessment of initial predictions. Both [7] and [12] require a person detection process as a preprocessing before detecting body joints, causing detection of joints cannot be done if the preprocessing failed to detect a person. Cao et al. (2017) [8] proposed the OpenPose that creates "Part Confidence Maps (PCM)" to detect joints and "Part Affinity Fields" to associate corresponding joints directly without detecting a person beforehand. OpenPose is capable to detect every joint of the human body in real time.

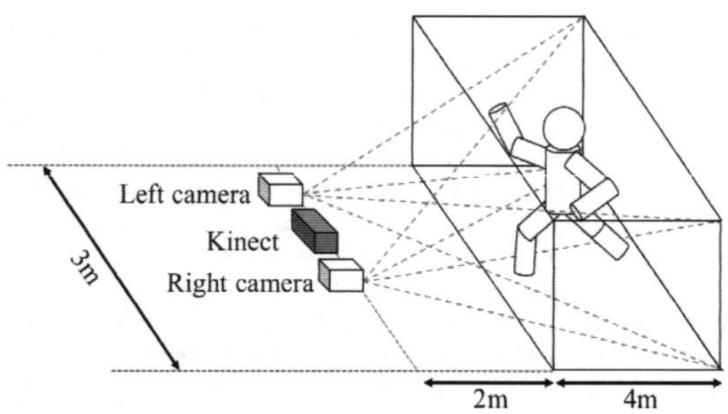

Figure 2. The experimental setup in this study.

Ohno et al. (2018) [9] applied OpenPose to a stereo camera and constructed 3D joints from 2D joints detected from each camera image using a stereo vision approach. Triangulation is used to find the optimal 3D joint by refining the location of joints from each image. The refining process uses the value of PCM to select a more reliable joint between two images and determine the corresponding joint from the counterpart image using template matching.

3D joint measurements based on stereo vision seem to be more promising on measuring ROM. It will be possible for patients to create self-reported ROM at home as long as dual cameras are available, and to send the report to clinicians to assess the ability to engage in ADL tasks.

III. Methods

This study evaluates two kinds of ROM measured by a stereo camera and the Kinect, respectively. Stereo camera and the Kinect are connected to a single computer to capture

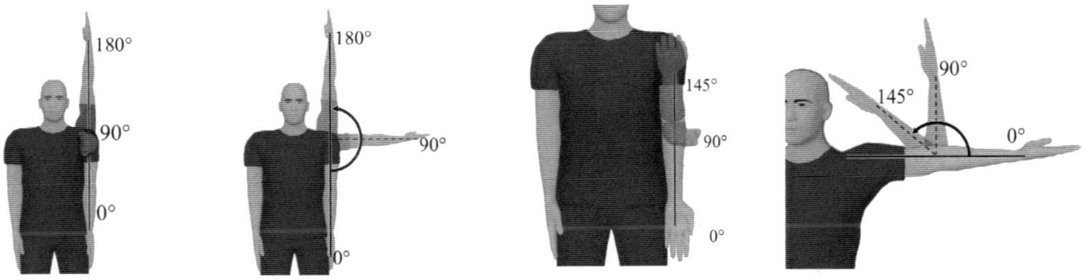

(a) Shoulder (Forward flexion) (b) Shoulder (Abduction) (c) Elbow (Frontal flexion) (d) Elbow (Side flexion)

Figure 3. Body postures measured in this study.

TABLE I. JOINTS MEASURED IN THIS STUDY

Joint	Pose	Incremental Measurement	Max. ROM
Shoulder	Forward flexion	30°, 60°, 90°, 120°, 150°	180°
	Abduction	30°, 60°, 90°, 120°, 150°	180°
Elbow	Frontal flexion	30°, 60°, 90°, 120°	145°
	Side flexion	30°, 60°, 90°, 120°	145°

the movement of the subject concurrently. The capturing speed is set to 30fps.

A. 3D Joint Measurement Using a Stereo Camera

The method of Ohno et al. (2018) [9] is adapted to measure 3D joints. Two cameras were calibrated and placed at intervals of 60cm. Each camera has a resolution of 1280px×720px. Errors were measured at 1,004 control points regularly placed in an area of 5m×10m. Here, Figure 1 shows distribution of errors on each control points. For all points, Root Mean Square Error (RMSE) was measured 8.12cm. For points located up to 6m against the cameras, the RMSE was measured 5.07cm. To get optimal results, subjects will be positioned between 2m and 6m from the center line between two cameras. This distance is considerably sufficient to measure ROM while engaging in ADL tasks.

B. 3D Joint Measurement Using Kinect

The Kinect is mounted on the middle between the stereo camera described above. Skeleton tracking function provided by the Kinect for Windows SDK 2.0 is used to measure 3D joints. No calibration is made on the resulted 3D joints. Here, Figure 2 shows the experimental setup for this study.

C. Data Extraction

With the stereo camera and the Kinect in frontal view, ROM for shoulder and elbow are measured to perform some ADL tasks. The measurements of shoulder and elbow, as shown in Figure 3, are conducted for subjects standing with upright postures for ROM measurements in this study. Values of angles between minimum and maximum angles on performing each posture are measured to be used as necessary data for the ADL [2]. Table I shows joints measured in this study whereas the maximum ROM for each joint movement was determined as in [11]. Here, we perform two types of measurements: incremental and continuous. The incremental measurement measures the absolute accuracy for a given posture. On the other hand, the continuous measurement will reveal how stable the values of angles are measured. Two healthy subjects (A and B) were participated in this study. Both subjects were required to wear ordinary clothes and to stand at 4m from the Kinect.

1) Incremental measurement: Subjects were asked to take each posture as shown in Figure 3, where angles for each pose is assigned as in Table I and set using a goniometer. Goniometric measurements were performed using standardized methods [13]. After the goniometer was aligned to the shoulder motion by the examiner, a second examiner read and recorded the measurement. Once the pose was set, measurements were recorded simultaneously using the Kinect and the stereo camera. For the maximum ROM, subjects were instructed to move each joint to their maximum capability. These procedures were repeated three times for both left and right joints. The agreement of two measurements against a goniometer were assessed by studying the mean bias and constructing Limits of Agreement (LOA) to determine validity [10][14]. Here, the 95% LOA were defined as the mean bias to ±1.96 Standard Deviation (SD). Since the goniometer has scale in 5° increments, the 95% LOA for the discrepancy exceeded ±5° can be defined as clinically significant.

2) Continuous measurement: During continuous measurement, subjects were asked to stand upright and slowly perform shoulder abduction and elbow frontal flexion as shown in Figure 3(b)-(c) and move back to the upright pose. During this movement, measurements were recorded using the Kinect and the stereo camera, concurrently. The resulted measurements were fitted separately using fourth-order polynomial regression models to investigate the stability of each measurement by each device. RMSE and R-squared (R^2) were calculated to evaluate the model. RMSE indicates the

TABLE II. MEAN BIAS AND LOA MEASUREMENTS OF SHOULDER AND ELBOW JOINT ANGLES OBTAINED USING THE KINECT AND THE STEREO CAMERA AGAINST A GONIOMETER

Subject	Joint	Pose	Kinect vs goniometer		Stereo camera vs goniometer	
			Mean bias	95% LOA	Mean bias	95% LOA
A	Shoulder	Forward flexion	3.39°	-16.62° to 23.39°	-1.65°	-26.74° to 23.44°
		Abduction	-2.82°	-32.41° to 26.78°	-3.90°	-17.51° to 9.71°
	Elbow	Frontal flexion	10.64°	-4.11° to 25.40°	9.91°	-14.29° to 34.11°
		Side flexion	-20.01°	-58.55° to 18.53°	-3.51°	-18.96° to 11.94°
B	Shoulder	Forward flexion	-10.95°	-75.50° to 53.59°	-0.22°	-20.80° to 20.35°
		Abduction	-4.96°	-17.93° to 8.01°	-6.29°	-20.88° to 8.31°
	Elbow	Frontal flexion	1.97°	-23.05° to 27.00°	1.27°	-26.28° to 28.82°
		Side flexion	-17.85°	-31.35° to -4.16°	-4.74°	-20.13° to 10.64°

TABLE III. MODEL FITTING AND ERROR ESTIMATION

Subject	Joint	Pose	RMSE (R^2)	
			Kinect	Stereo camera
A	Left shoulder	Abduction	8.86° (0.9704)	8.59° (0.9708)
	Right shoulder		7.14° (0.9084)	8.0° (0.9735)
	Left elbow	Frontal flexion	7.13° (0.9801)	9.06° (0.9672)
	Right elbow		6.85° (0.9784)	6.66° (0.9767)
B	Left shoulder	Abduction	2.78° (0.9971)	3.13° (0.9960)
	Right shoulder		3.81° (0.9947)	5.01° (0.9913)
	Left elbow	Frontal flexion	8.54° (0.9759)	4.79° (0.9886)
	Right elbow		5.13° (0.9917)	4.66° (0.9938)

absolute fit of the model to the data whereas R^2 is a relative measure of fit.

IV. RESULTS

1) Incremental measurement: Joints with poses shown in Table I were measured for left and right shoulders and elbows from each subject. Table II shows the mean bias and the 95% LOA of measurements of the Kinect and the stereo camera against those of the goniometer. The analysis results derived from each user are presented because the difference of the clothes is considerably affecting the measurement results. Subject A and B, as shown in Figure 4, wore short and long-sleeved shirts, respectively. The mean bias indicates that the stereo camera had relatively better accuracies than the Kinect. Especially, the measurement of poses from frontal view (forward flexion of shoulder and frontal flexion of elbow) where those joints were partly occluded from the devices. However, the 95% LOA for the discrepancy of both devices against the goniometer exceeded ±5°, which was defined as clinically significant. For the Kinect, this finding is consistent with [10].

2) Continuous measurement: Continuous movement during shoulder abduction and elbow frontal flexion were measured from each subject. Subjects were requested to perform these tasks within 20 seconds. No start or stop signs were given to the subjects because the necessary data can be extracted manually from the recorded scenes. Table III shows RMSE and R^2 for each model fitted to the measurement results from the Kinect and the stereo camera. RMSE values from data derived by the stereo camera and the Kinect are comparable. High values of R^2 were achieved from both model fitting. Here, Figure 5 and 6 visualize the measurement data and models fitted to the data.

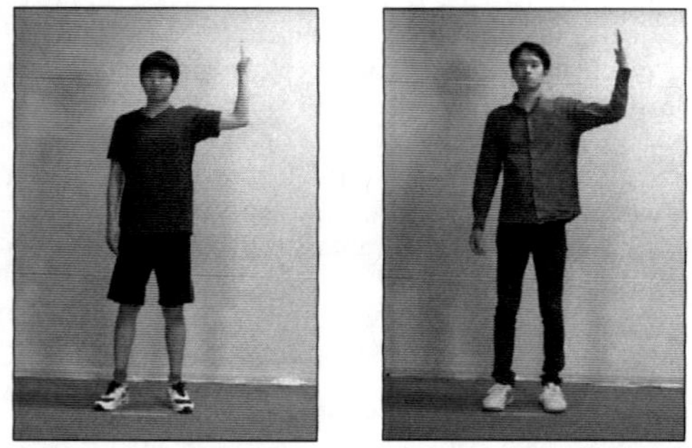

(a) Subject A (b) Subject B

Figure 4. The looks of subjects in this study.

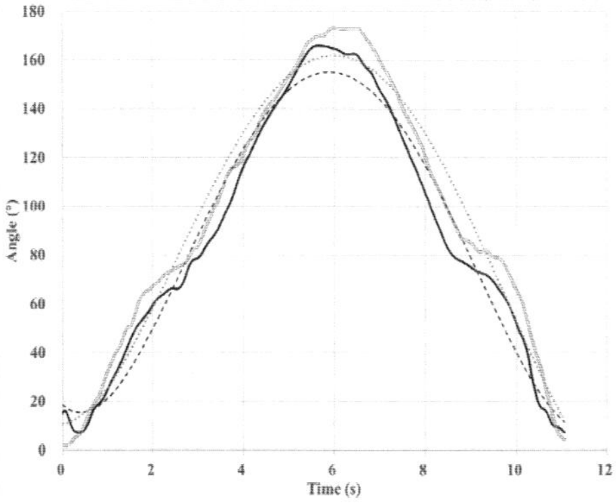

(a) Left shoulder (abduction)

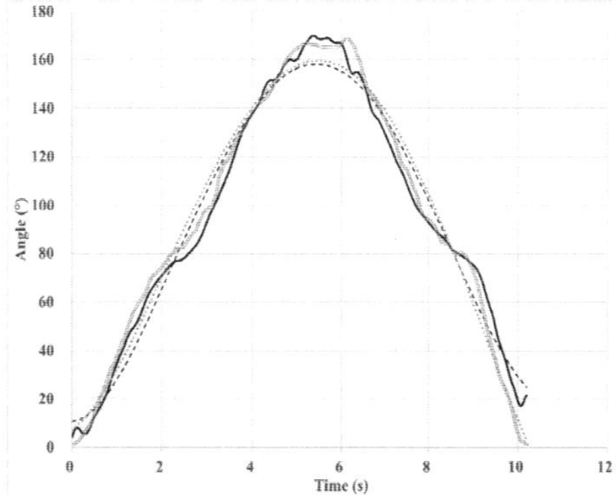

(b) Right shoulder (abduction)

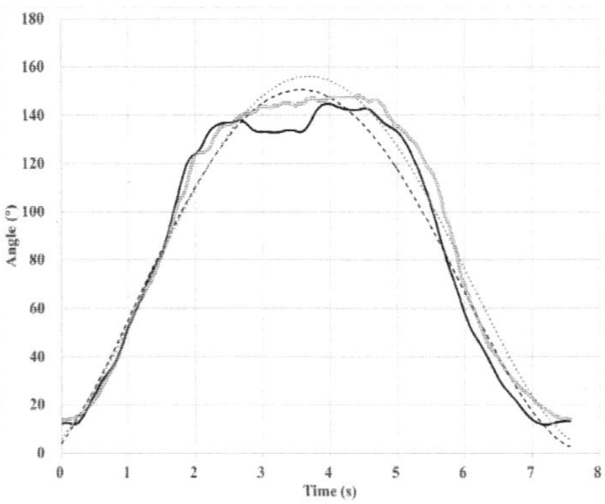

(c) Left elbow (frontal flexion)

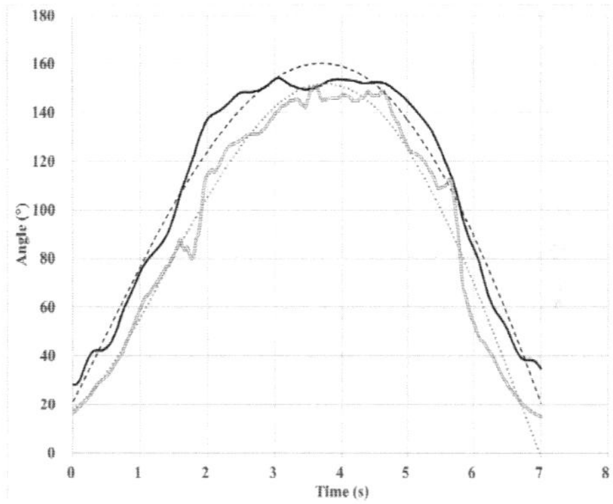

(d) Right elbow (frontal flexion)

—Stereo camera ---Stereo camera (fitted curve) —Kinect ·····Kinect (fitted curve)

Figure 5. Continuous measurement for abduction and frontal flexion (subject A).

V. CONCLUSION AND FUTURE WORK

In this study, we measured 3D joints remotely using a stereo camera and the Kinect. Our experiments show that the stereo camera had relatively better accuracies than the Kinect on measuring a pose at a given angle. On the other hand, the stereo camera was observed to be as stable as the Kinect on measuring joint angles continuously. However, before the stereo camera, as well as the Kinect can be used to measure ROM, it is important to understand their limitations in accuracy for the measurement of specific joint motions against a goniometer.

The stereo camera used in this study is superior to the Kinect because it was based on OpenPose that doesn't require the detection of full body posture to detect particular joints. Whereas, the Kinect may fail to detect joints when other part of the body is occluded. The stereo camera will enable us to observe various ADL tasks, such as dressing, eating, and bathing where a part of the body may be hidden easily.

ACKNOWLEDGMENT

We acknowledge the effort from the authors of OpenPose to make ROM measurements using a stereo camera possible. This work was supported by the MIC/SCOPE #181602007.

REFERENCES

[1] A. M. Oosterwijk, M. K. Nieuwenhuis, C. P. van der Schans, and L. J. Mouton, "Shoulder and elbow range of motion for the performance of activities of daily living: A systematic review," Physiotherapy Theory and Practice, vol. 34, no. 7, pp. 505-528, 2018.

[2] A. M. Oosterwijk, M. K. Nieuwenhuis, H. J. Schouten, C. P. van der Schans, and L. J. Mouton, "Rating scales for shoulder and elbow range of motion impairment: Call for a functional approach," PLOS ONE, vol. 13, no. 8, pp. 1-13, 2018, https://doi.org/10.1371/journal.pone.0200710.

[3] B. P. Pereira, A. Thambyah A, and T. Lee, "Limited forearm motion compensated by thoracohumeral kinematics when performing tasks requiring pronation and supination," Journal of Applied Biomechanics, vol.28, pp. 127–138, 2012.

[4] S. Aleesandro, C. Andrea, M. Matteo, and M. T. Lorenzo, "Kinect V2 Performance Assessment in Daily-Life Gestures: Cohort Study on Healthy Subjects for a Reference Database for Automated Instrumental Evaluations on Neurological Patients," Applied Bionics and Biometrics, pp. 1-16, 2018, https://doi.org/10.1155/2017/8567084.

[5] S. H. Lee et al. , "Measurement of Shoulder Range of Motion in Patients with Adhesive Capsulitis Using a Kinect," PLOS ONE vol. 10, no. 6, pp. 1-12, 2015, doi:10.1371/journal.pone.0129398.

[6] R. R. P. Kumar, S. Muknahallipatna, J. McInroy, M. McKenna, and L. Franc, "Real-time Range of Motion Measurement of Physical Therapeutic Exercise," Journal of Computer and Communications, vol. 5, pp. 19-42, 2017.

[7] A. Newell, K. Yang, and J. Deng, "Stacked Hourglass Networks for Human Pose Estimation," Computer Vision – ECCV 2016, pp. 483–499, Amsterdam, Netherlands, 2016.

[8] Z. Cao, T. Simon, S.E. Wei, and Y. Sheikh, "Realtime Multi-Person 2D Pose Estimation Using Part Affinity Fields," Computer Vision and Pattern Recognition 2017, pp. 7291-7299, 2017.

[9] Y. Ohno, O. D. A. Prima, and H. Ito, "3D Human Pose Estimation for Motion Analysis," The 80th Nation Convention of Information Processing Society of Japan, pp. 263-264, 2018. (in Japanese)

[10] M. E. Huber, A. L. Seitz, M. Leeser, and D. Sternad, "Validity and Reliability of Kinect Skeleton for Measuring Shoulder Joint Angles: a Feasibility Study," Physiotherapy, vol. 101, no. 4, pp. 389–393, 2015.

[11] K. Yonemoto, S. Ishigami, and T. Kondo, "The Method Guidelines for Range of Motion Measurement," The Japanese Journal of Rehabilitation Medicine, vol. 32, no. 4, pp. 207–217, 1995. (in Japanese)

[12] A. Toshev and C. Szegedy, "DeepPose: Human Pose Estimation via Deep Neural Networks," Computer Vision and Pattern Recognition (CVPR), 2014 IEEE Conference, pp. 1653–1660, 2014.

[13] N. B. Jain, R. B. Wilcox, J. N. Katz, and L. D. Higgins, "Clinical Examination of the Rotator Cuff," Physical Medicine and Rehabilitation, vol. 5, pp. 45–56, 2013.

[14] J. M. Bland, and D. G. Altman, "Statistical Method for Assessing Agreement between Two Methods of Clinical Measurement," Lancet, vol.327, pp. 307-310, http://dx.doi.org/10.1016/S0140-6736(86)90837-8, 1986.

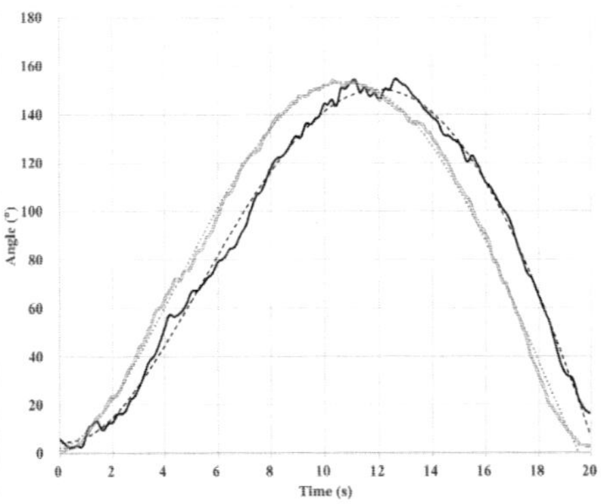

(a) Left shoulder (abduction)

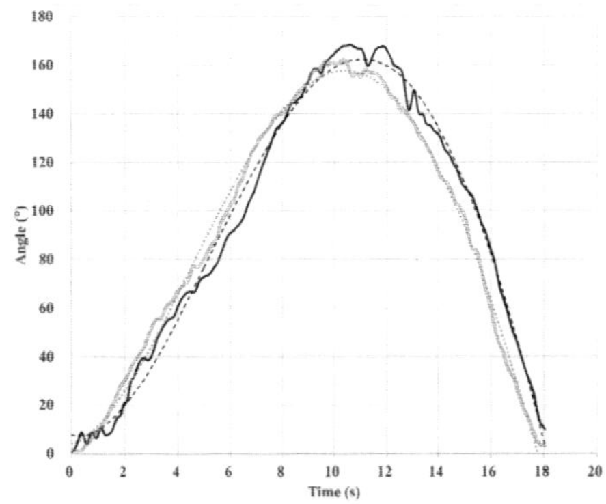

(b) Right shoulder (abduction)

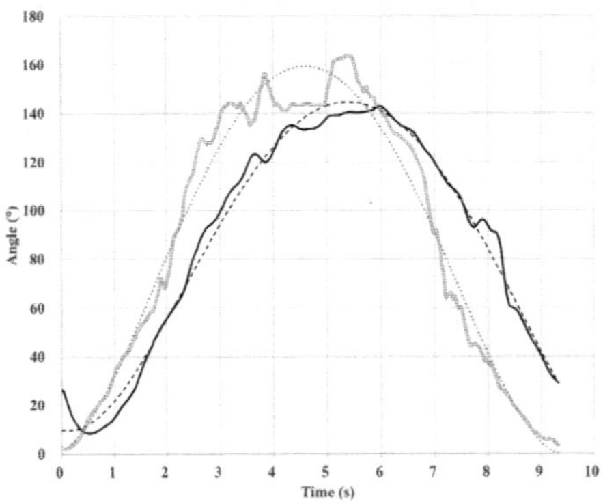

(c) Left elbow (frontal flexion).

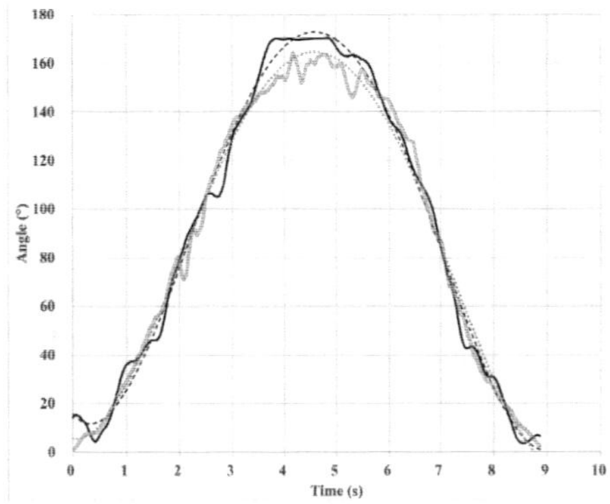

(d) Right elbow (frontal flexion)

—Stereo camera - - -Stereo camera (fitted curve) —Kinect ····Kinect (fitted curve)

Figure 6. Continuous measurement for abduction and frontal flexion (subject B).

Taking the Difference Between Leisure Time and Workdays Into Account to Improve Virtual Coaching

Talko Dijkhuis[1,2], Johan Blok[1], Hugo Velthuijsen[1], Koen Lemmink [2]
[1]Hanze University of Applied Sciences, Institute for Communication, Media & IT, Groningen, The Netherlands
[2]University of Groningen, Center for Human Movement Sciences, University Medical Center Groningen, Groningen, The Netherlands
[1]{t.b.dijkhuis, j.blok, h.velthuijsen}@pl.hanze.nl
[2]{t.b.dijkhuis, k.a.p.m.lemmink}@umcg.nl

Abstract—**A sedentary lifestyle is a cause for many health problems. To motivate employees to a healthier lifestyle and to physical activity, the Hanze University initiated a health promotion program, including a fortnightly coaching on lifestyle and activity. An activity tracker was used to monitor the participants' steps. However the used activity tracker doesn't provide timely personalized coaching. In this paper, we investigate the possibility of enhancing timely personalized virtual coaching. Therefore we investigated the manner in which the predictability of physical activity of a participant during the day can be improved. We focussed on the individual differences as well as the difference in activity and the circadian rhythm between free time and workdays. Exploring the data of the experiment the collected step count data was used to examine whether there was a significant difference between leisure time, such as weekends, holidays, and workdays. Taking the findings of this investigation into account, we augmented individual machine learning models. The training of algorithms per participant in combination with time sliced datasets improved the accuracy of the prediction during the day of one meeting his or her daily goal. The use of personalized prediction models, applied machine learning, and the consideration of the difference between leisure time and workdays, will become a valuable and viable addition to a virtual coaching system helping the participants in achieving a healthy lifestyle and enabling a virtual coaching system that intervenes at the appropriate points in time.**

Keywords–monitoring physical activity; circadian rhythm; inference statistics; machine learning; intervention.

I. Introduction

An unhealthy lifestyle with insufficient daily physical activity shortens life expectancy. Insufficient physical activity is associated with 5.3 million deceases globally in 2008 [1]. Contrarily sufficient physical activity is related to a reduced risk of metabolic syndrome [2], cardiovascular disease [3] and mortality [4]. To promote a healthy lifestyle and physical activity during the workday, the Hanze University of Applied Sciences (HUAS) initiated a health promotion program called -in Dutch- 'Het Gezonde Nieuwe Werken' (HGNW). This initiative contained a focus on the improvement of physical activity. Participants received an activity tracker to increase the awareness of their daily progress in achieving their goals in terms of numbers of steps. The daily feedback of the activity tracker was complemented with a fortnightly coaching session on lifestyle and physical activity. However, the feedback of the activity tracker and its platform didn't provide the participant with timely personalised feedback. Furthermore, current activity trackers do not provide a personalized probability of reaching the daily goal or take the difference between leisure time and workdays into account, therefore lacking an ability to adapt timing of coaching interventions to optimal points in time. Leisure time and workdays are known to show different patterns in activity [5] as well as a different circadian rhythm [6]. To enable a more personalized virtual coach these differences have to be taken into account.

In this paper, we investigate the influence of the circadian pattern of leisure time, workdays and holidays on the physical activity pattern during the day. In addition we investigate whether the level of activity depends on leisure and working time. Pattern differences due to these factors provide an opportunity to implement personalized automated intervention strategies for an individualized virtual coach.

The remainder of this paper is organized as follows. The first section introduces the state of the art on measuring activity levels, and the use of machine learning for monitoring. Subsequently, we describe the study on health promotion at HUAS, the collected dataset on daily physical activity of the participants, the method of statistical analysis of the difference between week, workweek, weekend, holiday and bank holiday, and measuring the performance of the trained machine learning models. In the third section we present the results of the statistical analysis and the performance of trained machine learning models. The conclusion on the results and a short discussion on future work completes the paper.

II. State of the Art

Activity trackers provide a measure for the number of steps and enable monitoring progress during the day and over time. Adding an activity tracker on steps to physical therapy or counselling was effective in some groups [7] [8]. The collection of step data is not only effective for therapy or counselling, it is also an intervention mechanism in itself [9]. Only the fact of using an activity tracker could motivate physical activity and improvement of health [10]. To improve on physical activity in combination with activity tracking monitoring, coaching is helpful. Effectiveness of (e)Coaching depends on timeliness and on personal contextual information in combination with actionable insights [11]. In other words the participant needs to receive the information and the advice when it is most relevant. To the best of our knowledge no studies exist on the use of activity trackers in combination with the circadian pattern and machine learning algorithms to establish *individualized* models or studies on *individualized* models used in virtual coach systems on monitoring activity helping the participant to improve his or her physical behaviour.

III. Methods

In this section, we present the study design of the HNGW, the data set we analysed, the statistical method to identify the differences in physical activity between leisure time and work week, and the training of algorithms with time sliced data and the methods used for analysis of the accuracy of the trained models.

A. Study design

The study data stems from the HNGW project. Forty eight healthy employees were recruited from the HUAS. The 48 participants were divided according to age, gender, BMI, and baseline self-reported health prior to being randomized into two groups. Group A followed a twelve-week health promotion intervention; the other group, group B, served first as a control group and thereafter also received the twelve-week health promotion intervention. Only the intervention period was used to study. The outcome measures included, among other values, the daily steps. The daily steps were measured with the Fitbit Flex, which is known to be a trustworthy and valid activity tracker for step count [12] and suitable for health promotion programs [13].

B. Data set

We prepared the individual participants available minute step data to investigate the difference between the activity patterns on the weekends, the holidays, the bank holidays and the workdays, and the corresponding activity patterns. We followed a step-by-step approach. First, we performed a data pre-processing step to remove the incomplete records from the data set. We also eliminated all records per day whenever no step was gathered during that day. Second, we augmented an hourly summarised data set per participant with new derived variables representing:

1) the year (2014-2015)
2) the week of the year (range 0-52)
3) the day of the week (range 0 - 6)
4) the hour of day (range 0 - 23)
5) the steps per hour
6) the cumulative sum of the steps per hour

Third, the work week is defined as the weekdays Monday till Friday. Fourth, a weekend is defined as Saturday and Sunday. Fifth, because all employees work on at an university of applied sciences, they have the same holidays and information on the holidays was added to the data set. Sixth, the bank holidays were identified and added. For the column the average number of steps per participant per day the amount of steps between 6:00AM and 11:00PM was considered. The average number of steps per participant per day the amount of steps column was regarded as a threshold per day in order to determine the binary outcome column. Finally, we constructed individual binary outcome variables based on the threshold of the time slices work week, weekend, holiday, bank holiday, and week.

C. Statistical Analysis

To determine whether there were differences in level of activity and moment of activity, we fitted a regression model and used the ANOVA analysis in conjunction with the Tukey HSD post-hoc analysis to identify possible correlations and differences between the hourly datasets whole week, work week, weekend, holiday and bank holiday. We reject the null hypothesis when the mean of the steps per hours is equal between the different time slices.

The Random Forest algorithm was chosen as the algorithm to enable prediction on whether a participant would meet his daily threshold. Random Forest is known as one of the most accurate algorithms predicting the participant meeting his or her goal [14]. The Random Forest algorithm was trained for the individual participant and hourly time sliced datasets. The techniques cross validation and parameter tuning were used to optimize the trained model. To investigate the performance of individualized hourly work week, weekend, holiday and bank holiday models, we compared these predictions with the predictions of a baseline model. The baseline model was trained per participant on the whole data set without distinction between the time slices work week, weekend, holiday or bank holiday and then the baseline model predicted adopting the four time sliced datasets leading to four sets of predictions. Next the four models for the work week, weekend, holiday and bank holiday predicted adopting the four time sliced datasets. The confusion matrix method was used to classify the difference between the predicted value and the actual value of both the baseline model and the time sliced models. A confusion matrix provides an overview of the true positives (TP; a predicted a 'true' and the actual data contained a 'true'), true negatives (TN; the model predicted a 'false' and the actual data was a 'false'), false positives (FP; the model predicted a 'true' label, but the actual data was a 'false'), and false negatives (FN; the model predicted a 'false' label, but the data was 'true') of a model. The confusion matrix served as a basis for the calculation of the performance measure F1-score and the accuracy [15].

The F1-score and accuracy are calculated for each model on group level, both metrics have a range of zero to one, where one is the best score. To calculate the F1-score, two other metrics known as the precision and the recall are used. Precision is the proportion of the true positives and the false negatives, and is calculated as:

$$\frac{TP}{(TP+FN)}$$

Recall is the true positive rate, which is calculated as follows:

$$\frac{TP}{(TP+FP)}$$

Using precision and recall, the F1-score is calculated as:

$$2 \cdot \frac{Precision \cdot Recall}{Precision + Recall}$$

To calculate the accuracy is a metric to determine the nearness of the prediction to the true value. A value of the accuracy close to one indicates the best performance. It calculates the ratio between the correctly classified cases and all cases as:

$$\frac{TP+TN}{TP+TN+FP+FN}$$

The personalized work week, weekend, holiday and bank holiday models were studied on the performance in comparison with the baseline week model.

IV. Results

The ANOVA test and the Tukey HSD post-hoc analysis identified significant differences between the hourly level of activity and the time sliced datasets. The ANOVA test found a significant difference in activity level ($p<0.001$) between the work week, the weekend, the holiday and the bank holiday at six, seven, eight o'clock. The participants were less active in the beginning of the day. Fig. 1 is an illustration of the difference of level of activity per hour of the day between the work week and the weekend.

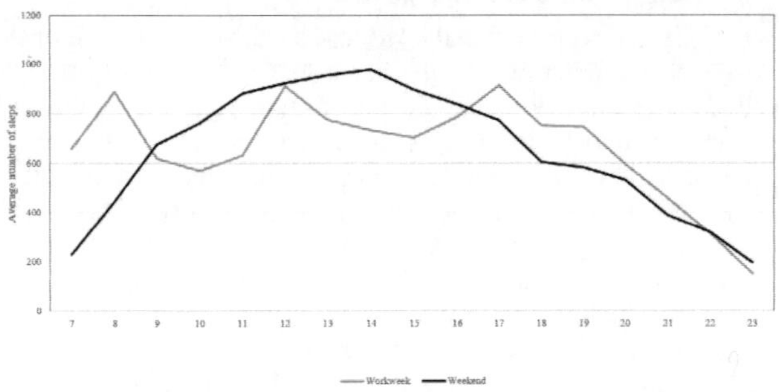

Figure 1. The difference in pattern of number of steps per hour during the work week versus the weekend.

The Tukey HSD post-hoc analysis showed where there are significant differences between the time sliced datasets. The majority of the datasets (21 out of 30) showed significant differences at six, seven, and eight o' clock. In the remainder of the day the differences between the different time sliced datasets were not significant.

Table I is an example of the Tukey HSD post-hoc analysis representing the seven o'clock results.

TABLE I. MULTIPLE COMPARISON OF MEANS - TUKEY HSD, ALFA=0.05 EXAMPLE

Dataset 1	*Dataset 2*	*Hour*	*Mean difference*	*Reject*
all week	bank holiday	7	-225	True
all week	holiday	7	-167	True
all week	weekend	7	-316	True
all week	work week	7	165	True
bank holiday	holiday	7	88	False
bank holiday	weekend	7	-60	False
bank holiday	work week	7	421	True
holiday	weekend	7	-148	True
holiday	work week	7	333	True
weekend	work week	7	481	True

The personalized Random Forest models based on the hourly datasets proved to show a better F1-score and accuracy then the baseline week model. Table II an Table III represent the improvement of the performance when using the work week, weekend, holiday and bank holiday model instead of the baseline weekmodel.

TABLE II. COMBINATIONS OF TIME SLICE BASED MODELS AND THEIR F1-SCORE.

Model 1	*F1-score*	*Model 2*	*F1-score*	*Difference*
week model	0.59	weekend model	0.86	48%
week model	0.58	holiday model	0.71	22%
week model	0.57	bank holiday model	0.68	19%
week model	0.81	work week model	0.96	18%

V. CONCLUSION

The level of activity in the morning is influenced by the difference between leisure time and work time. The individualized time sliced models improved the F1-score and accuracy in comparison to the non-time sliced week model. It

TABLE III. COMBINATIONS OF TIME SLICE BASED MODELS AND THEIR ACCURACY.

Model 1	*Accuracy*	*Model 2*	*Accuracy*	*Difference*
week model	0.89	weekend model	0.92	2.9%
week model	0.89	holiday model	0.89	0%
week model	0.76	bank holiday model	0.98	28%
week model	0.81	work week model	0.96	18%

is recommended to construct time sliced models per individual to improve the performance of the individualized models. In the future, contextual data that influences physical activity, like weekly non-workdays, sports, and celebrations, may be taken into account. Another possible future direction is to individualize interventions that allow for more personalized coaching. The individualization of the predictive models enables automated personalized, contextualized, and timely coaching. The results of this paper will be applied in the preventive eHealth virtual coach platform as suggested by Blok et al. [16].

ACKNOWLEDGMENT

The authors thank the Hanze University Health Program, especially M. van Ittersum, for providing the physical activity data of the Health Program, all the participants in the experiment, and M. Aiello for suggestions on the structure of the paper.

REFERENCES

[1] I. Min-Lee et al., "Effect of physical inactivity on major non-communicable diseases worldwide: An analysis of burden of disease and life expectancy," The Lancet, vol. 380, no. 9838, 2012, pp. 219–229. [Online]. Available: http://dx.doi.org/10.1016/S0140-6736(12)61031-9

[2] D. Zhang et al., "Leisure-time physical activity and incident metabolic syndrome: a systematic review and dose-response meta-analysis of cohort studies," pp. 36–44, oct 2017. [Online]. Available: http://www.ncbi.nlm.nih.gov/pubmed/28927737https://linkinghub.elsevier.com/retrieve/pii/S002604951730207X

[3] S. A. Lear et al., "The effect of physical activity on mortality and cardiovascular disease in 130 000 people from 17 high-income, middle-income, and low-income countries: the PURE study," www.thelancet.com, vol. 390, 2017. [Online]. Available: http://dx.doi.org/10.1016/

[4] D. D. Carter, K. Robinson, J. Forbes, and S. Hayes, "Experiences of mobile health in promoting physical activity: A qualitative systematic review and meta-ethnography," PLOS ONE, vol. 13, no. 12, dec 2018, p. e0208759. [Online]. Available: http://dx.plos.org/10.1371/journal.pone.0208759

[5] C. E. Matthews et al., "Sources of Variance in Daily Physical Activity Levels in the Seasonal Variation of Blood Cholesterol Study," American Journal of Epidemiology, vol. 153, no. 10, 2001, pp. 987–995.

[6] S.-J. Paine and P. H. Gander, "Explaining ethnic inequities in sleep duration: a cross-sectional survey of Mori and non-Mori adults in New Zealand." Sleep health, vol. 2, no. 2, jun 2016, pp. 109–115. [Online]. Available: http://www.ncbi.nlm.nih.gov/pubmed/28923252

[7] H. J. de Vries, T. J. Kooiman, M. W. van Ittersum, M. van Brussel, and M. de Groot, "Do activity monitors increase physical activity in adults with overweight or obesity? A systematic review and meta-analysis," Obesity, vol. 24, no. 10, oct 2016, pp. 2078–2091. [Online]. Available: http://doi.wiley.com/10.1002/oby.21619

[8] M. Miyauchi, M. Toyoda, N. Kaneyama, H. Miyatake, E. Tanaka, M. Kimura, T. Umezono, and M. Fukagawa, "Exercise Therapy for Management of Type 2 Diabetes Mellitus: Superior Efficacy of Activity Monitors over Pedometers," Journal of Diabetes Research, vol. 2016, sep 2016, pp. 1–7. [Online]. Available: https://www.hindawi.com/journals/jdr/2016/5043964/

[9] L. A. Cadmus-Bertram, B. H. Marcus, R. E. Patterson, B. A. Parker, and B. L. Morey, "Randomized Trial of a Fitbit-Based Physical Activity Intervention for Women." American journal of preventive medicine, vol. 49, no. 3, sep 2015, pp. 414–8. [Online]. Available: http://www.ncbi.nlm.nih.gov/pubmed/26071863http://www.pubmedcentral.nih.gov/articlerender.fcgi?artid=PMC4993151

[10] Z. H. Lewis, E. J. Lyons, J. M. Jarvis, and J. Baillargeon, "Using an electronic activity monitor system as an intervention modality: A systematic review," BMC Public Health, vol. 15, no. 1, dec 2015, p. 585. [Online]. Available: http://bmcpublichealth.biomedcentral.com/articles/10.1186/s12889-015-1947-3

[11] M. Gerdes, S. Martinez, and D. Tjondronegoro, "Conceptualization of a Personalized eCoach for Wellness Promotion," in Proceedings of ACM 11th EAI International Conference on Pervasive Computing. Barcelona: Association for Computing Machinery, 2017.

[12] T. J. M. Kooiman, M. L. Dontje, S. R. Sprenger, W. P. Krijnen, C. P. van der Schans, and M. de Groot, "Reliability and validity of ten consumer activity trackers," BMC Sports Science, Medicine and Rehabilitation, vol. 7, no. 1, dec 2015, p. 24. [Online]. Available: http://bmcsportsscimedrehabil.biomedcentral.com/articles/10.1186/s13102-015-0018-5

[13] J. Wang, R. Chen, X. Sun, M. F. H. She, and Y. Wu, "Recognizing human daily activities from accelerometer signal," Procedia Engineering, vol. 15, 2011, pp. 1780–1786. [Online]. Available: http://dx.doi.org/10.1016/j.proeng.2011.08.331

[14] T. B. Dijkhuis, F. J. Blaauw, M. W. van Ittersum, H. Velthuijsen, and M. Aiello, "Personalized Physical Activity Coaching: A Machine Learning Approach," Sensors, vol. 18, no. 2, feb 2018, p. 623. [Online]. Available: http://www.mdpi.com/1424-8220/18/2/623

[15] S. Schoeppe et al., "Efficacy of interventions that use apps to improve diet, physical activity and sedentary behaviour: a systematic review," International Journal of Behavioral Nutrition and Physical Activity, vol. 13, no. 1, dec 2016, p. 127. [Online]. Available: http://ijbnpa.biomedcentral.com/articles/10.1186/s12966-016-0454-y

[16] J. Blok, A. Dol, and T. Dijkhuis, "Toward a Generic Personalized Virtual Coach for Self-management : a Proposal for an Architecture," no. c, 2017, pp. 105–108.

Wearable and App-based Resilience Modelling in Employees (WearMe)

Herman de Vries, [1,2,3,6], Wim Kamphuis[2,7], Hilbrand Oldenhuis[1,8], Cees van der Schans[4,9] & Robbert Sanderman[3,10]

[1] Hanze University of Applied Sciences, Professorship Personalised Digital Health, Groningen, The Netherlands
[2] TNO, Department of Human Behaviour & Organisational Innovation, Soesterberg, The Netherlands
[3] Department of Health Psychology, University Medical Center Groningen, Groningen, The Netherlands
[4] Hanze University of Applied Sciences, Centre of Expertise Healthy Ageing, Groningen, The Netherlands
[5] Deptartment of Psychology, Health and Technology, University of Twente, Enschede, The Netherlands

E-mail: [6] h.j.de.vries@pl.hanze.nl; [7] wim.kamphuis@tno.nl; [8] h.k.e.oldenhuis@pl.hanze.nl; [9] c.p.van.der.schans@pl.hanze.nl; [10] r.sanderman@umcg.nl

***Abstract* – Occupational stress can cause all kinds of health problems. Resilience interventions that help employees deal with and adapt to adverse events can prevent these negative consequences. Due to advances in sensor technology and smartphone applications, relatively unobtrusive self-monitoring of resilience-related outcomes is possible. With models that can recognize intra-individual changes in these outcomes and relate them to causal factors within the employee's own context, an automated resilience intervention that gives personalized, just-in-time feedback (e.g., a virtual coach) can be developed. This 'Work in Progress' paper presents the study protocol for the Wearables and app-based resilience Modelling in employees (WearMe) project that aims to develop such models. A cyclical conceptual framework based on existing theories of stress and resilience is presented, as the basis for the WearMe project. The included concepts are operationalized and measured using sleep tracking (Fitbit Charge 2), heart rate variability measurements (Elite HRV + Polar H7) and Ecological Momentary Assessment (mobile app), administered in the morning (7 questions) and evening (12 questions). Analyses will target the development of both within-subject (n=1) models, as well as between-subjects models. If successful, future work will focus on further developing these models and eventually exploring the effectiveness of the envisioned personalized resilience system.**

Keywords* – *Occupational Stress; Personalized eHealth; Sensors; Wearables; Virtual Coaching

I. INTRODUCTION

Occupational stress can cause health problems, such as musculoskeletal disease, cardiovascular disease, depression and burnout [1], but also has financial consequences due to treatment costs, productivity loss and absenteeism [2]. Especially the cumulative wear and tear on bodily systems caused by stress or inefficient management of mechanisms that promote adaptation is detrimental for health and well-being [3]. This so-called 'allostatic load' increases the brain's sensitivity to appraise stimuli as threats and reduces resources to cope, which can result in a loss spiral.

There are all kinds of ways people cope with stress. The process of positively adapting to adverse events is also known as resilience [4]. In order to demonstrate positive adaptation (e.g., maintaining job performance and health), both individual (e.g., personality) and contextual (e.g., social support) resources can be used to cope with adversity [5]. By using these resources, resilient individuals are able to recover from the negative impact of stress relatively quickly and thus decrease their risk of negative long-term consequences.

Resilience interventions are often offered to a broad population. However, those that target employees with a higher risk of experiencing stress tend to have better long-term effects [6]. An even more personalized approach could be to monitor for early signs of the consequences of stress, relate these to causal factors in the employee's context, and provide personalized advice to better cope with the stressor. Due to advances in sensor technology and smartphone applications, relatively unobtrusive self-monitoring of changes in resilience related outcomes is possible [7]. What is needed are models that can recognize intra-individual changes in these outcomes and relate these to causal factors and future consequences. With such models, one can potentially create an automated resilience intervention that gives personalized, just-in-time feedback, for instance in a virtual coaching application.

In this 'Work in Progress' paper, we present the study protocol of the ongoing Wearables and app-based resilience Modelling in employees (WearMe) project. After introducing the WearMe project in Section I, a cyclical conceptual framework that is based on existing theories on stress and resilience is described in Section II. In Section III, we elaborate on how these concepts are operationalized in the first WearMe study. This includes the use of consumer-available wearables and an Ecological Momentary Assessment (EMA) app. In Section IV, some possible approaches for future work are described in our overarching goal to develop predictive models of employee resilience that can be used in personalized interventions such as virtual coaching applications.

II. CONCEPTUAL FRAMEWORK

The conceptual framework presented in Figure 1 has a cyclical nature and is based on several existing theories on stress and resilience.

When (job) *demands* such as time pressure or physical workload are interpreted as threats because the available *resources* to *adaptively cope* with the demands are perceived to be insufficient, it results in stress [8]. Depending on whether the person can utilize the available resources to adaptively cope with the demands, the short-term accumulated *stress* determines the individual's perceived *need for recovery* afterwards, which is

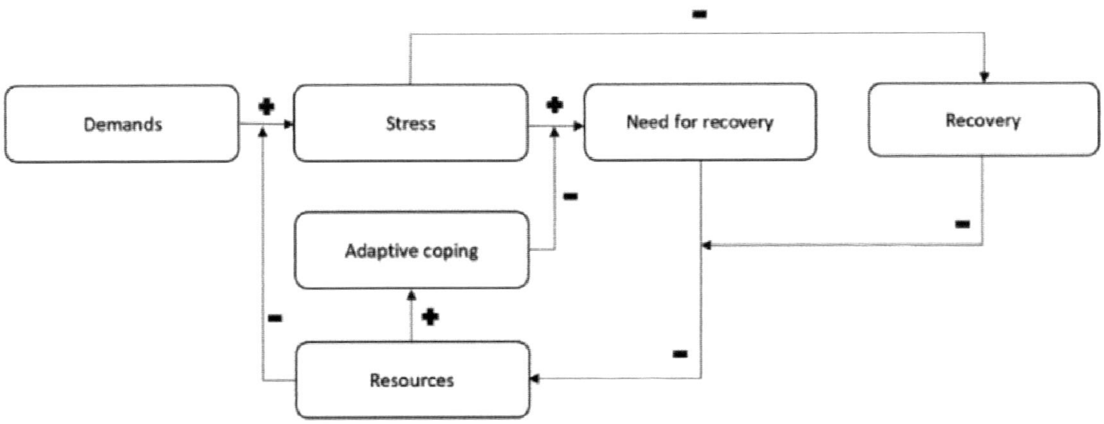

Figure 1. Conceptual framework for the WearMe study.

characterized by feelings of exhaustion and having less vigour to undertake new activities [9], [10]. The need for recovery therefore has a negative impact on the individual resources to appraise and cope with new demands – unless there is sufficient *recovery* present to alleviate this effect [9]. Besides causing a perceived need for recovery, stress also decreases sleep quality [11] and psychological detachment [12], which are aspects of recovery [13].

The cyclical nature of the conceptual framework is also supported by the Conservation of Resources theory [14] that states that resource loss leads to stress and that initial loss of resources may cause a loss spiral because resources are also used to prevent resource loss.

III. MEASUREMENT CYCLE

Based on the conceptual framework, a measurement cycle was developed that suggests how the described concepts may be operationalized using consumer-available wearables and EMA smartphone application to monitor for early signs of the consequences of stress. With exception of adaptive coping, which was not included because it is highly context-specific and thus difficult to quantify, all concepts are measured daily (Figure 2).

Although the *resources* that are needed to cope with demands are context-specific as well, several individual psychological and physiological resources were identified that are be relevant in a broad spectrum of situations, that may change on a day-to-day basis, and that can be measured using consumer-available wearables and apps. The first resource, Heart Rate Variability (HRV), is a measure for the variability in the intervals between two heartbeats and is considered to be a proxy for autonomous nervous system functioning [15]. While HRV is mostly known as a parameter that illustrates physiological changes during acute stress, the resting HRV can remain decreased during and afterwards acute stress [15] [16]. In contrast, having a lowered resting HRV has been associated with an increased sensitivity for stress [18], decreased emotion-regulation [19], a decrease in physical performance [20] and an increased risk of long-term physical or mental health problems [21]. In the WearMe study, resting HRV is therefore considered to be a potential indicator for the accumulation of stress, as well as an individual resource used in the appraisal of and coping with upcoming demands. Participants measure their resting HRV in the morning after waking up and before getting out of bed during 2 minutes in a supine position using the Elite HRV smartphone application [22] and a Polar H7 chest strap [23].

Besides HRV, perceived happiness, work engagement (vigour and dedication) and generalized self-efficacy are individual *resources* that are measured in a short EMA questionnaire in the morning and evening. Similarly, perceived *demands*, *stress* and *need for recovery* during the day are measured during the evening EMA questionnaire in a smartphone application.

The concept of *recovery* consists of two components that are known to limit the spill over of a perceived need for recovery during the next day; sleep and being able to psychologically detach from work during leisure time [24]. Therefore, psychological detachment is measured in the evening EMA questionnaire, while the morning EMA questionnaire includes an item on the perceived sleep quality. Furthermore, sleep onset latency, the number of awakenings, wake time after sleep onset, total sleep time and sleep efficiency are also objectively measured using the

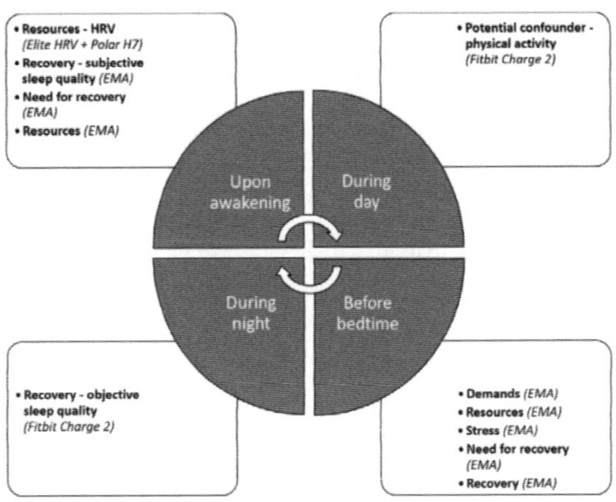

Figure 2. Measurement cycle of the WearMe study.

Fitbit Charge 2 wrist-worn tracker [25]. Since stress is also known to have a negative effect on sleep quality [11] and psychological detachment [12], it is also considered to be a potential indicator for the accumulation of stress.

Finally, alcohol intake [26] and physical activity [27] were measured due to potential confounding effects on sleep and/or HRV. Alcohol intake during the previous evening was measured in the morning EMA questionnaire and physical activity (steps, sedentary minutes, minutes of moderate-to-vigorous physical activity) was measured throughout the day using the Fitbit Charge 2.

III. PRESENT STUDY

The first WearMe study aims to explore the feasibility of the described measurement cycle to monitor for early signs of the accumulation of stress and to relate these to other factors based on the hypotheses described in the conceptual framework. Additionally, the development of population models will be explored. The study protocol was approved by the ethical committee of the Hanze University of Applied Sciences Groningen (heac.2018.008).

A. Population

For this ongoing 15-week study, twelve students in Applied Psychology (n=5) and Social Work (n=7) that are on their first full-time internship, are at least 18 years old and own an Android or iOS smartphone were recruited. Due to the potentially stressful nature of the context of these internships, as well as this being the participants' first full-time internship in their curriculum, we anticipate this population to be at risk of experiencing stress.

B. Data collection

Besides the daily measurements that were described in section two, several questionnaires are being administered to benefit the development of population models using between-subject analyses. Therefore, questionnaires on personality traits [28], coping strategies [29], burnout [30], work engagement [31] and symptoms of somatisation, distress, depression and anxiety [32] were administered at study onset. The questionnaires on burnout, work engagement and symptoms of somatization, distress, depression and anxiety will also be administered after 5, 10 and 15 weeks. Finally, participants will fill out a resources questionnaire to retrospectively assess the perceived personal and environmental resources throughout the internships after 15 weeks.

C. Data analysis

Data analyses will target the development of within-subject (n=1) models to predict changes in individual physiological and psychological resources (resting HRV, vigour, happiness, generalized self-efficacy) and recovery (sleep, psychological detachment) based on the hypotheses described in the conceptual framework. Furthermore, the development of population models will be explored. While no specific data-analysis techniques are pre-defined, both the use of traditional statistical analysis techniques and machine learning will be explored.

IV. CONCLUSION

If the results affirm that tracking sleep and resting HRV using consumer wearables is feasible and may be useful in resilience modelling, the current models could be expanded. Future studies will therefore focus on the development of predictive models that allow early detection of stress-related symptoms. In addition, expanding the current model by using additional consumer-available wearables or apps that can unobtrusively collect potentially relevant data (e.g., GPS location, calendar events) may be explored. When our conceptual framework that illustrates hypotheses based on deductive reasoning shows to be valid, a more inductive approach to data-analysis may be explored (e.g., using machine learning) to increase the explained variance of the individual models. If successful, these models can be implemented in applications that create personalized feedback on how to cope with demands.

Furthermore, it is possible that within-subject models can be formed but require a long period of data collection. If subgroups with similar outcome trajectories can be identified using between-subject analyses of baseline and first-week data in a larger sample in order to create a classification algorithm, it might be possible to develop a system that combines both methods [33]. In such a system, participants would first receive semi-personalized feedback based on their subgroup classification and start receiving fully personalized feedback when enough within-subject data are available.

REFERENCES

[1] T. W. Colligan and E. M. Higgins, 'Workplace stress: Etiology and consequences', *J. Workplace Behav. Health*, vol. 21, no. 2, pp. 89–97, 2006.

[2] S. Béjean and H. Sultan-Taïeb, 'Modeling the economic burden of diseases imputable to stress at work', *Eur. J. Health Econ.*, vol. 6, no. 1, pp. 16–23, 2005.

[3] B. S. McEwen, 'Stress, adaptation, and disease: Allostasis and allostatic load', *Ann. N. Y. Acad. Sci.*, vol. 840, no. 1, pp. 33–44, 1998.

[4] T. D. Cosco, A. Kaushal, R. Hardy, M. Richards, D. Kuh, and M. Stafford, 'Operationalising resilience in longitudinal studies: a systematic review of methodological approaches.', *J. Epidemiol. Community Health*, vol. 71, no. 1, pp. 98–104, Jan. 2017.

[5] T. W. Britt, W. Shen, R. R. Sinclair, M. R. Grossman, and D. M. Klieger, 'How Much Do We Really Know About Employee Resilience?', *Ind. Organ. Psychol.*, vol. 9, no. 02, pp. 378–404, 2016.

[6] A. J. Vanhove, M. N. Herian, A. L. U. Perez, P. D. Harms, and P. B. Lester, 'Can resilience be developed at work? A meta-analytic review of resilience-building programme effectiveness', *J. Occup. Organ. Psychol.*, vol. 89, no. 2, pp. 278–307, 2016.

[7] R. L. Drury, 'Wearable biosensor systems and resilience: a perfect storm in health care?', *Front. Psychol.*, vol. 5, p. 853, 2014.

[8] R. S. Lazarus and S. Folkman, 'Transactional theory and research on emotions and coping', *Eur. J. Personal.*, vol. 1, no. 3, pp. 141–169, 1987.

[9] M. Van Veldhoven, 'Need for recovery after work: An overview of construct, measurement and research', in

Occupational health psychology: European perspectives on research, education and practice, vol. 3, de J. Houdmont and S. Leka, Eds. Nottingham, 2008, pp. 1–25.

[10] A. B. Bakker and E. Demerouti, 'The Job Demands-Resources model: state of the art', *J. Manag. Psychol.*, vol. 22, no. 3, pp. 309–328, 2007.

[11] E.-J. Kim and J. E. Dimsdale, 'The effect of psychosocial stress on sleep: a review of polysomnographic evidence.', *Behav. Sleep. Med.*, vol. 5, no. 4, pp. 256–278, 2007.

[12] S. Sonnentag, I. Kuttler, and C. Fritz, 'Job stressors, emotional exhaustion, and need for recovery: A multi-source study on the benefits of psychological detachment', *J. Vocat. Behav.*, vol. 76, no. 3, pp. 355–365, 2010.

[13] S. Sonnentag and C. Fritz, 'The Recovery Experience Questionnaire: development and validation of a measure for assessing recuperation and unwinding from work.', *J. Occup. Health Psychol.*, vol. 12, no. 3, p. 204, 2007.

[14] S. E. Hobfoll, 'The Influence of Culture, Community, and the Nested-Self in the Stress Process: Advancing Conservation of Resources Theory', *Appl. Psychol.*, vol. 50, no. 3, pp. 337–421, 2001.

[15] J. F. Thayer, F. Ahs, M. Fredrikson, J. J. Sollers III, and T. D. Wager, 'A meta-analysis of heart rate variability and neuroimaging studies: implications for heart rate variability as a marker of stress and health.', *Neurosci. Biobehav. Rev.*, vol. 36, no. 2, pp. 747–756, Feb. 2012.

[16] M. Hall *et al.*, 'Acute stress affects heart rate variability during sleep.', *Psychosom. Med.*, vol. 66, no. 1, pp. 56–62, Feb. 2004.

[17] E. Hynynen, N. Konttinen, U. Kinnunen, H. Kyrolainen, and H. Rusko, 'The incidence of stress symptoms and heart rate variability during sleep and orthostatic test.', *Eur. J. Appl. Physiol.*, vol. 111, no. 5, pp. 733–741, May 2011.

[18] G. Park, J. J. Van Bavel, M. W. Vasey, and J. F. Thayer, 'Cardiac vagal tone predicts inhibited attention to fearful faces.', *Emot. Wash. DC*, vol. 12, no. 6, pp. 1292–1302, Dec. 2012.

[19] D. P. Williams, C. Cash, C. Rankin, A. Bernardi, J. Koenig, and J. F. Thayer, 'Resting heart rate variability predicts self-reported difficulties in emotion regulation: a focus on different facets of emotion regulation.', *Front. Psychol.*, vol. 6, p. 261, 2015.

[20] S. Jimenez Morgan and J. A. Molina Mora, 'Effect of Heart Rate Variability Biofeedback on Sport Performance, a Systematic Review.', *Appl. Psychophysiol. Biofeedback*, vol. 42, no. 3, pp. 235–245, Sep. 2017.

[21] D. Liao, M. Carnethon, G. W. Evans, W. E. Cascio, and G. Heiss, 'Lower heart rate variability is associated with the development of coronary heart disease in individuals with diabetes: the atherosclerosis risk in communities (ARIC) study', *Diabetes*, vol. 51, no. 12, pp. 3524–3531, 2002.

[22] A. S. Perrotta, A. T. Jeklin, B. A. Hives, L. E. Meanwell, and D. E. R. Warburton, 'Validity of the Elite HRV Smartphone Application for Examining Heart Rate Variability in a Field-Based Setting', *J. Strength Cond. Res.*, vol. 31, no. 8, pp. 2296–2302, 2017.

[23] D. J. Plews, B. Scott, M. Altini, M. Wood, A. E. Kilding, and P. B. Laursen, 'Comparison of heart-rate-variability recording with smartphone photoplethysmography, Polar H7 chest strap, and electrocardiography', *Int. J. Sports Physiol. Perform.*, vol. 12, no. 10, pp. 1324–1328, 2017.

[24] S. Sonnentag and C. Binnewies, 'Daily affect spillover from work to home: Detachment from work and sleep as moderators', *J. Vocat. Behav.*, vol. 83, no. 2, pp. 198–208, 2013.

[25] M. de Zambotti, A. Goldstone, S. Claudatos, I. M. Colrain, and F. C. Baker, 'A validation study of Fitbit Charge 2™ compared with polysomnography in adults', *Chronobiol. Int.*, vol. 35, no. 4, pp. 465–476, 2018.

[26] D. S. Quintana, I. S. McGregor, A. J. Guastella, G. S. Malhi, and A. H. Kemp, 'A meta-analysis on the impact of alcohol dependence on short-term resting-state heart rate variability: implications for cardiovascular risk.', *Alcohol. Clin. Exp. Res.*, vol. 37 Suppl 1, pp. E23-29, Jan. 2013.

[27] S. F. Smagula, K. L. Stone, A. Fabio, and J. A. Cauley, 'Risk factors for sleep disturbances in older adults: Evidence from prospective studies.', *Sleep Med. Rev.*, vol. 25, pp. 21–30, Feb. 2016.

[28] J. J. A. Denissen, R. Geenen, M. A. G. Van Aken, S. D. Gosling, and J. Potter, 'Development and validation of a Dutch translation of the Big Five Inventory (BFI)', *J. Pers. Assess.*, vol. 90, no. 2, pp. 152–157, 2008.

[29] W. C. Kleijn, G. L. Van Heck, and A. Van Waning, 'Ervaringen met een Nederlandse bewerking van de COPE copingvragenlijst: De COPE-Easy', *Gedrag Gezondh.*, vol. 28, pp. 213–226, 2000.

[30] J. R. B. Halbesleben and E. Demerouti, 'The construct validity of an alternative measure of burnout: Investigating the English translation of the Oldenburg Burnout Inventory', *Work Stress*, vol. 19, no. 3, pp. 208–220, 2005.

[31] W. B. Schaufeli, A. B. Bakker, and M. Salanova, 'The measurement of work engagement with a short questionnaire: A cross-national study', *Educ. Psychol. Meas.*, vol. 66, no. 4, pp. 701–716, 2006.

[32] B. Terluin *et al.*, 'The Four-Dimensional Symptom Questionnaire (4DSQ): a validation study of a multidimensional self-report questionnaire to assess distress, depression, anxiety and somatization', *Bmc Psychiatry*, vol. 6, no. 1, p. 34, 2006.

[33] G. Spanakis, G. Weiss, B. Boh, L. Lemmens, and A. Roefs, 'Machine learning techniques in eating behavior e-coaching: Balancing between generalization and personalization', *Pers. Ubiquitous Comput.*, vol. 21, no. 4, pp. 645–659, 2017.

Virtual Coach: Towards Personalized Mental Support

K. van Noppen – Kleist[1] , W. Mulder [1,2], T.B. Dijkhuis [2,3], M.R. Dam[2]

[1]KPN ICT Consulting, Data Innovation Lab, Groningen, The Netherlands

[2]Institute for Communication, Media & IT, Hanze University of Applied Sciences, Groningen

[3]Center for Human Movement Sciences, University Medical Center Groningen, Groningen, The Netherlands

[1] e-mail:{katharina.vannoppen-kleist, wico.mulder}@kpn.com

[2] e-mail:{m.r.dam, t.b.dijkhuis}@pl.hanze.nl

Abstract— **Self-management is widely seen as a viable contribution to sustainable health-care as it allows to promote physical and mental well-being. A promising approach to promoting a healthy lifestyle is the deployment of personalized virtual coaches, especially in combination with the latest developments in the fields of Data Science and Artificial Intelligence. This paper presents a framework for a virtual coaching system, as well as a use case in which parts of this framework are applied. The virtual coach in the use case aims to encourage customer contact center employees to protect their mental health. This article outlines one part of the use-case in particular, viz. how to promote employee autonomy and supervisor support by, inter alia, monitoring employees' levels of emotional exhaustion. Current systems focus on providing users with insight in their health status or behavior, the authors developed the functional architecture for a system that can be implemented for different goals and generates personalized, real-time advice based on the combination of user preferences, motivational success and predicted user behavior.**

Keywords- virtual coach; emotional exhaustion; artificial intelligence; mental support; data science; behavioral change.

I. INTRODUCTION

Most people struggle with protecting their physical and mental well-being in daily life. A promising approach to promoting a healthy lifestyle stems from the development of eHealth and telemedicine services. Particularly, services that aim at increasing users' health prevention and self-management efforts, are gaining momentum [16]. This development, together with the increasing availability of data from mobile devices and techniques for real-time data analysis, provides new opportunities for health self-management via personalized virtual coaching.

As described by Blok and Dijkhuis [4], we define a virtual coach as an automated system that assists the user, either as a stand-in for or supplementary to a human coach, to exhibit desired user behavior. Additionally, we expand this definition by a generic lifestyle intervention framework which stresses the application of virtual coaches for health benefits. This framework emphasizes three domains of health: physical health, mental health, and social health [10]. A healthy lifestyle is a balancing act between the load, the recovery and the capacity of the individual [12]. Specifically, striking a balance between load (physical, mental and social) and recovery, enables individuals to sustain a healthy capacity to cope with daily life situations. For instance, mental load and the ability to recover are dependent on the individual's capacity to deal with stressful situations [27]. Examples of sources of mental stress are pressure at work or major life changes that are perceived negatively [12][27].

The remainder of this paper is organized as follows: the first section introduces the state of the art on virtual coaching systems. Subsequently, we describe the general, functional architecture for the virtual coach proposed by us. In the third section we present the use case in which the first modules of the virtual coach architecture are being implemented at the time of writing this article. The conclusion of the current state of this implementation and description of future steps completes the paper.

II. STATE OF THE ART

Effectiveness of (e)-Coaching depends on timeliness and personal contextual information in combination with actionable insights [8]. In other words the user needs to receive relevant information and advice at the appropriate time. In order to achieve this goal, various types of virtual coaches have been developed, for example self-monitoring of chronic diseases such as diabetes [15], motivating increased activity levels [25] or making healthier choices in daily life [17]. Currently available systems provide insight in health and behavior, but do not facilitate long-term behavioral change [13]. Spanakis et al. [23] predict eating behavior for the near future using rules learned from a predictive algorithm, however, they do not use personalized motivation techniques.

The architecture proposed in this article is a general, functional architecture that can be implemented in various situations, both for coaching on physical and mental health. The architecture provides not only insights, but also personalized real-time advice. During the generation of the advice, personal preferences regarding activities as well as the success rates of different motivational techniques are taken into account in order to strengthen motivation and support behavioral change. Furthermore, advice is not only based on current behavior, but also on predicted behavior in the near future.

To the best of our knowledge no systems exist that can be implemented for multiple goals (e.g., mental and physical health) and generate personalized real-time advice based on the combination of activity preferences, motivational success and predicted user behavior.

III. VIRTUAL COACH: FUNCTIONAL ARCHITECTURE

On the basis of our definition of a virtual coach, we have designed a functional, modular architecture that is applicable to multiple domains. The architecture proposed in this paper focuses on providing relevant real-time advice and supporting behavioral change.

In order to increase the influence of a virtual coach, we propose to channel behavioral insights into concrete advice for the user. This advice needs to be personalized with regard to the user's context and (motivational) preferences. This is done in several steps, as shown in *Figure 1*. In Step 1, the virtual coach identifies the moments when the user is in need of intervention. When intervention is called for, an appropriate personalized intervention method is generated by performing steps two and three simultaneously, the function of Step 2) being to determine which action a user had best perform to support him/her in reaching his/her goal, and that of Step 3) to determine the motivation technique most likely to influence his/her behavior.

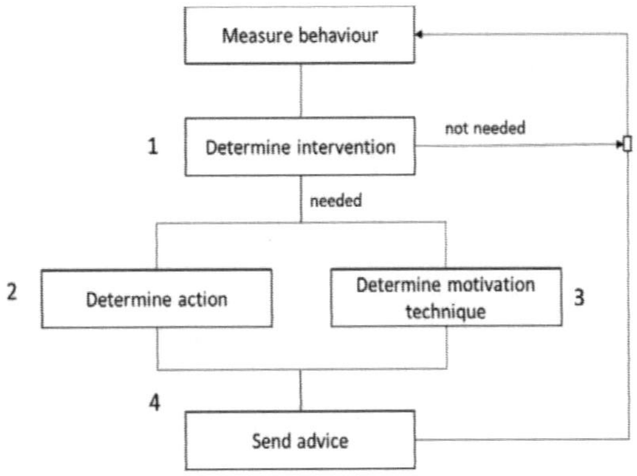

Figure 1. The steps in the functional architecture of the virtual coach.

All steps are geared to the user's specific context. In order to capture the user's context we combine behavioral data (generated by the user) with data from IoT-sensors and other sources (e.g., weather, location, time). This way, we take advantage of the variety of available data sources and enable a virtual coach to learn to interpret the user's context automatically. The final advice is then presented to the user, for example via a mobile application. In the following sections the above mentioned steps are described in more detail.

A. Determine if intervention is needed

The first step is the automatic identification of the moments when a user might need intervention. For this to be successful, a measurable and realistic goal must be set: the goals and needs of the user will have to be captured and objectively measured. Next, an appropriate amount of data must be collected in order to successfully establish whether or not a user reaches his goal. Patterns in the data can be learned using machine learning techniques, resulting in a predictive model [7]. The model has to be updated over time: the behavior of the user may change. Also, when the user frequently reaches his goal, the goal needs to be re-defined as well. The choice to intervene is context-aware and takes, for example, the hour of the day as well as the user's calendar into account.

B. Determine action

Once the virtual coach has determined that an intervention is necessary, the system provides the user with personalized suggestions for actions. These actions are tailored to the user's interests and preferences. Providing the user with a high variety of recommended actions is expected to increase user experience [14]. For the system to be able to recommend an appropriate and personalized action to the user, a list of possible actions must be created first. In order to overcome the cold start problems in the steps of the architecture, a new user completes an intake profile. This intake profile consists of, inter alia, the user's demographics and preferences for a random subset of items on the action list. Also, the intake profile is used to assess the user's ability. The virtual coach considers the feasibility of an action and links it to the user's ability level using a Rasch model in order to select appropriate action items [19]. Based on the user's ability, the probable willingness to engage in a particular action can be determined. When this probability is high, the action is seen as feasible. Note that a user's ability might change over time (for better or worse, depending on fluctuations in his/her health and fitness). These changes in ability need to be recomputed regularly in order to recommend appropriate actions. Furthermore, all action recommendations must be context-aware (e.g., don't send a user outside when it is raining).

C. Determine motivation techniques

There are many theories on how to motivate humans and they attempt to link individual motivation techniques to, inter alia, personality factors and locus of control. We are aiming for an approach that compares different motivational theories and systematically explores their effectiveness among users. Ultimately, we investigate whether tailoring actions by creating personality and preference profiles for motivation techniques stimulates behavioral change. First attempts to generate messages aimed at changing health-related behavior (e.g., handle potentially conflicting goals and motives) were introduced by Baskar et al. [3] and Op den Akker et al. [16]. Along with the above mentioned intake profile, a test battery, as well as a few examples of motivation messages may be given to the user right at the beginning, allowing him to rate the messages. Furthermore, clustering can be used to identify groups of users with a similar profile based on a questionnaire at the intake of the virtual coach. Individuals within the same group could initially be treated similarly [23]. The proposed model can be applied to several situations as long as the users' behavior can be measured frequently and the probability of users reaching their goal can be predicted accurately.

IV. USE CASE: SERVICE WITH A SMILE

Research suggests that people who report high levels of work stress run a greater risk of developing a range of mental and physical health conditions such as depression, anxiety, hypertension and heart disease [11]. Work-related issues, like the combination of high demand and low control, have been shown to require sustained physical and psychological effort. These sustained efforts typically come at significant physiological and psychological costs [6][22] and have been linked to increased employee burnout, turnover, absenteeism and decreased performance [2]. Unfortunately, after recovering from these experiences, individual employees are often assigned less responsibility, they work fewer hours and are paid less than before [18], turning work-related mental health issues into the cause of serious long-term problems for both employees and employers.

Especially in highly demanding service environments, where employees are expected to deliver 'service with a smile', many of the above mentioned negative health consequences are very common. In addition, the subsequent high rates of employee turnover entail enormous direct and indirect costs because of the expenditure of hiring and training new employees and the financial impact of diminished service quality [1]. It is crucial that employers create environments where employees can be themselves, because exhausted employees can neither enjoy their job nor deliver good service.

Customer service centers are typically such highly demanding service environments. Hence, most customer service centers struggle to sustain their employees' wellbeing; exhaustion, high stress levels and burnouts are common among call center employees [5].

While employees need to take into account that a customer's perception of a company is determined by the quality of their interactions [21] and hence have to maintain a friendly disposition, they are also expected to quickly solve the issues of their customers. Matching these job demands has been shown to negatively influence job satisfaction and job performance because of, inter alia, emotional dissonance [28] and emotional exhaustion [24]. The above mentioned research outlines how high levels of psychological stress in call center employees lead to health impairment (particularly emotional exhaustion) and demotivation [26].

Both, job autonomy and supervisor support are the most commonly found work-related factors to negate the above mentioned health risks for customer service center employees (particularly emotional exhaustion)[28]. Providing customer service center employees with autonomy means providing them with a degree of freedom to make at least some decisions on their own. Autonomy has been shown to improve coping with emotionality of customers, to relate significantly to job satisfaction and performance [20] and to alleviate negative emotions felt by customer service employees doing stressful work. When autonomy in the workplace is absent, however, customer service center employees have been found to be more stressed, less satisfied and to report poorer mental and physical health [9].

V. DIGITAL CUSTOMER CONTACT CENTER ASSISTANT

In collaboration with the customer contact center of Van Lanschot, a private Dutch bank, we set out to develop a virtual coach or 'digital customer contact center assistant' (hereafter: DC3A) with the aim of diminishing negative mental health influences and increasing the motivation of employees. As outlined in the virtual coach framework, the DC3A assesses the mental state of the employees (e.g., emotional exhaustion) on the basis of both direct and indirect factors. After each call, employees indicate, inter alia, the perceived tone of the call (positive, neutral or negative). This feedback serves as a direct indicator of their emotional exhaustion (i.e. the higher the number of negative customer interactions, the higher the level of emotional exhaustion). When employees are emotionally exhausted, both key performance indicators and service quality indicators are expected to decrease. Hence, the DC3A extracts relevant key performance indicators and service quality indicators from customer interaction records and links the employees' feedback to these two sets of indicators. Machine learning techniques (such as classification and clustering), but also automated text analysis of call scripts are used to find patterns and assess the level of emotional exhaustion. On the basis of this assessment, the DC3A determines whether an intervention is needed and, if so, which type of action needs to be recommended (e.g., taking a break or seeking supervisor support). As described in the general framework for the virtual coach, the type of motivation that is paired with the recommended action is determined by using hybrid recommender systems. This allows us to optimize the action with a personalized motivational approach, based either on facts extracted from calls or on an informed selection of the type of action. We postulate that this added motivational component enhances a) the action's level of persuasion and therefore b) the employee's level of compliance. Both action and motivation are then combined to formulate the intervention message for the employee, aiming at initiating behavior that decreases levels of emotional exhaustion. The effect of this intervention, and whether or not the message is understood, is measured through key performance indicators and service quality. On the basis of the results of our first prototype we will adjust algorithms for both content and timing of the messages for the employees, and incorporate the necessary changes.

VI. CONCLUSION

We have designed a functional architecture for a virtual coach that addresses the mental health and operational support of customer contact center employees. Our design and approach are based on research that addresses mental health problems of people working in highly demanding service environments. A first prototype, aiming at supporting the mental health and well-being of customer contact employees, is currently being tested in an organization whose customer contact center is a key instrument in achieving customer satisfaction.

During the development of the presented architecture, as well as during the building process of the first prototype, the authors encountered a few challenges. One of these challenges was bridging the various, necessary fields of expertise (e.g., find a common understanding of the problem and agree on vocabulary and methods). Another challenge was the business driven use-case; the authors were asked to work with mostly predetermined information and minimize the collection of new data points. Both challenges could be addressed successfully by forming a dedicated, multidisciplinary research team with a common goal. This dedicated team, consisting of experts in the areas of Computer Science, Software Engineering, Psychology, Applied Mathematics and Artificial Intelligence, formed the basis for a new ecosystem. This ecosystem is sponsored by the involved private businesses as well as academic institutions and enables the team to take the next steps in the development of the virtual coach. These steps are: 1) development and implementation of the first prototype in the use case, and 2) combine this work with exploratory research at academic institutions to fill gaps and further interdisciplinary research in other related areas.

ACKNOWLEDGMENT

The authors thank van Lanschot for permitting us to describe their use case in this paper, the Data & Analytics department of KPN ICT Consulting for the financial means to attend the eTelemed 2019 in order to present this paper, the Hanze University of Applied Sciences for their collaboration and partnership and L.M. van Noppen for text correction.

REFERENCES

[1] A. Alexandrov, E. Babakus, and U. Yavas, "The effects of perceived management concern for frontline employees and customers on turnover intentions: moderating role of employment status," Journal of service research, vol. 9, no. 4, pp. 356-371, 2007.

[2] A. B. Bakker, and E. Demerouti, "The job demandsresources model: State of the art," Journal of managerial psychology, vol. 22, no. 3, pp. 309-328, 2007.

[3] J. Baskar, R. Janols, E. Guerrero, J.C. Nieves and H. Lindgren, "A Multipurpose Goal Model for Personalised Digital Coaching," In: Agents and Multi-Agent Systems for Health Care A2HC 2017, AHEALTH 2017. Lecture Notes in Computer Science, vol. 10685, S. Montagna, P. Abreu, S. Giroux and M. Schumacher, Eds. Springer, Cham, 2017.

[4] J. Blok, T. Dijkhuis and A. Dol, "Toward a generic personalized virtual coach for self-management: A proposal for an architecture," The Ninth International Conference on eHealth, Telemedicine, and Social Medicine (eTelemed 2017), Nice, France.

[5] F. Castanheira and M. J. Chambel, "Reducing burnout in call centers through HR practices," Human Resource Management, vol. 49, no. 6, pp. 1047-1065, 2010.

[6] E. R. Crawford, J. A. LePine and B. L. Rich, "Linking job demands and resources to employee engagement and burnout: A theoretical extension and meta-analytic test," Journal of applied psychology, vol. 95, no. 5, pp. 834, 2010.

[7] T. Dijkhuis, F. Blaauw, M. van Ittersum, H. Velthuijsen and M. Aiello, "Personalized Physical Activity Coaching: A Machine Learning Approach" Sensors, vol. 18, no. 2, pp. 623, 2018.

[8] M. Gerdes, S. Martinez, and D. Tjondronegoro, "Conceptualization of a Personalized eCoach for Wellness Promotion," in Proceedings of ACM 11th EAI International Conference on Pervasive Computing. Barcelona: Association for Computing Machinery, 2017.

[9] L. Holdsworth and S. Cartwright, "Empowerment, stress and satisfaction: An exploratory study of a call centre," Leadership & Organization Development Journal, vol. 24, no. 3, pp. 131-140, 2013.

[10] M. Huber et al., "How should we define health?," BMJ, vol. 343, no. 9666, p. 41-63, 2011.

[11] International Labour Office (ILO) (2016). Workplace Stress: A collective challenge. International Labour Office – Geneva: ILO, 2016.

[12] G. Kentta and P. Hassmen, "Overtraining and recovery: A conceptual model," Sport Medicine, vol. 26, no. 1, pp. 1–16, 1998.

[13] R. Klaassen, K. Bul, R. op den Akker, G. J. van der Burg, P. M. Kato and P. Di Bitonto, "Design and Evaluation of a Pervasive Coaching and Gamification Platform for Young Diabetes Patients," Sensors, vol. 18, no. 2, pp. 402-429, 2018.

[14] B. P. Knijnenburg et al., "Explaining the user experience of recommender systems," User Modeling and User-Adapted Interaction, vol. 22, no. 4-5, pp. 441-504, 2012.

[15] L. Piwek, D. A. Ellis, S. Andrews and A. Joinson, "The rise of consumer health wearables: Promises and barriers," PLoS Med., vol. 13, no. 2, 2016.

[16] H. op den Akker, M. Cabrita, R. op den Akker, V. M. Jones and H. J. Hermens, "Tailored motivational message generation: A model and practical framework for real-time physical activity coaching," Journal of Biomedical Information, vol. 55, pp. 104–115, 2015.

[17] H. J. A. op den Akker, R Klaassen and A Nijholt, "Virtual Coaches for Healthy Lifestyle" In: Toward Robotic Socially Believable Behaving Systems, Volume II., Intelligent Systems Reference Library, vol. 106, A. Esposito and L. Jain, Eds. Springer, Cham, 2016.

[18] Organisation for Economic Co-operation and Development (OECD 2012) "Sick on the job? Myths and realities about mental health and work", ISSN: 22257985.

[19] M. Radha, M. C. Willemsen, M. Boerhof and W. A. IJsselsteijn, "Lifestyle Recommendations for Hypertension through Rasch-based Feasibility Modeling", Conference on User Modeling Adaptation and Personalization (ACM, 2016), New York, NY, USA, pp. 239-247.

[20] S. Saragih, "The effects of job autonomy on work outcomes: Self-efficacy as an intervening variable," International Research Journal of Business Studies, vol. 4, no. 3, pp. 203-215, 2015.

[21] O. Sawyerr, S. Srinivas and S. Wang, "Call center employee personalityfactors and service performance," Journal of Services Marketing, vol. 23, no. 5, pp301-317, 2009.

[22] W. B. Schaufeli, A. B. Bakker and W. van Rhenen, "How changes in job demands and resources predict burnout, work engagement, and sickness absenteeism," Journal of Organizational Behavior: The International Journal of Industrial, Occupational and Organizational Psychology and Behavior, vol. 30, no. 7, pp. 893-917, 2009.

[23] G. Spanakis, G. Weiss, B. Boh, L. Lemmens, and A. Roefs, "Machine learning techniques in eating behavior e-coaching", Pers Ubiquit Comput, vol. 21, no. 4, pp. 645-659, 2017.

[24] D. D. van Jaarsveld, D. D. Walker and D. P. Skarlicki, "The role of job demands and emotional exhaustion in the relationship between customer and employee incivility," Journal of Management, vol. 36, no. 6, pp. 1486-1504, 2010.

[25] A. Watson, T. Bickmore, A. Cange, A. Kulshreshtha, A. and J. Kvedar, "An internet-based virtual coach to promote physical activity adherence in overweight adults: randomized controlled trial," Journal of medical Internet research, vol. 14, no. 1, p. e1, 2012.

[26] D. Wheatley, "Autonomy in paid work and employee subjective well-being," Work and Occupations, vol. 44, no. 3, pp. 296-328, 2017.

[27] J. M. Williams and M. B. Andersen, "Psychosocial antecedents of sport injury: Review and critique of the stress and injury model'," Journal of Applied Sport Psychology, vol. 10, no. 1, pp. 5–25, 1998.

[28] M. Zito, F. Emanuel, M. Molino, C. G. Cortese, C. Ghislieri and L. Colombo, "Turnover intentions in a call center: The role of emotional dissonance, job resources, and job satisfaction," PloS one, vol. 13, no. 2, 2018.

Constructing Healthcare Ontologies of any Data Format

Athanasios Kiourtis, Argyro Mavrogiorgou, Dimosthenis Kyriazis
Department of Digital Systems
University of Piraeus
Piraeus, Greece
e-mail: {kiourtis, margy, dimos}@unipi.gr

***Abstract*—Current health systems are becoming stronger and more efficient, with promises for better and healthier life quality. However, one of the prerequisites for this promise is that health systems are able to connect and interact with each other, by quickly and seamlessly exchanging data, using open standards and bypassing interoperability constraints. For that purpose, several researches have been proposed focusing on the field of interoperability, dealing with specific one-to-one scenarios of data transformation. Among these solutions, the translation of healthcare data into ontologies is considered as a solution towards interoperability. However, healthcare data can be found in multiple formats, while most of the current approaches are dealing with specific data formats and designs. To address this gap, in this paper, we are proposing an extended approach to create an ontology model from any data format, taking into account complex cases arising from multiple data design styles, by transforming the healthcare data into XML Schema Graphs (XSG) for providing it as an input to the Apache Jena, and finally generating Resource Description Framework (RDF) entities.**

Keywords—healthcare; interoperability; ontologies; heterogeneous data

I. INTRODUCTION

Worldwide, healthcare is at the intersection of the introduction of disruptive digital innovations, with digital health revolutions playing a significant role in decreasing long-term healthcare costs, enabling better healthcare outcomes, and empowering both the patient and the healthcare provider with real-time data and connections with each other [1]. However, for this promise it is considered crucial that health systems are able to communicate with each other, by quickly and seamlessly exchanging data using open standards [2]. Nevertheless, one of the biggest issues in healthcare research is the ability of researchers to obtain healthcare record data, as information silos of healthcare data exist across both private and public sectors, affecting both patient care and medical research [3]. In order to overcome these information silos, there is a great need for legislative directives that allow and encourage data sharing across both federal and commercial healthcare environments, including security requirements to protect personally identifiable information and protected health information [4]. Even with these directives, issues will remain regarding information exchange, since data has to be shared with as many stakeholders as possible. To this domain, interoperability is the only sustainable way for letting systems to talk with one another and getting the complete image of a patient [5], considered of great importance to overcome today's fragmented and proprietary global health systems, allowing stakeholders to make the most of health data, including extending, upgrading, and preserving it. According to [6][7], while global annual health spending reached $7.077 trillion dollars in 2015, this metric should increase to $8.734 trillion dollars by 2020. Moreover, according to [8], researchers predict that the healthcare market will have a value of $18.7 billion by 2020, up from a size of $5.8 billion in 2015. What is more, the global healthcare middleware market is expected to reach $3.07 billion by 2023 from $1.90 billion in 2018, in order to overcome healthcare interoperability issues [9]. Such issues include exchanging healthcare data between disparate systems to support care coordination, population health management initiatives and most importantly to stay Health Insurance Portability and Accountability (HIPAA) compliant to support meaningful use of data and protect the privacy of individuals [10]. Based on [11], 86% of the mistakes made in healthcare are due to administrative reasons, where 30% of clinical tests have to be re-ordered since the results cannot be found. Moreover, the same report has showed that patient charts cannot be found on 30% of visits, while about 80% of all serious medical errors involve miscommunication during care transitions.

Henceforth, it is undeniable that healthcare provisioning and research require healthcare data to be restructured into a common format and standard terminologies. In general, healthcare data can be either structured such as in relational databases, or semi-structured such as in eXtensible Markup Language (XML) data sources [12]. Therefore, data sources can be heterogeneous in syntax, schema, or semantics, thus making data interoperation a difficult task [13]. Consequently, syntactic heterogeneity is caused by the use of different models or languages, schematic heterogeneity is caused by structural differences, while semantic heterogeneity is caused by different meanings or interpretations of the data in various contexts. In all these cases, interoperability can be achieved by manufacturing medical domain ontologies for representing the different concepts, relationships, and axioms among the different healthcare datasets, and for representing medical terminology systems [14]. More specifically, ontologies provide a promised technology to solve especially the semantic heterogeneity problem, as a formal, explicit specification of a shared conceptualization [15], referring to an abstract model of phenomena in the world, having identified the relevant concepts of those phenomena. 'Explicit' means that the type of concepts used and the constraints on their use are explicitly defined. 'Formal' refers to the fact that the ontology should be machine readable, while 'Shared' means that the ontology should capture consensual knowledge accepted by the communities.

In all the aforementioned cases, healthcare interoperability is currently delivered through the HL7 Fast Healthcare Interoperability (HL7 FHIR) standard [16], which despite its wide adoption, it still needs much time to become a global healthcare data exchange standard, as there exist systems that still produce data that are not related to it. Taking into consideration these challenges, in [17] a holistic approach has been presented for achieving interoperability through the transformation of healthcare data into its corresponding HL7 FHIR structure. Shortly, the provided mechanism was building the healthcare ontologies that were primarily stored into a triplestore, in order to identify and compare their syntactic and semantic similarity with the HL7 FHIR Resources. Consequently, according to the aggregation of the syntactic and semantic similarity results, the matching and translation to the HL7 FHIR was taking place. In this paper, we are going to describe the mechanism of the automatic ontology creation of any data format that is currently in [17], taking into consideration that dozens of ontology creation mechanisms have been developed over the last decade, and nearly all of them are able to perform transformations only on specific data formats and standards (e.g., csv or xml files).

The remainder of the paper is organized as follows. In Section II, the related work is illustrated, while Section III depicts the overall approach with regards to the automatic ontology creation of any data format. Section IV includes the results of the evaluation of the proposed approach, while Section V presents our concluding remarks and future directions.

II. Related Work

A. Ontologies in Healthcare

Current healthcare systems need to have the ability to communicate complex and detailed medical data in a secure, automated, and effective way. This can be achieved by building medical domain ontologies [18], in order to represent medical terminology systems. In general, an ontology represents a common, shareable, and reusable view of a particular application domain that gives meaning to information structures that are exchanged among different information systems. As a result, ontology's aim is to achieve shareable knowledge for transmitting it between different stakeholders and application systems. Henceforth, the main goal is to create semantics in a generic way, providing the basis for agreement within a domain. Generally, the advantage of wrapping each information source to a local ontology is to allow the development of a source ontology independently of other sources or ontologies. Engineering new ontologies is not a deterministic process, since many design decisions must be made, while the designers' backgrounds and the target application, influence their decisions in different ways.

In the context of healthcare, ontologies may bring the indispensable integration of knowledge and data [19]. They are composed of existing medical knowledge, providing a way for building systems that help healthcare providers to perform clinical actions in a more efficient and effective way. Ontologies also specify semantic-based criteria that support different statistical aggregations for different purposes in healthcare, improving the accuracy of diagnoses by providing real time correlations of symptoms, test results and individual medical histories through standards-based systems for systematic crosschecking diagnoses. To this context, the authors in [20] developed a healthcare information system based on ontology methods, where healthcare practitioners were able to construct ontologies about new diseases, being easier to predict upcoming diseases. Furthermore, the authors in [21] built an ontology-based healthcare context information model to implement a ubiquitous environment. In that case, contextual information was extracted and classified to implement the healthcare services using the context information model that could be defined through ontologies. As a result, application and healthcare service developers could use the sensed information in various environments by authoring device- and space-specific ontologies based on this common ontology. The research of [22] should be mentioned, where the authors proposed a framework and toolset that could provide a secure single point of access to a client's full picture of her personal health information, by proposing an ontology-based framework. This framework was an independent tool that could automatically gather and combine a client's health information from the various providers in their circle of care and provide the information securely and electronically without inconveniencing the client with multiple requests and sharing agreements. Finally, the researchers in [23] developed an ontology-based toolkit for improving the field of semantic interoperability in healthcare data. Shortly, the authors identified and specified the potential data sources, they conceptualized their semantic meaning and defined to what extent routine data could be used as a measure of the process or outcome of care required, in order to finally formalize and validate the final ontology.

B. Ontology creation from different data sources

In the field of ontology creation, several strategies and researchers have been developed for deriving ontologies mainly from heterogeneous XML data sources. Some of these approaches target on a general mapping between XML and Resource Description Framework (RDF) [24], while others aim at mapping XML Schema [25] to Web Ontology Language (OWL) [26]. In this domain, the authors in [27] proposed direct mappings from XML Schema to OWL, describing mappings from XML to RDF graphs, where the generated instances did not necessarily respect the ontology created from the XML Schema. The XML Schema to OWL mapping process was based on a set of interpretation and transformation rules from XML Schema to OWL. What is more, the authors in [28] developed a tool to create an OWL ontology from an XML Schema, and convert XML data to OWL instances compliant to the created ontology, through the development of four Extensible Style Sheet Language Transformation (XSLT) [29] instances to transform XML files to OWL, without any other intervention on semantics and structures during the transformation. The research in [30] proposed a tool that aimed at building an OWL ontology from an XML data source. This method was based on XML Schema to automatically generate the ontology structure, as well as a set of mapping bridges between the entities of the XML data source and the created ontology, contributing into query translation between OWL and XML. The authors in [31] proposed a framework that aimed to generate an ontology from a large source of XML Schemas. They presented a set of patterns that enabled the direct, and automatic transformation from XML Schema into OWL, allowing the integration of huge amounts of XML data in the Semantic Web. They focused on an advanced logical representation of XML Schema components and presented an implementation that was possible to mine XML Schema sources in order to extract enough knowledge to build semantically correct ontologies with considerable expressivity. Moreover, the research in [32] proposed an approach to construct OWL ontology from XML document with the help of entity relation model. The authors proposed an XML-to-Relational (XTR) mapping approach to map an XML document to an entity-relation model, and then a Relational-to-Ontology (RTO) mapping approach to map an entity-relation model to an OWL ontology. What is more, the authors in [33] proposed an approach to integrate heterogeneous XML sources using an ontology-based mediation architecture. The ontology integration process contained the schema transformation and ontology merging steps, used for modelling each XML source as a local RDF ontology. Finally, the authors in [34] were based on XML Schema to build an ontology. In the case that the schema did not exist, it could be automatically generated from the source XML document, coping with all the possible complex cases that could arise from different XML Schema design styles.

C. *Key contributions of the proposed approach*

Several researches have been proposed so far in the literature focusing on the healthcare interoperability field, whereas dealing with specific one-to-one scenarios of data transformation. Among these solutions, the translation of healthcare data into ontologies is considered as a solution towards interoperability. However, healthcare data can be found in multiple data formats, while during ontology transformations, different terms are produced for the same concept. Most of the developed approaches are dealing with specific data format (i.e., mostly XML data format), while they do not tackle the question on how to create the ontology model, if there is not any XML Schema available. To address this gap, in this paper we are proposing an extended approach to create an ontology model from any data format, taking into account even the most complex cases arising from different data formats and design styles.

III. PROPOSED APPROACH

This study is based on an existing approach [17] that can transform any healthcare dataset of any format and nature, into HL7 FHIR through the translation of the latter into ontologies, and their matching through syntactic and semantic similarities. In our case, we are going to propose a mechanism for describing the automatic creation of the healthcare ontologies for the ingested healthcare dataset. Fig. 1 illustrates the overall architecture of the mechanism as depicted in [17], where the components that are highlighted with grey color are going to be ignored since they have already been discussed. Therefore, in the proposed mechanism only the Ontology Building System is going to be described in deep details.

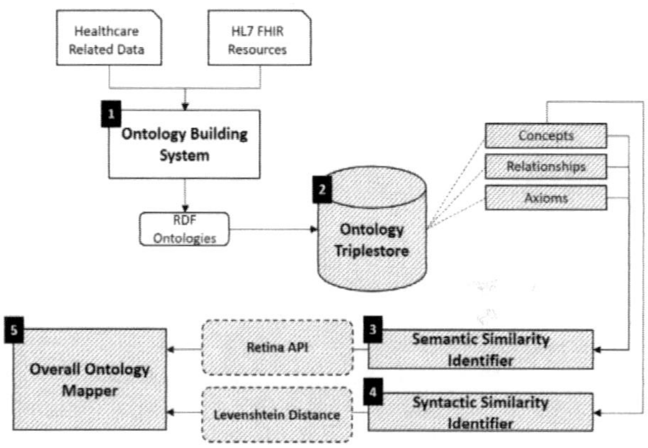

Figure 1. Overall architecture.

A. *Ontology Building System*

The objective of the Ontology Building System is to gather the XML Schema of the gathered healthcare datasets, and conclude in building the corresponding ontologies as described in the following five steps (Fig. 2):

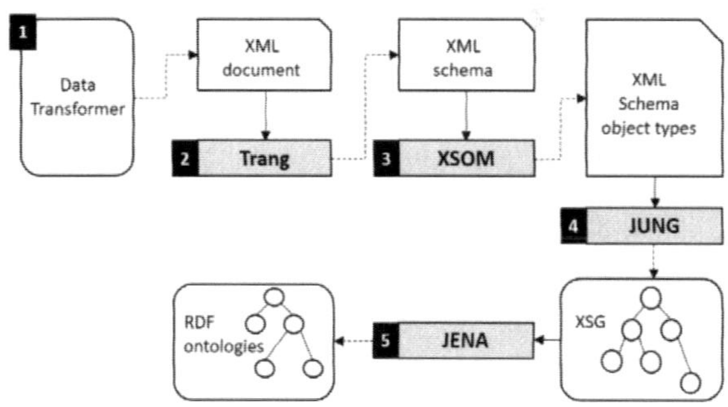

Figure 2. Ontology Building System.

1. Each healthcare dataset is being preprocessed through a document parser (Data Transformer) in order to identify the type of the document in which the dataset is stored (e.g., JSON, CSV, etc.). In any case, the Data Transformer converts the type of the healthcare dataset into XML format, by identifying the different elements of the document, as well as the parent and the child nodes, along

with their attributes. Henceforth, all the converted data is being stored into an XML document.

2. The created XML document is transformed into XML Schema using the Trang API [35]. Shortly, Trang performs conversions between different schema languages for XML, and is constructed around a RELAX NG object model [36] designed to support schema conversion, taking as input a file written in XML syntax and producing an XML Schema.

3. Afterwards, the XML Schema is analyzed using the XML Schema Object Model (XSOM) [37]. The SOM provides a set of classes in the System.Xml.Schema namespace that allows the reading of a schema from a file or the creation of a schema in-memory. The schema can then be traversed, edited, compiled, validated, or written to a fileXSOM. In this context, XSOM allows applications to easily parse XML Schema documents and inspect information into them.

4. Sequentially, the output of the XSOM is used as an input for the Java Universal Network/Graph framework (JUNG) [38], which is used for graph-based manipulations. The latter, generates an XML Schema Graph (XSG) [39] that describes the schema in the same way whatever its design style is.

5. Finally, the XSG is used as an input to the Apache Jena [40], in order to generate ontologies in the form of RDF entities [41]. It should be mentioned that RDF entities emerge from complex types, element group declarations, and attribute-group declarations according to specific matching rules. Apart from these, ontologies contain different object properties that emerge from relationships that are of type element-subelement relationships, while they contain datatype properties that emerge from attributes and from simple types.

IV. EVALUATION

A. Use Case description

The use case exploits a healthcare dataset structured in CSV format, covering all the previous steps. The dataset that has been used to evaluate the efficiency of the proposed approach (Fig. 3) was a sub-dataset of anonymized citizens' personal information, derived from Karolinska Institute [42].

```
subject;gender;dateOfBirth;dateOfDeath;causeOfDeath
ID0054619;1;1/6/1954;NA;NA
ID1012175;0;1/6/1968;NA;NA
ID0332172;0;1/11/1961;NA;NA
ID0186244;0;1/7/1963;NA;NA
ID0782623;1;1/5/1957;NA;NA
ID0944842;1;1/5/1961;NA;NA
ID0131218;1;1/7/1976;NA;NA
ID0131218;1;1/12/1951;NA;NA
ID0911095;1;1/11/1945;NA;NA
ID0951273;0;1/5/1928;20060212;C2299
ID0695153;1;1/4/1980;NA;NA
ID0695153;0;1/4/1977;NA;NA
ID0695153;1;1/4/1959;NA;NA
ID0695153;1;1/6/1951;NA;NA
ID1164801;0;1/1/1978;NA;NA
ID1409311;0;1/9/1960;NA;NA
ID0097871;0;1/8/1950;NA;NA
ID1095873;1;1/6/1924;NA;NA
ID0250381;0;1/10/1954;NA;NA
```

Figure 3. Use case dataset sample.

In more details, it consists of 5000 different instances of the personal information of certain citizens about their:

(i) subject: string that depicts the personal identifier
(ii) gender: integer that varies between 0 (male) and 1 female
(iii) date of birth: date/time that depicts the birth date
(iv) date of death: date/time that depicts the death date
(v) cause of death: string that depicts the cause of death

B. Application of the Ontology Building System

In order to depict the proposed ontology creation process, after providing as an input the described dataset, we gather the following results, for each different step of the described process of Section III.

Initially, after the preprocessing of the CSV use case dataset through the Data Transformer, the results of the XML creation of this dataset are represented in Fig. 4.

```
<root>
    <entry id="1">
        <subject>ID0054619</subject>
        <gender>1</gender>
        <dateOfBirth>1/6/1954</dateOfBirth>
        <dateOfDeath>NA</dateOfDeath>
        <causeOfDeath>NA</causeOfDeath>
    </entry>
    <entry id="2">
        <subject>ID1012175</subject>
        <gender>0</gender>
        <dateOfBirth>1/6/1968</dateOfBirth>
        <dateOfDeath>NA</dateOfDeath>
        <causeOfDeath>NA</causeOfDeath>
    </entry>
    <entry id="3">
        <subject>ID0332172</subject>
        <gender>0</gender>
        <dateOfBirth>1/11/1961</dateOfBirth>
        <dateOfDeath>NA</dateOfDeath>
        <causeOfDeath>NA</causeOfDeath>
    </entry>
    <entry id="4">
        <subject>ID0186244</subject>
        <gender>0</gender>
        <dateOfBirth>1/7/1963</dateOfBirth>
        <dateOfDeath>NA</dateOfDeath>
        <causeOfDeath>NA</causeOfDeath>
    </entry>
```

Figure 4. XML document of the use case dataset.

In sequence, the XML Schema that is created through the derived XML document is represented in Fig. 5.

```
<?xml version="1.0" encoding="UTF-8"?>
<xsd:schema targetNamespace=""
elementFormDefault="qualified"
xmlns:xsd="http://www.w3.org/2001/XMLSchema">
    <xsd:element name="root">
        <xsd:complexType mixed="true">
            <xsd:sequence>
                <xsd:element name="entry" maxOccurs="unbounded">
                    <xsd:complexType mixed="true">
                        <xsd:sequence>
                            <xsd:element name="subject" minOccurs="0"
                            type="xsd:normalizedString"/>
                            <xsd:element name="gender" minOccurs="0"
                            type="xsd:int"/>
                            <xsd:element name="dateOfBirth" minOccurs="0"
                            type="xsd:normalizedString"/>
                            <xsd:element name="dateOfDeath" minOccurs="0"
                            type="xsd:normalizedString"/>
                            <xsd:element name="causeOfDeath" minOccurs="0"
                            type="xsd:normalizedString"/>
                        </xsd:sequence>
                        <xsd:attribute name="id" type="xsd:int"
                        use="required"/>
                    </xsd:complexType>
                </xsd:element>
            </xsd:sequence>
        </xsd:complexType>
    </xsd:element>
</xsd:schema>
```

Figure 5. XML Schema of the use case dataset.

The next step of the proposed mechanism deals with the analysis of the XML Schema through the XSOM, whose results are provided as an input to the JUNG in order to finally generate the XSG. The latter is represented in Fig. 6.

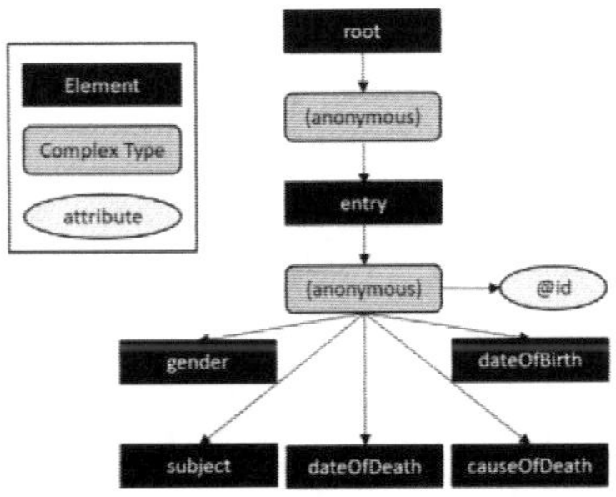

Figure 6. XSG of the use case dataset.

To this end, the XSG is used as an input to the Apache Jena to generate the RDF entities that are depicted in Fig. 7.

```
<rdf:Description rdf:nodeID="A14">
  <owl:allValuesFrom
  rdf:resource="http://uri-not-absolute.com#entry"/>
  <owl:onProperty rdf:resource="http://uri-not-absolute.com#entry"/>
  <rdf:type rdf:resource="http://www.w3.org/2002/07/owl#Restriction"/>
</rdf:Description>
<rdf:Description rdf:nodeID="A6">
  <owl:allValuesFrom
  rdf:resource="http://www.w3.org/2001/XMLSchema#normalizedString"/>
  <owl:onProperty rdf:resource="http://uri-not-absolute.com#subject"/>
  <rdf:type rdf:resource="http://www.w3.org/2002/07/owl#Restriction"/>
</rdf:Description>
<rdf:Description rdf:nodeID="A11">
  <owl:allValuesFrom
  rdf:resource="http://www.w3.org/2001/XMLSchema#normalizedString"/>
  <owl:onProperty rdf:resource="http://uri-not-absolute.com#dateOfBirth"/>
  <rdf:type rdf:resource="http://www.w3.org/2002/07/owl#Restriction"/>
</rdf:Description>
<rdf:Description rdf:nodeID="A0">
  <owl:maxCardinality
  rdf:datatype="http://www.w3.org/2001/XMLSchema#int">1</owl:maxCardinality>
  <owl:onProperty rdf:resource="http://uri-not-absolute.com#dateOfBirth"/>
  <rdf:type rdf:resource="http://www.w3.org/2002/07/owl#Restriction"/>
</rdf:Description>
```

Figure 7. RDF entities of the use case dataset.

Finally, Fig. 8 represents the final output of the mechanism, depicting the ontological hierarchical tree of the use case dataset, as it is visualized by WebVOWL [43].

Figure 8. Ontological hierarchical tree of the use case dataset.

V. Conclusions

Currently, health systems are becoming stronger and more efficient, with promises for better and healthier life quality. However, one of the prerequisites for this promise is that health systems are able to interact with each other, by seamlessly exchanging data using open standards, thus bypassing interoperability constraints. In this domain, healthcare ontologies have been developed to resolve data heterogeneity issues, resulting in interoperability by knowledge sharing and reuse. Most of the developed approaches that aim in ontology building are dealing with specific data format, while they do not tackle the question on how to create the ontology model, if there is not any XML Schema available. In order to address the gap of automated ontology creation, a previous research was considered about a healthcare ontology matching mechanism that has the ability of transforming healthcare data to HL7 FHIR format, by building healthcare ontologies, and finding syntactic and semantic similarities among these ontologies and the ontologies of the HL7 FHIR Resources. In that case, an extended approach was presented that is able to create an ontology model from any data format, taking into account complex cases arising from different data formats and design styles. According to the results of the evaluation of the proposed approach, the overall ontology creation mechanism provided reliable results that are compatible with the manually derived results.

Generally, writing an automated ontology building mechanism that is able to cover multiple data formats is still a challenging research task, since there exist several mechanisms that provide the same functionalities. In this context, we will work on the evaluation of the Ontology Building System with the similar mechanisms that were studied and are currently used, in order to identify potential errors and gaps that could be addressed in the current approach. What is more, we want to perform a similar comparison and evaluation concerning the other sub-mechanisms of the current approach for healthcare interoperability, in order to conclude to more reliable and efficient results. To this end, we will continue on the evaluation of the Ontology Building Systems with datasets of different sizes, standards and formats, respecting privacy issues, based on the mechanism developed in [44].

Acknowledgment

The CrowdHEALTH project has received funding from the European Union's Horizon 2020 research and innovation programme under grant agreement No 727560.

References

[1] Is digital health the future of healthcare?, https://www.forbes.com/sites/quora/2018/04/04/is-digital-health-the-future-of-healthcare/#201d7f592556 [last accessed: 07/01/2019]

[2] Data Integration and Interoperability: ISO/OGC, https://www.directionsmag.com/article/3396 [last accessed: 07/01/2019]

[3] Why Health Systems Must Embrace Interoperability Now, https://www.webpt.com/blog/post/so-long-to-silos-why-

health-systems-must-embrace-interoperability-now [last accessed: 07/01/2019]

[4] Data Silos, https://www.himss.org/news/data-silos-together-we-can-bust-them [last accessed: 07/01/2019]

[5] Moving Forward with Interoperability, https://www.philips.com/a-w/about/news/archive/blogs/innovation-matters/moving-forward-with-interoperability-how-patients-will-drive-change.html [last accessed: 07/01/2019]

[6] The Economist Intelligence Unit (EIU) is the research and analysis division of The Economist Group and the world leader in global business intelligence, http://www.eiu.com/home.aspx [last accessed: 07/01/2019]

[7] Life Sciences and Healthcare, https://www2.deloitte.com/global/en/pages/life-sciences-and-healthcare/articles/global-health-care-sector-outlook.html [last accessed: 07/01/2019]

[8] Healthcare analytics market worth $18.7 billion by 2020, https://www.whatech.com/market-research/medical/archive/192126-healthcare-analytics-market-worth-18-7-billion-by-2020 [last accessed: 07/01/2019]

[9] Healthy Growth Forecast for Medical Devices Global Market, https://www.bccresearch.com/pressroom/hlc/healthy-growth-forecast-for-medical-devices-global-market [last accessed: 07/01/2019]

[10] Healthcare Security: Understanding HIPAA Compliance and its Role in Patient Data Protection, https://digitalguardian.com/blog/healthcare-security-understanding-hipaa-compliance-and-its-role-patient-data-protection [last accessed: 07/01/2019]

[11] 3 Signs You Should Invest in Healthcare Automation, https://www.formstack.com/blog/2017/invest-in-healthcare-automation/ [last accessed: 07/01/2019]

[12] R. Arnon, L. Seligman, and R. Costello, "XML, Databases, and Interoperability." Federal Database Colloquium, AFCEA, San Diego, 1999.

[13] I. F. Cruz, and X. Huiyong, "The role of ontologies in data integration." Engineering intelligent systems for electrical engineering and communications, vol. 13(4), p.245, 2005.

[14] B. A. Gylys, and M. E. Wedding, "Medical terminology systems: a body systems approach", FA Davis 2017.

[15] E. M. Gontier, "Web Semantic and Ontology", 2015.

[16] HL7 FHIR, https://www.hl7.org/fhir/ [last accessed: 07/01/2019]

[17] A. Kiourtis, A. Mavrogiorgou, D. Kyriazis, and S. Nifakos, "Aggregating the Syntactic and Semantic Similarity of Healthcare Data towards their Transformation to HL7 FHIR through Ontology Matching", Int. Journal of Medical Informatics (IJMI), 2018. under review

[18] P. Wennerberg, "Aligning medical domain ontologies for clinical query extraction", 12th conference of the EU chapter of the association for computational linguistics: student research workshop. Association for Comp. Linguistics, 2009.

[19] Ontologies in healthcare supporting the integration of knowledge, http://www.dataversity.net/ontologies-in-healthcare-supporting-the-integration-of-knowledge/ [last accessed: 07/01/2019]

[20] V. Kumar, "Ontology based public healthcare system in Internet of Things (IoT)", Procedia Computer Science, vol. 50, pp. 99-102, 2015.

[21] J. Kim, and K. Y. Chung, "Ontology-based healthcare context information model to implement ubiquitous environment", Multimedia Tools and Applications, vol. 71(2), pp. 873-888, 2014.

[22] M. Azarm, and L. Peyton, "An ontology for a patient-centric healthcare interoperability framework", IEEE/ACM International Workshop on Software Engineering in Healthcare Systems (SEHS), pp. 34-41, 2018.

[23] H. Liyanage, P. Krause, and S. de Lusignan, "Using ontologies to improve semantic interoperability in health data". Journal of innovation in health informatics, vol. 22(2), pp. 309-315, 2015.

[24] Resource Description Framework (RDF), https://www.w3.org/RDF/ [last accessed: 07/01/2019]

[25] Xml schema, https://www.w3.org/standards/xml/schema [last accessed: 07/01/2019]

[26] Web Ontology Language (OWL), https://www.w3.org/OWL/ [last accessed: 07/01/2019]

[27] M. Ferdinand, C. Zirpins, and D. Trastour, "Lifting XML Schema to OWL", In Web Engineering - 4th International Conference, Munich, Germany, pp. 354– 358, 2004.

[28] H. Bohring, and S. Auer, "Mapping XML to OWL Ontologies", In Leipziger Informatik-Tage, vol. 72, pp. 147–156. Society, Washington, DC, USA, 2007.

[29] Extensible Stylesheet Lagnuage (XSLT), https://www.w3schools.com/xml/xsl_intro.asp [last accessed: 07/01/2019]

[30] R. Ghawi, and N. Cullot, "Building Ontologies from XML Data Sources", 1st Workshop on Modelling & Visualization of XML and Semantic Web Data, pp. 480-484, 2009

[31] I. Bedini, G. Gardarin, and B. Nguyen, "Transforming XML Schema to OWL Using Patterns", 5th IEEE Int. Conference on Semantic Computing, Palo Alto: US, pp.1-8, 2011.

[32] J. Xu, and W. Li, "Using Relational Database to Build OWL Ontology from XML Data Sources", International Conference on Computational Intelligence and Security Workshops, pp. 124– 127, 2007.

[33] I. F. Cruz, X. Huiyong, and H. Feihong, "An ontology-based framework for XML semantic integration", pp. 217-226, 2004.

[34] N. Yahia, S. A. Mokhtar, and A. Ahmed, "Automatic generation of OWL ontology from XML data source", arXiv preprint arXiv:1206.0570, 2012.

[35] Trang, http://www.thaiopensource.com/relaxng/trang.html [last accessed: 07/01/2019]

[36] Relax NG, http://www.kohsuke.org/relaxng/ [last accessed: 07/01/2019]

[37] XML Schema Object Model (XSOM), https://directory.fsf.org/wiki/XSOM [last accessed: 07/01/2019]

[38] Jung, http://jung.sourceforge.net/ [last accessed: 07/01/2019]

[39] M. Hacherouf, S. N. Bahloul, and C. Cruz, "Transforming XML documents to OWL ontologies: A survey", Journal of Information Science, vol. 41(2), pp. 242-259, 2015.

[40] Apache Jena, https://jena.apache.org/ [last accessed: 07/01/2019]

[41] RDF entities, https://www.infowebml.ws/rdf-owl/ [last accessed: 07/01/2019]

[42] Karolinska Institute, https://ki.se/ [last accessed: 07/01/2019]

[43] Web Ontology Visualizer, http://vowl.visualdataweb.org/webvowl.html [last accessed: 07/01/2019]

[44] A. Kiourtis, A, Mavrogiorgou, and D. Kyriazis, "Towards a Secure Semantic Knowledge of Healthcare Data Through Structural Ontological Transformations", In Joint Conference on Knowledge-Based Software Engineering, pp. 178-188, 2018.

The Patient Summary Case: Challenges in Archetypes Terminology Binding Using SNOMED-CT Compositional Grammar

Luis Marco-Ruiz
Norwegian Centre for e-healthresearch
Tromsø, Norway
Luis.marco.ruiz@ehealthresearch.no

Rune Pedersen
Norwegian Centre for e-healthresearch
Tromsø, Norway
Rune.pedersen@ehealthresearch.no

Abstract— **In order to cover the requirements of interoperability in the Norwegian context, we studied the adequacy of expressing the clinical semantics contained in archetypes as terminology expressions using SNOMED-CT's compositional grammar. As a result, we identified important challenges categorized as technical, expressivity, human, and models mismatch related. Technical challenges include the binding of archetype elements to sections of the SNOMED-CT expressions that are semantically equivalent, lack of tooling for performing guided binding based on pre-defined semantic patterns, and lack of guidance about the infrastructure to use ontology-based terminologies. Expressivity challenges include variations in the precision of the semantics expressed by the archetype and the SNOMED-CT models, challenges expressing temporal semantics in SNOMED-CT, and lack of mechanisms for specifying expressions whose values are only known at runtime. Human challenges include lack of guidance to discern what to leave represented at archetype level and what to project as a terminology expression depending on the expressivity requirements, and the need for more clarity about which hierarchies and attributes to use when several options are available for expressing the same concept. Model mismatch issues were related to the issue of grounding (referencing) the sections of one model to the other and clarify the role of the context model and in which situations it makes sense to annotate archetypes using its verbose expressions. The challenges detected show a pressing need for the collaboration between the openEHR community and SNOMED International for providing better guidelines about terminology binding, better tooling for facilitating the binding process, and developing mechanisms that allow for extending SNOMED-CT with other biomedical ontologies in order to increase the coverage of archetypes semantics for scenarios with high expressivity requirements such as data reuse ones.**

Keywords-openEHR; SNOMED-CT; archetype; terminology binding; biomedical ontology.

I. INTRODUCTION

Over more than a decade, health authorities worldwide have faced a growing pressure to accomplish smooth information flow between systems of health organizations. This has led to the adoption of clinical information standards and their use in combination with terminologies. Terminologies are used both to specify the meaning of the sections inside clinical information models (semantic binding), and to specify the content of the information recorded inside these sections (content binding). However, the effort in annotating clinical information models with terminologies is often underestimated [1]. In fact, few guidelines are available with clear instructions on how terminologies should be used in combination with clinical information models [2]. For example, when one faces the terminology binding of an openEHR archetype, it is not clear what elements of the archetype should be coded and which should not; when to use a pre- or a post-coordinated expression; or how to proceed when several terminology concepts are available. In fact, terminology binding guidelines often focus on the annotation of just the main sections of clinical documents, but they do not concern aspects related to the consequences of choosing an ontology-based terminology, the granularity that should be pursued when annotating archetypes, or the requirements that advanced phenotyping queries used in clinical research may pose in the terminology binding of archetypes [1][2]. These are issues that need to be carefully addressed since they will have a major impact on the interoperability of clinical data extracts, thus determining the limitations of eHealth infrastructures. For example, the needs for harmonizing primary care and secondary care health information if only openEHR is used differ from the needs when SNOMED-CT is also used as reference terminology. One could only use SNOMED-CT for semantic tagging, i.e., indicate the meaning of each section in a clinical document (e.g., "this section contains vital signs"). Or one could decide to use the terminology to actually specify content abstracting the information shared from the syntactic representation format of one of the systems. For example, an archetype measurement of 155 mmHg systolic blood pressure could be shared just as the SNOMED-CT concept "On examination - blood pressure reading very high (finding) 163028000". Also, depending on the type of terminology selected, it may be possible to perform filtering operations based on its structure and hierarchy (e.g. which patients in treatment with anticoagulants suffered from a major bleeding event last year?).

Adopting openEHR in combination with SNOMED-CT raises questions such as: which sections of archetypes should be bound to the reference terminology? Is there any benefit in fine-grained terminology binding of archetypes items? Should we use SNOMED-CT as a vocabulary or do we need to deal with its formal ontological nature to maximize the benefits?

These research questions have been a matter of discussion among the Norwegian Centre for E-health Research, the Norwegian Directorate of eHealth (sub-ordinate institution of our Ministry of Health and Care

Services), and the national and regional comities for standardization (Nasjonal IKT). In addition, the Norwegian Agency for eHealth considers the adoption of a common reference terminology a long-term commitment.

In the Norwegian context there has been a growing interest about the implications derived from ontology-based terminologies (e.g. SNOMED-CT) and the impact on present and future health information systems. As a result, the Directorate of eHealth decided the enrollment in SNOMED International and advised the acquisition and use of SNOMED CT at national level for a three-year period.
In November 2015 several projects started exploring the mapping of SNOMED-CT towards the most commonly used Electronic Medical Record (EMR) functionalities (e.g., critical information, clinical findings, the patient pathway for chronically ill patients, and the patient summary). Among the suggestions of the directorate is the elicitation of clinical information models (archetypes) and terminology value sets for building the patient summary for continuity of care.

A. The national work with openEHR archetypes in Norway

Throughout the last four years OpenEHR has grown to gain a national anchorage in Norwegian healthcare. From 2008 to 2012, the National ICT Health Trust conducted projects using archetypes for clinical chart systems, which resulted in a recommendation to build an openEHR-based infrastructure for specialized healthcare. In 2013 the National ICT Health Trust, decided to establish the National Editorial Group for Archetypes (NRUA) for coordinating the development of the national repository of clinical models (archetypes). NRUA also established a close collaboration with the openEHR foundation and the international openEHR Clinical Knowledge Manager (CKM). Through this collaboration, NRUA has translated and reviewed existing archetypes from the international CKM into the Norwegian national CKM [27]. Currently, the Norwegian CKM includes circa 400 reviewers of which around 30% are clinicians. The review process is being conducted at a national level [8] with the collaboration of contributors from all health regions. The number of archetypes approved in Norway in October 2018 was 80, and over 100 more are in process. The most immediate goal is to have 200 nationally approved archetypes by the end of 2019 to represent the EMR core content.

B. Archetypes binding to SNOMED-CT

Archetypes terminology binding is a complex, time-consuming task. Such task is even more challenging when the terminology used is an ontology-based terminology such as SNOMED-CT. In that case, both the semantics contained in the archetype and the terminology conceptual models need to be carefully analyzed. For example, when an archetype element is bound to a terminology code, clinical modelers need to consider not only the name of the code but also the hierarchy of SNOMED-CT that contains that code to determine if it is appropriate to use it or not. For example, concepts to represent the blood pressure are available in the Observable Entity and Clinical Finding hierarchies. The different choices available when binding archetypes to terminologies need to be carefully addressed since they will impact interoperability [5]. Actually, several levels of interoperability are possible: syntactic interoperability using common data schemas, partial semantic interoperability (SIOp) where some sections of clinical statements are bound to terminologies, or full SIOp based on machine-understandable models [2]. Therefore, it is important to determine several factors to design a roadmap towards interoperability. The first factor is to determine why interoperability is desired, i.e., what are the requirements that need interoperable clinical data. The second factor is to determine what level of interoperability is needed to cover those requirements. The third factor is to determine what are the tools needed to grant the level of interoperability desired. These determinants may vary from one application context to another, and from one country to another. In the case of Norway the interoperability of clinical data pursues three main objectives:

- Develop semantically enriched clinical information models that implementers can use to drive the development of health information systems.
- Facilitate mappings across different information models based on the archetypes provided.
- Facilitate the secondary use of clinical data by providing a standard format for clinical information representation and expressive queries based on standard terminologies.

In most cases the approach followed to bind archetypes to terminologies relies on assigning a terminology code to the main sections of the archetype. For example, a SNOMED-CT binding allows for specifying that the number set in the element at0078.13 of the archetype openEHR-EHR-OBSERVATION.lab_test-full_blood_count.v1 refers to the SNOMED-CT concept 250271003 /White blood cell finding. This approach endows the archetype with certain level of semantics by attaching a code that refers to a concept in the terminology. However, these semantics correspond to concepts framed within an external ontology (i.e. SNOMED-CT concept model) and may not suffice to cover the requirements previously described.

In the Norwegian case, the first requirement described (i.e. provide semantically enriched clinical information models) can be covered with that approach, however the second and third requirements cannot. When it comes to facilitating mapping tasks across different information models, the approach of binding independent terminology codes has been related to SIOp challenges. For example, Dixon et al. documented problems when establishing SIOp across organizational boundaries caused by issues interpreting the meaning of the terminology codes even when the same terminology was used by all organizations [9]. Clinical information models define a syntactic schema with rich data constraints definitions. However, the information model does not specify the semantics of concepts and relationships explicitly beyond the description associated to the code. When SNOMED-CT is used, the ontology that unambiguously defines the semantics of the concept is detached and external to the conceptual model that the archetype conveys. This also affects the third requirement since secondary use of clinical data requires running expressive queries. These queries require the machine-interpretation of semantic relations such as subsumptive or equivalence ones to manage the complexity of clinical information. Markwell documented the problems of interdependencies among the semantics of the archetype elements when they are annotated using independent terminology concepts [1]. For example, let us consider the archetype family_history whose root element is coded as 57177007 /Family history with explicit context and one of the internal nodes is coded as 73211009/Diabetes mellitus. The intended semantics are that a family member of the patient had diabetes. However, that is not unambiguously described following any formal model. Therefore, if a query returns the diabetes value one may interpret that the patient has the disease. This mismatch may cause wrong inferences and erroneous (or at least confusing) queries results if both the archetype and the terminology bindings are not analyzed carefully [10]. Another problem related to this mismatch between models is not purely technical but related to the human interpretation. Sometimes organizations using the same terminology decide to use different codes to identify the same concept, thus hampering SIOp [5][9].

Aiming to overcome these challenges and to cover the requirements set by NRUA, we determined that we needed a machine-interpretable representation of the clinical semantics carried by archetypes. We considered that the full expression of archetypes (including data constraints) as ontologies is not scalable due to computational restrictions [11], nor necessary to accomplish the requirements presented. Therefore, we opted for using SNOMED-CT compositional grammar to build semantic models that distilled the clinical semantics contained in archetypes (leaving data constraint aside). This way, each of the elements of the archetype can be represented within an ontology that provides unambiguous semantics and reasoning.

Previous studies have covered these issues for a specific set of clinical models [1], [3]-[5]. However, these studies focused mainly on other standards than openEHR and they did not consider the expressivity requirements that data reuse scenarios introduce. Many of the scenarios for data reuse and decision support requires a more complete terminological projection in order to ensure that the meaning of clinical information is preserved across systems. That is actually the case of several projects in Norway that aim to enable data reuse [6][7]. The objective of this paper is to report about the main challenges found in our evaluation for introducing SNOMED-CT in combination with the openEHR-based infrastructure for fulfilling the requirements of the Norwegian context.

II. Method

In order to specify the archetype clinical semantics as an unambiguous ontology model, the SNOMED-CT compositional grammar was used. Together with NRUA we selected the most representative archetypes of each of the sections contained in the Norwegian patient summary. The archetypes selected are archetypes reviewed nationally and published enabling interoperability at a national level. The sections of the patient summary correspond to those specified in epSOS [28]. Archetypes were downloaded from the Norwegian CKM and their estate was either published or in review. For each archetype, terminology binding was attempted by creating a projection of the archetype clinical semantics using the SNOMED-CT compositional grammar. We tried to maximize the coverage of the elements represented. The binding was performed from the root to the leaves. When some element/section of the archetype could not be represented using a SNOMED-CT expression for the whole archetype, we defined a new expression and tagged the reason for using several expressions to represent one archetype. Additionally, when a blocker caused by the complexity of the process, lack of tooling etc. was found, we tagged it. When we had doubts about the specific meaning of an archetype element we asked NRUA for clarification. Afterwards we reviewed all the tags and classified them into categories of challenges. Table 1 contains the set of archetypes used in the study.

TABLE I. SET OF ARCHETYPES COVERED.

Archetypes analyzed
openEHR-EHR-OBSERVATION.blood_pressure.v1
openEHR-EHR-EVALUATION.tobacco_smoking_summary.v0
openEHR-EHR-EVALUATION.immunisation_summary.v0
openEHR-EHR-EVALUATION.problem_diagnosis.v1
openEHR-EHR- openEHR-EHR-EVALUATION.adverse_reaction_risk.v1
openEHR-EHR-INSTRUCTION.medication_order.v0

III. Results

After reviewing all tags four categories of challenges were identified: 1) technical, 2) SNOMED-CT expressivity related, 3) human, and 4) mismatch between models.

Figure 1. Adverse reaction archetypes and their projection into SNOMED-CT.

A. *Technical challenges*

Technical challenges are associated to limitations or barriers found when using current technologies. The first technical challenge is associated with limitations to link the archetype and the SNOMED-CT expression due to a lack of available tooling. The main technical challenge found was the lack of support of archetypes to include verbose post-coordinated expressions in their term_bindings section. This could be solved by using proper URLs pointing to an external server that contains the expression provided that URIs have been recently accepted as the best approach for annotating archetypes. However, in that case, another limitation appears related to SNOMED-CT tooling. We did not find any mechanism to reference different sections of the expression. Therefore if the entire archetype is linked to a terminology expression that represents its meaning at an ontology level, it is not possible to specify to which concept or attribute of the expression the archetype element refers to. For example, in Figure 1 black arrows shows the need for establishing references from the archetype adverse_reaction_risk elements to the parts of the expression on the right. Other studies have proposed the use of linked data principles and W3C standards for referencing different ontology sections [12][13]. However, appropriate tooling to make that process transparent to clinical modelers is necessary. These referencing mechanisms are something that, to our knowledge, have not been covered in the medical informatics literature and, in many cases, have been considered too complex to achieve. However, it is a common and necessary operation named "grounding" in semantic web programming [12]-[14]. Such operation is necessary for establishing mappings and transformation mechanisms between the syntactic and ontological layers in semantic web applications. In fact, recent projects such as the Yosemite Project [15] and HL7 FHIR RDF [16] are pursuing the serialization of information models into RDF(S) for facilitating clinical information mapping, retrieval, and management.

The second limitation is related to the lack of tooling to assist the binding process. The SNOMED-CT compositional grammar contains a huge variety of attributes and hierarchies. The correct specification of semantics depends on choosing the correct hierarchy and attribute. However, at the moment, this is an extremely difficult task since one needs to check continuously the compositional grammar and the semantics of each attribute. Tooling to support this process by offering guidance in choosing the appropriate elements/attributes is needed. Those tools should guide clinical modelers not only by restricting the set of values that can be assigned to an attribute, but also by informing about the exact meaning and provide examples during the coding workflow.

The last technical challenge identified is the information about the infrastructure to process SNOMED-CT expressions. Although some examples of reasoners are mentioned, exploiting the expressivity of SNOMED-CT in large deployments may impose high demands on the hardware infrastructure required. Guidelines for software architects should be provided to allow them planning enterprise deployments.

B. *Expressivity challenges*

Challenges in expressivity relate to those concepts which semantics cannot be fully expressed with the compositional grammar. Some occur because the archetype element coded has a candidate in SNOMED-CT that introduces a slight variation in the semantics originally intended in the archetype. That is the case of "reaction mechanism" (archetype element) vs. "Immune hypersensitivity reaction by mechanism (disorder)" (SNOMED-CT concept). This problem has been approached elsewhere by extending SNOMED-CT [14]; however, since nationally approved archetypes intend to be as generic as possible, divergences between the SNOMED-CT implementation used for coding and the standard release should be avoided.

Sometimes the variation in semantics occurred as a consequence of relying on a SNOMED-CT attribute that has a broader meaning. An example is displayed in Figure 1 where the element substance can be mapped to the attribute "due to" of the SNOMED-CT expression, which has a wider range than substances (any ClinicalFinding; any Procedure etc.) Another loss of meaning occurs in the example where the archetype element "onset of reaction", with an openEHR data type Date/time, acquires the meaning "date and time of the onset of reaction". However, the SNOMED-CT candidate for matching is "Date of last episode". One should note that it is unknown if it refers to the onset, end etc. Other expressivity problems occur for not being able to express contextual semantics. The SNOMED-CT context model does not allow expressing temporal orders. This was expected since archetypes typically provide much more contextual information (epistemology) than biomedical ontologies [18]. This can be seen in Figure 1 where the section Reaction event (with meaning previous adverse events) is mapped the expression section with the Event that wraps the finding adverse reaction. We did not find the way to specify this kind of semantics that often occur in patient summaries of drugs, allergies etc. The time context SNOMED-CT attribute allows expressing if the situation occurred in the present or past but not order of events. Besides, in the compositional grammar, the time context attribute cannot be used as a valid attribute of an event. Other type of expressivity problems are those related to the restrictions that the compositional grammar imposes. For some nodes there exists a valid candidate in SNOMED-CT. However, it is not possible to include it using the compositional grammar or the expression needs to be overcomplexified for including it. In some cases, we found elements that cannot be coded due to a total lack of candidates available. Examples are "Duration of exposure" or "initial exposure" element in Figure 1. This is usual in concepts whose semantics are purely contextual such as time related properties.

Other expressivity challenge concerns the need for leaving the terminology projection incomplete. Marked with green italics in Figure 1. This situation is common when the expression should contain a Qualifier Value after an attribute. In Figure 1 it is seen how the substance that causes the adverse reaction should map to the attribute "due to" that may contain a substance such as penicillin. However, this value is unknown at archetype design time and will only be determined in runtime for each instance. In order to overcome these situations a mechanism to indicate that the value is available at instance level would be needed. Although SNOMED-CT is not intended to manage instances, it would be appropriate to provide guidelines for determining how expressions that have a dependence on the archetype instance values should be managed. Finally, we detected problems related to differences between the semantics of the conceptual model carried by the archetype and the semantics that can be expressed with SNOMED-CT. This becomes evident in the archetype openEHR-EHR-EVALUATION.problem_diagnosis.v1 that uses the same clinical model for expressing both the Problem (Disorder) and the Diagnosis, whereas in SNOMED-CT these are separated concepts. A solution for this could be to use the compositional grammar to specify a concept that inherits from two similar concepts such as "Clinical finding+Disorder" (or in this case clinical finding only). However, the archetype semantics have a wider meaning that includes concepts such as Diagnosis (439401001) that may belong to a different hierarchy in SNOMED-CT. In that case it is not possible to build concepts such as "Clinical finding+Diagnosis" since the compositional grammar discourages the combination of concepts that come from different hierarchies [19].

C. *Human challenges*

Human challenges relate to the difficulties found by clinical modelers inherent to the complexity of the process. First, we had many doubts in determining what to represent in the SNOMED-CT expression and what to leave unrepresented. Although we had determined that we should aim for a maximum representativeness to grant SIOp and enable expressive queries for clinical research, it became clear that we had not specified use cases with the appropriate level of detail to determine what sections should be represented with the terminology expression. To avoid this challenge it is needed to have a clear set of representative use cases related to the scenarios that will exploit the terminology. We acknowledge that anticipating that use is somehow idealistic at a national level. However, a set of use cases related to each of the requirements explained would help to narrow down the binding task.

The second human challenge is related to the one just described and is the lack of guidance about what can be expressed in the terminology; i.e. guidance on what to represent. This has been already reported by studies using independent codes for the sections of the clinical model [1]. However, it is needed that both SNOMED-CT implementers and archetype editors define general directives on what will be the minimum set to code and provide guidelines with examples for that. This is extremely important when the SNOMED-CT context model is used since it opens the door to terminology projections of a high percentage of the information model elements that previously were maintained only at the syntactic level. Finally, another aspect linked to the need of guidance is how to choose the appropriate attributes. Previously it was mentioned the need of tooling, but also guidelines are needed. An example is found when deciding on expressing the blood pressure using the Observable Entity hierarchy or the Clinical Finding hierarchy. Some studies have analyzed this issue [5], but this knowledge needs to be translated into pragmatic terminology binding guidelines.

D. *Model mismatch*

This category identifies the challenges related to structural differences between the archetype and the expression created with the compositional grammar. A first structural challenge comes from the fact that the context model sets many of the contextual information wrapping clinical statements. However, in the archetype this is in many cases the opposite complicating to establish equivalences between the archetype and the SNOMED-CT representation of the clinical concept (i.e. grounding operation). Again SNOMED-CT and archetype editors should provide guidelines about the alignment of both models and the way of referencing one to the other. Finally, we found that the attributes of the protocol section had a very low coverage. Examples appear, for example, in the blood_pressure archetype for cuff size or systolic pressure formula. This was expected since they refer mainly to contextual information, but we found that it is necessary to clarify the use of SNOMED-CT context model with regards to clinical information models in general, and archetypes in particular.

IV. DISCUSSION

The long term needs of semantic interoperability across EHRs and secondary use of data in the Norwegian context need the representation of archetype semantics as expressions that allow for defining them unambiguously and performing expressive queries over data sets. Terminology binding of only some archetype sections does not provide that level of expressivity. Expressing them by using the ontology of the terminology becomes necessary. We have coded terminological expressions using SNOMED-CT's compositional grammar using the archetypes that shape the Norwegian patient summary. Technical, expressivity, human and models mismatch challenges have been identified. The first challenge found is that currently SNOMED-CT expressions cannot be referenced from archetypes. On the one hand, it is not possible to include long post-coordinated expressions. Thus, the use of URLs to reference the expression should be recommended. This is a recommendation that has currently be taken into the openEHR specifications. The second challenge found is the lack of tooling available to guide the definition of expressions, not only validating post-coordinated expressions, but also assisting modelers while using the compositional grammar. In many cases, the need for continuously checking SNOMED-CT guidelines made us lose track of our own work. Nowadays, the availability of published archetypes is growing [6]. Tools that define general use cases to represent different types of archetypes would be very useful. For example, there are tools to guide clinical modelers in coding archetypes with graphical representations and powerful matching techniques [20]. This kind of tools may consider incorporating the functionality to guide users for building expressions using SNOMED-CT´s compositional grammar. Another challenge is related to the expressivity of SNOMED-CT. For simple archetypes that relate to readings such as blood pressure, it is possible to create terminology projections of archetypes´ semantics using the compositional grammar. However, for more complex archetypes such as the adverse_reaction or medication_order ones, the context model does not allow defining the relations that are needed to express them at a semantic level. This problem varied from issues related to the inability to include a concept in the compositional grammar, loss of semantics due to the variation of meaning and, in the worst case, not been able to express relations between archetype sections. This may influence the patterns that archetypes follow. We are aware that previous experiences show that the best way to approach interoperability is to annotate just some sections of the clinical information models (archetypes) [3]. However, for the national interoperability program it is important to determine until what extent the expressivity of SNOMED-CT can be used to define the semantics covered by archetypes and the implications of adopting ontology-based terminologies. This was part of the national project to study the use of formal ontologies for healthcare and the impact and possibilities regarding their adoption [2]. In that work we determined that there is not only one way to approach this challenge since determining the level of semantic enrichment of archetypes depends highly on each use case. For EHR interoperability it may suffice binding just the main sessions of the EHR, whereas for performing deep phenotyping [21], it would be necessary to define fine-grained annotations with several biomedical ontologies [2].

Both terminology and archetype experts should agree on a minimal set of recommendations for different purposes and agree on some design patterns for terminology binding that facilitate the scalability of semantic infrastructures for healthcare and research. This scalability principle should allow for starting with basic semantic annotations and progressing towards more complex semantic infrastructures on "as-needed" basis as recommended in other domains[22]-[24]. At the moment, when we reviewed the guidelines for SNOMED-CT adoption and compared them with the development of archetypes we detected an important lack of coordination between the guidelines from the archetype design point of view and the terminology point of view (e.g., SNOMED-CT guidelines). For example, the SNOMED-CT context model clearly overlaps with the function of the archetype but does not have enough expressivity to allow for projecting archetypes at a semantic level. Both openEHR and SNOMED International should agree on a minimal set of patterns as mentioned above. We do understand that the compositional grammar (and expressions like those created using the context model) may be necessary for some scenarios, but then they need to provide means for aligning SNOMED-CT expressions with archetypes. In order to enable a more complete representation of the semantics included in the archetypes at a terminology level, a more complete context model would be necessary. For example, approximately half of the concepts in the protocol section of archetypes could not be represented. However, extending the context model would lead to a mix of ontological and epistemological aspects. Even when both aspects are represented at a semantic level, it is appropriate to keep the models of meaning and information separated to enable separate scalability and reasoning [25]. An option to avoid mixing different models may be to maintain a context model ontology outside the medical terminology (SNOMED-CT). Approaches to develop parallel context models have been defined by relying on more expressive logics (OWL DL) [26]. However, these kind of logic leads to models that are not tractable [11], thus they may jeopardize large enterprise deployments. Experiences in Semantic Web development have shown that in many cases it is better to sacrifice expressiveness in order to gain scalability [12][23]. If those context models evolve into lighter models and define clear extension mechanisms to use other ontologies, they may be very useful to provide machine-interpretable representations of archetypes clinical semantics.

Finally, implementers must be aware that projecting most of the archetype as a formal ontology is expensive in terms of computational demands and models definition. Therefore, as mentioned before, semantic infrastructures should aim to follow a bottom-up approach starting by adopting a reference ontology and slowly evolving their semantic infrastructure as new requirements for higher expressivity arise.

V. Conclusion

The long-term requirements of the national interoperability frameworks in Norway need explicit representations of archetype semantics. We studied if SNOMED-CT´s compositional grammar can provide the level of expressiveness needed for covering these requirements. Technical, expressivity, human, and models mismatch challenges are present when defining archetype semantics as SNOMED-CT expressions. These challenges show a pressing need for the collaboration between archetypes and SNOMED-CT editors so both become aware of each other´s challenges. Future collaboration should: provide better guidelines for archetypes terminology binding assessing the best approach to follow depending on the interoperability needs; set the requirements for better tooling to ease the binding process; and work towards developing mechanisms that allow for improving expressivity, alignment between models, and extensibility of the SNOMED-CT concept model with other biomedical ontologies.

ACKNOWLEDGMENT

This work was funded by the Norwegian Centre for E-health Research.

REFERENCES

[1] D. Markwell, L. Sato, and E. Cheetham, "Representing Clinical Information using SNOMED Clinical Terms with Different Structural Information Models.," in *Proceedings of the 3rd International Conference on Knowledge Representation in Medicine*, Phoenix, Arizona, 2008, vol. 410, pp. 72–79.

[2] L. Marco-Ruiz, K. Nicolaisen, R. Pedersen, A. Makhlysheva, and P. A. Bakkevoll, Ontology-based terminologies for healthcare - Impact assessment and transitional consequences for implementation, 1st ed., vol. 1. Tromsø: Norwegian Centre for E-health Research, 2017.

[3] "HL7 TermInfo Project." [Online]. Available: http://wiki.hl7.org/index.php?title=TermInfo_Project#. [Accessed: 01-Jan-2019].

[4] T. A. Oniki, J. F. Coyle, C. G. Parker, and S. M. Huff, "Lessons learned in detailed clinical modeling at Intermountain Healthcare.," J Am Med Inform Assoc, vol. 21, no. 6, pp. 1076–1081, Nov. 2014.

[5] A. R. Rasmussen and K. Rosenbeck, "SNOMED CT implementation: implications of choosing clinical findings or observable entities," *Stud Health Technol Inform*, vol. 169, pp. 809–813, 2011.

[6] A. Budrionis , l. Marco-Ruiz , KY. Yigzaw, and JG. Bellika . Building a Learning Healthcare System in North Norway. Proceedings from The 14th Scandinavian Conference on Health Informatics 2016, Gothenburg, Sweden, April 6-7 2016, Linköping University Electronic Press; 2016, p. 1–5.

[7] L. Marco Ruiz, JA. Maldonado, R Karlsen, and JG. Bellika. Multidisciplinary Modelling of Symptoms and Signs with Archetypes and SNOMEDCT for Clinical Decision Support. Stud Health Technol Inform., Madrid: IOS press; 2015.

[8] H. Lærum, S. L. Bakke, R. Pedersen, and J. T. Valand, "An update on OpenEHR archetypes in Norway: Response to article Christensen B & Ellingsen G: 'Evaluating Model-Driven Development for large-scale EHRs through the openEHR approach' IJMI May 2016, Volume 89, pages 43-54," *Int J Med Inform*, vol. 93, p. 1, Sep. 2016.

[9] B. E. Dixon *et al.*, "A pilot study of distributed knowledge management and clinical decision support in the cloud," *Artificial Intelligence in Medicine*, vol. 59, no. 1, pp. 45–53, Sep. 2013.

[10] A. L. Rector, "The Interface between Information, Terminology, and Inference Models," in *Proceedings of the 10th World Congress on Medical Informatics*, London, UK, 2001, vol. 84, pp. 246–250.

[11] D. Karlsson, M. Berzell, and S. Schulz, "Information Models and Ontologies for Representing the Electronic Health Record.," in *International Conference on Biomedical Ontology*, Buffalo, NY, 2011, p. :153–157.

[12] L. Marco-Ruiz, C. Pedrinaci, J. A. Maldonado, L. Panziera, R. Chen, and J. G. Bellika, "Publication, discovery and interoperability of Clinical Decision Support Systems: A Linked Data approach," *Journal of Biomedical Informatics*, vol. 62, pp. 243–264, Aug. 2016.

[13] C. Pedrinaci, J. Domingue, and A. P. Sheth, "Semantic Web Services," in *Handbook of Semantic Web Technologies*, J. Domingue, D. Fensel, and J. A. Hendler, Eds. Springer Berlin Heidelberg, 2011, pp. 977–1035.

[14] P. D. Fensel, D. F. M. Facca, D. E. Simperl, and I. Toma, "Web Service Modeling Ontology," in *Semantic Web Services*, Springer Berlin Heidelberg, 2011, pp. 107–129.

[15] D. Booth, "Yosemite Project | Comparing the Yosemite Project and ONC Roadmaps for Healthcare Information Interoperability," *Yosemite Project*, 02-Apr-2015. [Online]. Available: http://yosemiteproject.org/comparing-the-yosemite-project/. [Accessed: 01-Jan-2019].

[16] "Rdf - FHIR v3.0.1." [Online]. Available: https://www.hl7.org/fhir/rdf.html. [Accessed: 01-Jan-2019].

[17] C. Soguero-Ruiz *et al.*, "Ontology for heart rate turbulence domain from the conceptual model of SNOMED-CT," *IEEE Trans Biomed Eng*, vol. 60, no. 7, pp. 1825–1833, Jul. 2013.

[18] C. Martínez-Costa, R. Cornet, D. Karlsson, S. Schulz, and D. Kalra, "Semantic enrichment of clinical models towards semantic interoperability. The heart failure summary use case," *J Am Med Inform Assoc*, vol. 22, no. 3, pp. 565–576, May 2015.

[19] IHTSDO, "SNOMED CT Compositional Grammar Specification and Guide." [Online]. Available: https://confluence.ihtsdotools.org/display/DOCSCG. [Accessed: 01-Jan-2019].

[20] E. Sundvall *et al.*, "Integration of tools for binding archetypes to SNOMED CT," *BMC Med Inform Decis Mak*, vol. 8, no. Suppl 1, p. S7, Oct. 2008.

[21] P. N. Robinson, "Deep phenotyping for precision medicine," *Human Mutation*, vol. 33, no. 5, pp. 777–780.

[22] "Semantic Annotations for WSDL and XML Schema." [Online]. Available: http://www.w3.org/TR/sawsdl/. [Accessed: 01-Jan-2019].

[23] J. Kopecký and T. Vitvar, "MicroWSMO," *WSMO Working Draft - D38v0.1 MicroWSMO*, 19-Aug-2008. [Online]. Available: http://www.wsmo.org/TR/d38/v0.1/. [Accessed: 01-Jan-2019].

[24] K. Gomadam, A. Ranabahu, and A. Sheth, "SA-REST: Semantic Annotation of Web Resources," *SA-REST: Semantic Annotation of Web Resources*. [Online]. Available: https://www.w3.org/Submission/SA-REST/. [Accessed: 01-Jan-2019].

[25] A. L. Rector, R. Qamar, and T. Marley, "Binding ontologies and coding systems to electronic health records and messages," *Applied Ontology*, vol. 4, no. 1, pp. 51–69, Jan. 2009.

[26] C. Martínez-Costa and S. Schulz, "Ontology-based reinterpretation of the SNOMED CT context model.," in *Proceedings of the 4th International Conference on Biomedical Ontology*, Montreal, Canada, 2013, pp. 90–95.

[27] https://arketyper.no/ckm/ (15.02.2019)

[28] https://ec.europa.eu/digital-single-market/en/news/cross-border-health-project-epsos-what-has-it-achieved (15.02.2019)

TCM Named Entity Recognition Based On Character Vector With Bidirectional LSTM-CRF

Jigen LUO, Jianqiang DU*, Bin NIE, Wangping XIONG, Jia HE, Yanyun YANG
School of Computer
Jiangxi University of Traditional Chinese Medicine
Nanchang, China
E-mail: jianqiang_du@163.com

Abstract—**In order to better solve the problem of low accuracy caused by inaccurate word segmentation in the task of extracting Chinese medicine named entities, the extraction technology relies heavily on the characteristics of manual development, and needs guidance of domain knowledge. This paper proposes a named entity recognition in the field of Traditional Chinese Medicine (TCM) Based on character vectors for Bidirectional Long Short Term Memory with a Conditional Random Field (Bidirectional LSTM-CRF). First of all, the model uses the word2vec tool to convert the corpus into a character vector, which can avoid the influence of inaccurate word segmentation in Chinese medicine field on entity recognition; then use Bidirectional LSTM neural network to extract deep features of sentence level, and reduce the workload of manual feature setting in the traditional method; Finally access to the CRF layer, and the Viterbi algorithm is used to dynamically plan the most reasonable tag output of the sentence, and the correlation between the output tags is considered. We use different models to conduct experiments on the TCM corpus. The results show that the model proposed in this paper has a good effect. The F value of the evaluation index on the three types of Chinese medicine, prescription and syndrome type has reached 90%.**

Keywords-Named Entity Recognition; Character vector; Bidirectional LSTM-CRF; Chinese Medicine Informatics.

I. INTRODUCTION

The knowledge of TCM diagnosis is a historical treasure left by the Chinese nation for thousands of years and has a strong guiding role in the clinical treatment. In the field of traditional Chinese medicine, the forms of expression between syndromes, prescriptions and traditional Chinese medicines are diverse and complex, and they are used throughout the treatment process of Chinese medicine [1][2]. In order to construct a knowledge graph of the Chinese medicine field and form a structured knowledge, the extraction of three types of entities such as syndrome type, prescription and traditional Chinese medicine is particularly important and a very meaningful step.

Named entity recognition is a type of problem of sequence labeling. It is the basic work of information extraction, information retrieval, machine translation and other tasks [3]. The current mainstream methods for named entity recognition are based on statistical and rule-based language model methods [4]. The statistical based method automatically extracts the composition law of the named entity directly from the text, and identifies the entity through the model-trained language model [5]. The statistical method for the named entity recognition has a hidden Markov model [6] and the conditional random field Model [7]. The rule-based approach [8] uses the rational knowledge of linguists to identify named entities through rules written by linguists. Traditional named entity recognition methods require a wealth of domain expert knowledge to extract a large number of artificial features [8].

With the rapid development of deep learning, the advantages of neural networks in dealing with natural language processing problems have gradually emerged. The core technology of deep learning is to express text features in the form of word vectors. Zheng et al. [9] proposed using neural networks for named entity recognition tasks, and using perceptron algorithm for accelerated training models; later, Recurrent Neural Network applications appeared in sequence labeling problems. Chowdhury et al. [10] The Recurrent Neural Network is used for sequence tasks, achieving better results and avoiding feature engineering. However, it is difficult for recurrent neural networks to obtain long distance dependent information, and long distance information is lost. In order to solve the problem of long distance information dependence. Xu et al.[11] used a Bidirectional LSTM neural network and a conditional random field method to identify medical names. This method takes advantage of LSTM's feature extraction and considers the relationship between output tags in combination with CRF, making Long Short-Term Memory with a Conditional Random Field (LSTM-CRF) model is widely used in Chinese named entity recognition tasks [12][13]. However, due to the lack of obvious space between Chinese characters, word segmentation processing is required, but the current word segmentation software is aimed at the general field and rarely involves special fields. In order to solve the erroneous influence of inaccurate word segmentation on entity recognition, Dong et al. [14] used character-level vectors as input for deep learning to perform entity recognition, but these are all considered in the general field, without considering the particularity of the Chinese medicine field.

There are deficiencies in Chinese word segmentation in the field of Chinese medicine. Many Chinese medical terms are misclassified. For example, "Yin and Yang deficiency syndrome" is a word, and the commonly used word segmentation tool will divide it into: "Yin and Yang", "Two" and "Deficiency syndrome", this wrong participle will seriously affect the entity recognition effect.

This paper provides a Bidirectional LSTM-CRF named entity recognition method based on character vector, which lays a foundation for the knowledge graph construction in the field of traditional Chinese medicine. By continuously adjusting the parameters of the model, until the neural network parameter combination suitable for entity recognition in the Chinese medicine field is found, the effect of the entity recognition is optimized.

Section II introduces the model proposed in this paper; the model is based on character vector with Bidirectional LSTM entity recognition method in the field of traditional Chinese medicine. In Section III, the contrast experiment and adjustment model parameters are given. It is proved that the proposed model in this paper has certain advantages. The parameters of LSTM neural network model suitable for traditional Chinese medicine are found. Finally, the conclusion and future work of this research are found in Section IV.

II. Bidirectional LSTM-CRF Model Based On Character Vector

For the identification of entities in the field of traditional Chinese medicine, this paper identifies the three types of entities: syndrome type, prescription and traditional Chinese medicine, which are marked by two words. B-XX indicates the first type of entity, while I-XX indicates other words of the entity. The specific label forms of the three types of entities are shown in TABLE I.

As shown in Figure 1, the schematic diagram of the character vector based Bidirectional LSTM-CRF TCM domain named entity recognition model designed for this paper. The model is divided into three parts: the character vector layer, the Bidirectional LSTM neural network layer, and the CRF tag inference layer. The character vector layer converts the input text into a recognizable numerical form of the neural network, and then calculates the output feature vector h_t of each character through the bidirectional LSTM neural network. The CRF inference layer finds the most suitable output tag sequence of the sentence through dynamic programming to make up Subsequent to the defect that the Softmax output tags are independent of each other.

TABLE I. Entity Two Word Position Label Representation

Sign	Significance
B-SYN	The first Chinese character of the syndrome type entity
I-SYN	All Chinese characters remaining in the first characters of the syndrome type entity
B-PRE	The first Chinese character of the prescription type entity
I-PRE	All Chinese characters remaining in the first characters of the prescription type entity
B-MED	The first Chinese character of the Chinese medicine type entity
I-MED	All Chinese characters remaining in the first characters of the Chinese medicine type entity
O	Irrelevant words

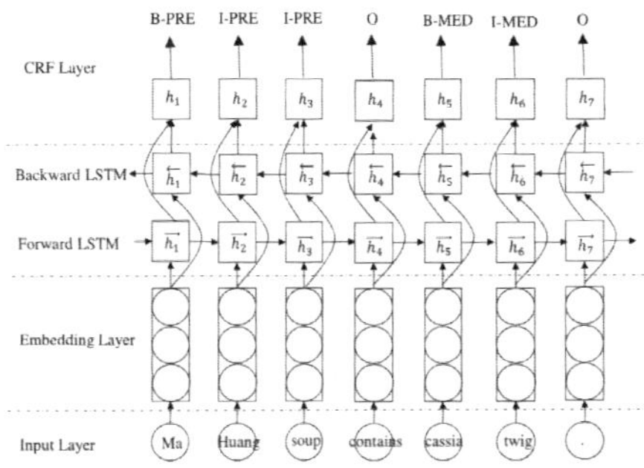

Figure 1. The model structure of BLSTM-CRF

A. Characters Vector Layer

The structure of the Bidirectional LSTM-CRF model is shown in Figure 1. The bottom end of the input is the sequence to be labeled "Ma Huang soup contains cassia twig", which is the input layer in the figure. Since the neural network algorithm cannot directly process Chinese characters, it is necessary to convert the text into a vector form. There are two ways to vector the text: one-hot and distributed representation. Since the distributed vector representation can reduce the dimension of the vector and can effectively represent the association between semantics. This paper adopts the distributed representation method for text vector. The character embedding operation is required before entering the model. This article use Google's opening source tool word2vec to convert the text into a character vector. Suppose the input sentence is S , and the set of characters contained is $W(w_1, w_2, w_3, ..., w_m)$, m is the length of the sentence, where the $t-th$ character vector is $w_t^* \in R^d$, where the dimension of the word vector is in the above formula, the input text is expressed as:

$$S = \left[w_1^*, w_2^*, ..., w_m^* \right] \in R^{T \times d} \quad (1)$$

B. Bidirectional LSTM Neural Network Layer

The LSTM neural network is an improvement to the common RNN neural network. Its main purpose is to solve the problem of gradient disappearance or gradient explosion. The biggest difference from the RNN neural network is the neural unit. The neural unit of the LSTM neural network joins the gate structure, including the input gate, the output gate and the forgetting gate. The neural unit of the LSTM neural network is shown in Figure 2.

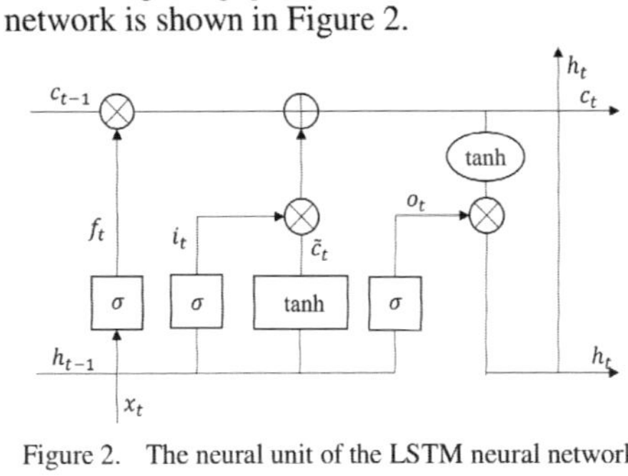

Figure 2. The neural unit of the LSTM neural network

At t time, the LSTM unit components are updated as follows.

$$f_t = \sigma(W_f \cdot [h_{t-1}, x_t] + b_f) \quad (2)$$

$$i_t = \sigma(W_f \cdot [h_{t-1}, x_t] + b_i) \quad (3)$$

$$\tilde{c}_t = \tanh(W_c \cdot [h_{t-1}, x_t] + b_c) \quad (4)$$

$$c_t = f_t \circ c_{t-1} + i_t \circ \tilde{c}_t \quad (5)$$

$$o_t = \sigma(W_o \cdot [h_{t-1}, x_t] + b_o) \quad (6)$$

$$h_t = o_t \circ \tanh(c_t) \quad (7)$$

Among them, represents the sigmoid activation function, which is the element multiplication, x_t is the input vector of LSTM at t time, h_t represents the hidden state; W_f , W_i , W_c , W_o represents the weight matrix of the forgotten gate, input gate, memory cell, and output gate; b_f , b_i , b_c , b_o represents bias of the forgotten gates, input gates, memory cells, output gate. f_t, i_t, c_t, o_t represents forgotten gate, input gate, memory cell status, and output gate.

The output feature vector calculation method of unidirectional LSTM neural network is introduced above. In order to make full use of context information and mine more hidden features, and effectively solve the problem of new word discovery in TCM named entity recognition task, this paper adds on top of LSTM neural network. A layer of reverse LSTM neural network structure forms a bidirectional LSTM neural network. The assumption $\overrightarrow{h_t}$ is that the output of the forward LSTM neural network at the moment, $\overleftarrow{h_t}$ is the output of the backward LSTM unit at the moment, and the output of the time t is the splicing of the preceding and succeeding vectors, that is $h_t = [\overrightarrow{h_t}, \overleftarrow{h_t}]$.

C. CRF Layer

The CRF algorithm is a machine learning model specifically designed for sequence labeling tasks. Suppose the input sequence of the model is $W = \{w_1, w_2, w_3, ..., w_m\}$, which w_i represents the $i-th$ Character in input sentence, output sequence is $Y = \{y_1, y_2, y_3, ..., y_m\}$, and y_i is the output sequence label of w_i. In this paper, P is the scoring matrix of the bidirectional LSTM layer, and P is a $m \cdot k$ matrix, where m is the length of sentences and k is the number of custom tags. P_{ij} Represents the $j-th$ label score value of the $i-th$ Character in the sentence. For example, P_{12} refers to the probability that the first Character in a sentence is labeled as the second label in the label set. In order to better calculate the sentence path, you need to add the start tag at the beginning of the sentence and the end tag at the end of the sentence. Therefore, the parameter of the CRF layer is a state transition matrix of $(k+2)*(k+2)$, then the whole model is marked as $Y = \{y_1, y_2, y_3, ..., y_m\}$ for the input sequence $W = \{w_1, w_2, w_3, ..., w_m\}$ .The final score is calculated as follows.

$$score(W, Y) = \sum_{i=0}^{m} A_{y_i, y_{i+1}} + \sum_{i=1}^{m} P_{i, y_i} \quad (8)$$

The final score of the sentence can be divided into two parts, where P_{i,y_i} is the scoring matrix of the bidirectional LSTM neural network model and A is the transfer scoring matrix between adjacent tags. In order to get the sentence probability, the score matrix s needs to be normalized. The probability formula is as follows:

$$p(Y \mid W) = \frac{\exp(score(W, Y))}{\sum_{y} \exp(score(W, y))} \quad (9)$$

In order to get the final prediction label, it can be obtained by the Viterbi decoding algorithm. The dynamic programming can find the best path. The calculation formula is as follows.

$$Y^* = \underset{Y}{argmax}\, score(W, Y) \quad (10)$$

III. Experiment And Analysis

In this section, we mainly verify the validity of the application of the Bidirectional LSTM neural network based on character vector on the entity extraction task in the field of traditional Chinese medicine.

A. Experiment Data And Evaluation Indicators

The corpus used in the experiment is the Chinese medicine diagnosis text and classic Chinese medicine books provided by Jiangxi University of Traditional Chinese Medicine, such as the TCM syndrome differential diagnosis book, a total of 32,700 sentences related to syndromes and prescriptions, and Chinese medicine, totaling more than 2 million character. After many years of clinical experience, the corpus has been comprehensively covered with information on syndromes, prescriptions, Chinese medicine and other related entities.

In order to display the experimental results comprehensively and intuitively, this paper adopts the Precision rate (P), Recall rate (R) and F value as model evaluation indicators.

B. Experimental Design And Results

Some parameter settings of the LSTM neural network are shown in TABLE II.

TABLE II. Initial Setting of Partial Parameters of Bidirectional LSTM Neural Network

Hyper parameter	Initial value
Learning rate	0.001
Dropout	0.5
Gradient clipping	5.0
Embedding-dim	300
Optimizer	Adam
Batch-size	64
Hidden-dim	300
Epoch	30

In order to verify the effect of the proposed character vector based Bidirectional LSTM-CRF on TCM entity recognition, this paper sets up three sets of comparative experiments.

Experiment 1: The first set of experiments was to verify the impact of different algorithms on entity recognition in the Chinese medicine field. TABLE III shows the experimental results of the CRF model, the Character vector based bidirectional LSTM-Softmax model (Char-BLSTM-Softmax) and the Character vector based Bidirectional LSTM-CRF model (Char-BLSTM-CRF).

As we saw from the test results in TABLE III, The Char-BLSTM-CRF model has the best experimental results, which is more than one percentage point higher than the F-value of the Char-BLSTM-Softmax model. Due to the Softmax classifier in the Char-BLSTM-Softmax model does not consider the dependencies between the outputs tags, CRF classifiers use Viterbi algorithm for dynamic programming, the output label path is the most scientific. Therefore, the Char-BLSTM-CRF model has a certain improvement compared to Char-BLSTM-Softmax. The CRF model has the worst experimental results because the model requires manual extraction of high quality features. This indicates that the character vector based on BLSTM-CRF model proposed in this paper has a good effect in entity recognition in the field of Chinese medicine.

Experiment 2: This paper constructs the embedding vector with character and words. Experiment to verify the effect of character and word vector on Bidirectional LSTM-CRF model. It is further proved that the Bidirectional LSTM-CRF entity recognition model based on character vector can avoid the influence of inaccurate word segmentation on entity recognition in TCM domain. The Bidirectional LSTM-CRF entity recognition model based on word vector is abbreviated as Word-BLSTM-CRF. The experimental results are shown in TABLE IV.

The experimental performance of the Char-BLSTM-CRF model is 6 percentage points higher than the Word-BLSTM-CRF in F value. Because of the Word-BLSTM-CRF model needs to segment the corpus before labeling, the existing word segmentation tools are inaccurate for the terminology of the Chinese medicine field, resulting in poor experimental results. However, Char-BLSTM-CRF model does not need to word segmentation. It labels each character to avoid errors caused by inaccurate word segmentation.

Experiment 3: In experiment 1 and experiment 2, we fixed the parameters of the LSTM neural network model. In this experiment, by adjusting the parameters, the best parameters suitable for the identification of named entities in the field of Chinese medicine were found. Parameters that can be adjusted include Dropout values, optimizer and learning rate.

TABLE III. COMPARISON OF DIFFERENT MODELS' RECOGNITION OF TCM ENTITIES

Model	P（%）	R（%）	F（%）
CRF	79.26	78.56	78.91
Char-BLSTM-Softmax	89.92	90.70	90.31
Char-BLSTM-CRF	**91.47**	**91.37**	**91.42**

TABLE IV. COMPARISON OF CHARACTER VECTOR AND WORD VECTOR RECOGNITION

Model	P（%）	R（%）	F（%）
Word-BLSTM-CRF	84.05	86.19	85.11
Char-BLSTM-CRF	**91.47**	**91.37**	**91.42**

The Dropout parameter adjustment experiment means that the fixed optimizer is Adam; the learning rate is 0.001; and the Dropout value ranges from 0.1 to 0.5, and is incremented by an integral multiple of 0.1. The experimental results are shown in TABLE V. It can be clearly seen that when the Dropout value is 0.5, the model has the best effect and the F value is the highest.

The optimizer parameter adjustment experiment refers to the fixed Dropout of 0.5, the learning rate is 0.001, and the experimental results of the optimizer can select Adam, Adagrad, SGD, and Momentum. The experimental results are shown in TABLE VI. It can be clearly seen that the fixed Dropout and learning rate, the Adam optimizer have the fastest convergence and the best results.

The experimental results of the learning rate parameter adjustment are shown in TABLE VII. At this time, the fixed Dropout is 0.5, the optimizer is Adam, and the learning rate is set to 0.01, 0.001 and 0.0001. According to the experimental results, when the optimizer and Dropout are fixed, learning rate is 0.001, the effect of the model is optimal at this time.

According to the results of the above three sets of parameter experiments, the parameter combination of the Bidirectional LSTM-CRF model which is most suitable for TCM entity identification is Dropout=0.5, the learning rate is 0.001, and the optimizer is Adam. At this time, the comprehensive performance F value of the model reaches 90% above, the identification of named entities in the field of Chinese medicine has a good effect.

TABLE V. THE DROPOUT PARAMETER ADJUSTMENT EXPERIMENT

Dropout, optimizer=Adam, learning rate=0.001	P（%）	R（%）	F（%）
0.5	**91.47**	**91.37**	**91.42**
0.4	91.31	91.18	91.25
0.3	90.12	90.59	90.35
0.2	88.50	89.26	88.88
0.1	85.48	87.14	86.30

TABLE VI. THE OPTIMIZER PARAMETER ADJUSTMENT EXPERIMENT

Dropout=0.5, optimizer, learning rate=0.001	P（%）	R（%）	F（%）
sgd	81.71	76.45	78.99
Momentum	87.23	82.50	84.80
Adam	**91.47**	**91.37**	**91.42**
Adagrad	77.91	77.32	77.61

TABLE VII. THE LEARNING RATE PARAMETER ADJUSTMENT EXPERIMENT

Dropout=0.5, optimizer=Adam, learning rate	P（%）	R（%）	F（%）
0.0001	87.02	87.44	87.23
0.001	**91.47**	**91.37**	**91.42**
0.01	80.07	87.83	83.77

IV. Conclusion and Future Work

The identification of TCM entities is a basic work in the construction of knowledge graph in this field. The main work of the article is to focus on this topic. This paper proposes a Bidirectional LSTM-CRF model based on character vector, which uses character vector as input to replace the word vector in traditional deep learning, to avoid the influence of inaccurate TCM segmentation on entity recognition. The bidirectional LSTM with context information is used as the input. The hidden layer of the neural network solves the problem of dependence on long text input and mitigates the gradient explosion. Finally, it accesses the CRF tag inference layer to solve the dependency problem between output tags. In this paper, a lot of experiments have been done on the TCM entity corpus. The results show that the Bidirectional LSTM-CRF model based on character vector is better than other algorithms, and the parameters that are most suitable for TCM entity recognition are found through experiments.

However, from the results, the F value of the best experimental results in the article is only 91.42%, and there is still much room for improvement. In order to seek a bigger breakthrough, the future work will focus on the establishment of a high-quality corpus. In this process, more scholars in the field of Chinese medicine will be introduced to participate in the establishment of the corpus, so that the work has a better effect.

Acknowledgment

The research was financially support by three National Natural Science Foundations (61562045 & 61762051). The Postgraduate innovation fund of Jiangxi Province (YC2017-S349). Jiangxi Province key research and development program key projects (20171ACE50021, 20171BBG70108).

References

[1] N.Sager, C.Friedman and M.S.Lyman . "Review of medical language processing: computer management of narrative data". Computational Linguistics, vol.15, pp.195–198, 1989.

[2] P.Lu, K.H.Lin and C.L.Zhou . "Research of TCM face diagnosis and symptom factor based on NN". Application Research of Computers, vol.25, pp.2655-2657, 2008.

[3] S.Qiu et al. "Chinese Named Entity Recognition Based on Part of Speech Feature with Edges". Computer Engineering, vol.38, pp.128-130, 2012.

[4] J.P.Ju, W.W.Zhang, J.J.Ning and G.D.Zhou. "Geospatial Named Entities Recognition Using Combination of CRF and Rules". Computer Engineering, vol.37, pp.210-212, 2011.

[5] Z.H.Wang. "Name entity recognition using language models". IEEE Workshop on Automatic Speech Recognition & Understanding. IEEE, pp. 554-559 , 2003.

[6] G.D.Zheng and J.Su. "Named entity recognition using an HMM-based chunk tagger". In Proceedings of the 40th Annual Meeting of the Association for Computational Linguistics, pp. 209–219, 2002.

[7] Y.J.Zhang, Z.T.Xu and T.Zhang. "Fusion of Multiple Features for Chinese Named Entity Recognition Based on CRF Model". Journal of Computer Research & Development, vol.45, pp.1004-1010, 2008.

[8] L.Chiticariu, F.Reiss,Y.Y.Li,R.Krishnamurthy and S. Vaithyanathan. "Domain Adaptation of Rule-based Annotators for Named-Entity Recognition Tasks". EMNLP, pp.1002-1012, 2010.

[9] X.Q.Zheng, H.Y.Chen and T.Y.Xu. "Deep learning for Chinese word segmentation and POS tagging". Proceedings of the 2013 Conference on Empirical Methods in Natural Language Processing, pp.647-657, 2013.

[10] S.Chowdhury et al. "A multitask bi-directional RNN model for named entity recognition on Chinese electronic medical records". BMC Bioinformatics, vol.19, pp.75-84, 2018.

[11] K. Xu, Z.F.Zhou,T.Y.Hao and W.Y.Liu. "A Bidirectional LSTM and Conditional Random Fields Approach to Medical Named Entity Recognition".International Conference on Advanced Intelligent Systems & Informatics, pp.355-365, 2017.

[12] L.Sun,Y.Rao,Y.Lu and X.Li. "A method of Chinese named entity recognition based on CNN-BILSTM-CRF model". Communications in Computer and Information Science, vol.902 , pp.161-175, 2017.

[13] Z.Q.Xu , S. Li and W.H. Deng . "Learning temporal features using LSTM-CNN architecture for face anti-spoofing".2015 3rd IAPR Asian Conference on Pattern Recognition (ACPR). IEEE,pp.141-145, 2015.

[14] C.H.Dong,J.J.Zhang, C.Q.Zong, M.Hattori and H.Di. "Character-Based LSTM-CRF with Radical-Level Features for Chinese Named Entity Recognition". Natural Language Understanding and Intelligent Applications. Springer International Publishing, vol.10102, pp.239-250, 2016.

Enabling Employee Co-Creation in eHealth

Propositions for a Methodology

Migle Helmersen
Society dept.
NORCE research
Kristiansand, Norway
e-mail: mihe@norceresearch.no

Tom Roar Eikebrokk
Dept. of Information Systems
University of Agder
Kristiansand, Norway
e-mail: tom.eikebrokk@uia.no

Niels F. Garmann-Johnsen
Society dept.
NORCE research
Kristiansand, Norway
e-mail: ngar@norceresearch.com

Abstract—**This article is a reflection by a team of researchers. After visiting and doing action and evaluation research in a municipal eHealth Living Lab project, the authors find that there is a need for a more formalized approach to the social aspect and fundaments for employee driven innovation. We find a need for combining the fields of organizational learning and technology innovation. Based on our research, we propose a model for value creation based on new eHealth technology where employee co-creation is stimulated as a resource for the learning organization. Here, employees evaluate existing services as the basis for both designing totally new services and evaluate proposed new solutions. This basis in employee co-creation creates a broad basis for implementing changes that will benefit service users through increased ability and speed of organizational change in such innovation ecosystems. In addition, to the usual agile implementation phase, we induct the need for prior employee-involving mobilization and ideation phases.**

Keywords- eHealth; employee; co-creation; Design Thinking; LEAN; Agile, learning organization; innovation.

I. INTRODUCTION

Healthcare is a labor-intensive profession and is likely to remain so in any foreseeable future. At the same time, productivity in this sector needs to grow with radical rates each year to meet future demands, due to an aging population in Norway and many other industrialized countries. eHealth – using Information and Communication Technology (ICT) in healthcare - is expected to be an important factor in achieving needed value innovation. The simultaneous reduction of cost or use of scarce resources, and increased productivity and quality [1].

Balancing life-critical operations and the introduction of new technologies requires a sound fundament of continuous improvement of work, high involvement by all concerned employees in the innovation and information system design processes, learning and developing new clinical practices, and a culture for sharing knowledge and experience [2]. This article will refer to this fundament as the learning organization.

Innovation is by nature experimental and requires risk willingness to find a better next practice. Innovation is a concept for change and is first used when the solution in the form of new services and products is put into use or implemented. The process from the ideas to implementation is a resource-intensive and risky process and means that the organization is in a continuous change. How does the organization master this?

Ergonomics (of Latin ergon, work, and nomos, law) is defined as the science of adaptation between the working environment, technique and human beings so that work can be done as effectively as possible, and without any adverse impact on health [3]. But whether the results of this science are practiced is often a matter of, inter alia, politics and economy in the organization. To avoid illness and strain injuries, both employer and employee must consider what is good ergonomics. Good ergonomics also mean that work processes are adapted to the employees' requirements and ambitions. Therefore, ergonomics also has psychological and social aspects and is closely related to how the work is organized [4]. Modern ergonomics are extended to the overall adaptation of human beings to work processes, the working environment and changes in society at large.

Neubauer and Stary [5] describe ergonomics as a recognition of the role of employees in innovation, leading to both improvements and economic benefits through human-centered design. Human-centered design for interactive systems is something that promotes the following main principles [5]:

• The design is based on an explicit understanding of users [also read employees], tasks and environments

• Users are involved through all parts of design and development (not just testing – authors red).

• The design is driven and refined by user-centered evaluation

• The process is iterative - it is repeated as many times as necessary

• The design addresses the entire user experience (the "travel trip")

• The design team includes interdisciplinary skills and perspectives (authors´ translation).

To promote a strategy (organizational policy) that can improve this, López-Gómez et al. [6] suggest that one should:

• Access highly qualified personnel to develop new concepts and service innovations internally

• Develop training methods for personnel to adapt innovations, innovative ideas, sourced from external sources

• Develop better adapted schemes in education and training to fit the requirements of a service-economy

• Recognizing the value of informal learning to increase the attractiveness of continuous training for employees

• Promote modern innovation management approaches that better support creativity and autonomy of service-executing staff (authors translation).

Innovation management is crucial for creating an innovation culture. The research shows that the top management's conscious role in the organization is crucial, as both banner bearers for new ideas and for their development, where follow-up through systematic work with innovation and innovation management is important [7]. Salaman and his colleagues have defined criteria for innovation management: focus on networking, developing the creative talent, promoting learning, and mobilizing through a clear vision of the target image, and creating innovative processes [8]. The ideal leader has thus a direct impact on the development of employee competencies, create space for participation, and helps to make the business more innovative. Such management will not only be able to reduce resistance to new solutions and changed work routines but will influence employees to learn to design new solutions continuously and create arenas for dialogue [9]. This process creates a continuous learning and self-improving organization. The material on this is based on Argyris and others' research on learning, organization and action research [10].

What happens if the process described above is not prioritized in the organization? What are the consequences for employees' working environment and health? Frameworks for employees for doing a good job in the learning organization include that they interact with actors from different levels. These actors can act as support or provide resistance: in management, among product and service users, suppliers, and the media. Incorrect organization of ICT, resistance among employees in the introduction of new ways of working, and resistance to ICT in general, can be decisive in how companies tackle development processes.

Positive results have been achieved for physical work environments in many places, by introducing robots and similar ICT-supported technology that helps in physically demanding tasks. In this way, musculoskeletal disorders can be prevented. But the use of ICT can also be seen more often in connection with the influence on mental health. Through several research studies, it was defined how the implementation of welfare technology, in general in the market and use in everyday life, influenced employee attitudes and health. When introducing new ICT tools and innovation processes in companies, several factors were reported which had a negative impact on mental health.

It was found that the requirements for accessibility, communication, control, and repetitive technical errors, employee monitoring, unmet need for increased training, expectation of increased productivity, increased responsibilities and workload were associated with work pressure, stress and burnout [11][12]. Only two facets of the welfare technology introduction were perceived positively: individual assistance and customer guidance. In a study among nurses, it has emerged that the fear of "dehumanizing" of human care was dominating [13].

It appears that health personnel are initially concerned that basic values in care can be lost by "technologizing" relationships between people. Care represents for us closeness, while the technology appears to be cold and insensitive [14]. This leads us up to the research problem in this study: What are the necessary conditions for enabling employee co-creation in eHealth? How do we build up the learning organization, and what are the benefits of doing so?

The rest of this article is laid out as follows: In the Section 2 we go through our method and present an action and evaluation study into an eHealth Living Lab. Then in section 3 we disseminate the recommendations given to the principal behind the action research mission after the case-study. Finally, in section 4 we generalize these findings by questioning whether they represent a more structured and formalized approach to the social aspect and fundaments for employee driven innovation.

II. METHOD

During 2017 and 2018 a joint research team did a study of eHealth Living Lab [15]. The project was initiated by the city of Grimstad in Norway, as the municipality hosting the living lab.

Agder Living Lab (ALL) is a collaborative project. The Norwegian Directorate of Health has provided The Centre for Development of Institutional and Home Care Services (USHT) in Aust Agder with a contribution to the development of a Living Lab methodology in the welfare technology field. USHT in Grimstad Municipality is supposed to function as a living test laboratory. Here are nurses, patients and relatives involved in finding tomorrow's welfare technology. The University of Agder is a main partner. In addition, the Norwegian Housing Bank has contributed to dissemination.

The methodology for Agder Living Lab was given by the project as a progressive, step model illustrated in Figure 1:

1. Define user demands
2. Regulatory compliance testing
3. Lab-testing of usability

4. Testing in living environment
5. Piloting improved services

These steps belong to the pre-procurement phase, a matter that raises questions regarding transparency in public procurement. The model basically describes a classic agile information systems development plan. What criteria should be followed and who should govern what technology, and which vendor to invite to participate in this development (and gain potential lock-in advantages in the pursuing procurement phase)? If there is a lack of transparency, this may arguably also lead to added stress for all employees, conflicts and potentially political issues, as healthcare is a matter of great public concern. The authors make it clear that this comment is general and does not apply to the municipality named in Section 2.

The project plan for the ALL project states that ALL will contribute to demand-driven innovation and development of health and care services. Needs-driven innovation is about understanding the user's existing and future needs to ensure the development of solutions that are rooted in real needs. The sampling methodology is important in the development and implementation of welfare technology and is also central to this project.

Users are the best experts on their own, and all their knowledge is very valuable in an innovation process. Information from the user should therefore be used systematically for the development of the rich solutions.

Innovation and development through Living Lab must be based on five key principles:

• Value for the users
• User involvement (how can users influence the process)
• Quality with robust, durable solutions that meet tomorrow's needs
• Openness and accessibility
• Real life situations.

ALL, according to the client, has a two-sided purpose: ALL must both be a venue for suppliers testing new eHealth solutions. At the same time, municipalities like Grimstad have a great need to move forward with service innovation in eHealth, to meet future needs.

The assignment that the research team received from Grimstad municipality was as follows:

"Through the follow-up research we (ALL, clients) want to answer how we can best achieve the ALL project goals.

1. We want answers to how we can best cooperate with the supplier industry. Several technology vendors believe that they have the solution-but this may not be the need the service and users experience.

2. How can Living Lab methodology ensure good solutions and meet user needs? The user is very central in the living lab methodology. How can we best get users to test and develop new solutions?

3. We want the method we work out in the project to be easily transferred to other municipalities and interested parties. What is needed to ensure spread? (Citations from the tender, translated) [15]"

The way the following results were achieved, were through discussions in workshops with participants from the Agder Living Labs project group, addressing these challenges. As background for these workshops, the action research team's members had performed independent literature reviews searching for state of art knowledge in the field.

III. RESULTS

In this section the authors disseminate the findings from the case-study. Basically, the research team advocates the need for an ideation phase (combining the methodologies of Design Thinking and LEAN) [16] before entering into the implementation phase of eHealth development.

In conclusion in the case-study report, the short answers to the questions listed in the previous section were found to be as follows: The following quotation is from project-report from the Agder Living Lab follow-up-research and translated by the authors (Norwegian) [15].

"1. The research team generally do not recommend the municipalities to start here. In the short term, ALL has focused on a combined product and user focus, and that must be respected based on the framework ALL has had as a project. It also has its advantages. Having the focus on concrete product solutions, according to a project manager in ALL, has been necessary as a starting point. It must be concrete, credible and recognizable to be clear to employees. There must be a delineation around the work.

The research team looked the most at the conceptual model, as depicted in Figure 1, the future and how to scale up ALL from serving one municipality to becoming a National or at least important regional center for eHealth innovation. Although not all the ideas we contribute from the follow-up research team in retrospect proved to be equally good, it can form a starting point for further work with frameworks and methodology.

We therefore believe that in the future and in the long run, it is most appropriate to start with users and their needs - not the technology. We encourage municipalities to keep up to date with changes in the technical possibilities room, and we like to see the municipalities participate in technology and eHealth fairs and other venues for professional refills. But start with a service design process instead" [15].

Why should you start according to the model for service design, "Double Diamond" [17]; user, needs and problem solving (also called the "Ideation" phase), before going into solution exploration (the Implementation phase)?

• It creates commitment and mobilizes all system service users (Internal; employees, and external service users; patients and next-of-kin, relatives, or partners).

• It provides a better offer to ICT providers; offers an open innovation [18] knowledge capital around needs. It focuses the efforts smarter.

• It saves a long time in solution exploration; towards comprehensive digital (computer-driven) management, smarter health systems, with more accurate priorities and decisions.

• You also save time, money and human resources in solution research (which becomes more "LEAN") at all stages of the supply chain [16]. Innovation processes are also a cost carrier, which should be affected in a value innovation perspective [1].

• It all becomes a more open and transparent process (contributing to solving the transparency issue, touched above). Service design explorations, where the results are published, in front of technical (trial) acquisitions places potential suppliers on a more similar line.

• It reduces the inherent risks of technical procurement and the entire innovation process.

• Service design methods [16] helps identify drivers for desired changes. It can provide input to a quality and goal management system (Performance Indicators, Key Objectives), which can be followed up throughout the entire process of innovation. Thus, ALL, in the future, can offer better services to all the stakeholders, including the ICT providers.

2. The research team recommend it would be best to start by ensuring good solutions through understanding and covering the needs of the user. The ALL methodology has elements of this in the use of the user panel [15][19], but has lacked a description of the steps needed to arrive at the correct problem definition:

"As follow-up researches, our role is to give constructive criticism to what we observe. We have sought to remedy shortcomings and advise on possible improvements through our follow-up research project. We recommend that a future ALL concept starts with users, both residents and employees, and their needs, before defining today's and tomorrow's services. Only then will you see what is missing from technology and how this should be specified to new ICT suppliers and other stakeholders. ALL will then also be able to add value to the supplier by providing them with knowledge of the really rooted needs for new solutions and the requirements for these. We come in this report with suggestions that fill the gaps in the methodology, based on, among other things, the International Design Thinking [20] methodology (...)" [15] (translated).

3. The research team outline and discuss different strategic scenarios for how ALL can be scaled up and become a CenterPoint of a vivid eHealth innovation ecosystem and what conditions needs to be met for it to succeed. Generally, these advises are also disseminated in discussion and conclusion in this article.

The municipality of Grimstad has done a pioneering work that potentially has an interest far beyond its own municipal boundary. Those involved have learned a lot of the process so that they are geared better for new rounds later.

IV. Discussion and Conclusion

Based on the literature and workshops we have held together with ALL's project group and the glimpses we have received in ALL as a project, we have launched the following ideas about what we believe may be necessary if ALL will become a central focal point for "eHealth-Norway":

The methodology must be further developed and expanded, especially "backwards" so that skills mapping and involvement of employees and service users have been stepped up for problem definition and not just afterwards. Figure 2 seeks to illustrate the missing "steps". Before you can define user needs you need to work on:

- On-boarding employees
- Joint challenges and desired values
- Discovering Service user demands
- Analyzing and designing user journeys

An innovation culture must be created, and new knowledge will be built in the municipalities that will play an active role in a Living Lab concept like ALL.

Becoming "The learning organization" should be the goal of all municipalities and other healthcare organizations. Such learning takes place through active involvement and participation from the planning phase of change processes, transformation management and prioritization and choice of measures. It is this management work that can be systematized, with an overall process management and quality assurance system. It also contributes to an easier "rollout" later. Are everyone on board in the beginning, everyone is included in the scaling too. Here, we answer the questions raised in the introduction: "What are the necessary conditions for enabling employee co-creation in eHealth? How do we build up the learning organization, and what are the benefits of doing so?". Not only do we need an ideation phase, the first diamond in the "Double diamond-model [17], we may also need a "point zero" diamond to mobilize the workforce and achieve all desired benefits and value innovation performance [1].

Involvement of all participant groups in the early stages of the processes can help prevent work conflicts and provide a background for a health-promoting, productive and long-term working method, which should lead to efficient innovation of new processes. The most important group to anchor a new service in, besides the patients themselves, contains the employees in the municipality. In particular, the employees represent the first line; those with whom the patients always interact. They constitute the most important persons since they are resources in connection with the introduction of a new eHealth technology-supported service. Organizational development often lacks focus on welfare technology development. An innovation process in an enduring organization, for example within a care organization that introduce welfare technology, is also a learning process that includes the entire organization. A managerial responsibility here is to provide good frameworks for organizational learning.

Organizational learning is something else and more than individual-oriented learning. Individual-oriented learning can be both positive and negative for the whole. As individuals, we can add both good and bad habits and attitudes, based on our own experiences.

To see past experiences and to see the whole picture, effective mechanisms and processes are needed to share information and knowledge in an organization. It again requires a plan. It is only when people learn effectively that the organization can change.

The development towards the learning organization is about exceeding the habit of thinking and opposition to thinking new and openly. There are individuals who trade and learn, but the organization provides a framework that can support or inhibit the interaction between individual and organizational learning.

Getting real changes to existing work processes is a complex process. Cooperation on the development and testing of new welfare technology in practice means collaboration on smarter work processes and managerial arrangements related to these, thus it entails both individual and organizational learning. Collaboration provides experience and expertise on how to work together across user groups for continuous improvement in the company, and externally between partners in a value chain.

In this way, motors for mobilizing for development and change (see Figure 3.) are created. Training should be perceived as an aid and not as yet an additional burden. The management and the employees get concrete experiences about the importance of participation and arenas for dialogue. Although, we have shown a need for combining the fields of organizational learning and technology innovation.

REFERENCES

[1] W. C. Kim and R. Mauborgne, Blue ocean strategy, expanded edition: How to create uncontested market space and make the competition irrelevant: Harvard business review Press; 2014.

[2] P. R. Oeij, S. Dhondt, R. Žiauberytė-Jakštienė, A. Corral and P. Totterdill, "Workplace innovation as social innovation in European companies," Paper for The ISPIM Innovation Summit, Kuala Lumpur, Malaysia on, pp. 4-7, 2016.

[3] C. Adams, An Introduction to Ergonomics ThougtCo.2019 [accessed 2019 Jan.]. Available from: https://www.thoughtco.com/what-is-ergonomics-1206379.

[4] T. Vavik and T. A. Øritsland, Menneskligе aspekter i design (Human aspects in design). Tapir Uttrykk, NTNU, Norway. 1999.

[5] M. Neubauer and C. Stary, S-BPM in the Production Industry: Springer; 2017.

[6] C. López-Gómez, D. Leal-Ayala, M. Palladino and E. O'Sullivan, "Emerging trends in global advanced manufacturing," Policy links: University of Cambridge; 2013.

[7] R. D. Fitjar, M. Gjelsvik and A. Rodríguez-Pose, "The combined impact of managerial and relational capabilities on innovation in firms," Entrepreneurship & Regional Development, vol. 25, no. 5-6, pp. 500-520, 2013.

[8] G. Salaman, J. Storey and J. Billsberry, Strategic human resource management: Theory and practice: Sage; 2005.

[9] H. Enehaug, M. Helmersen and S-E. Mamelund, Individual and organizational well-being when workplace conflicts are on the agenda. A mixed methods study. Nordic journal of working life studies, Vol 6, No 1, pp. 83-104, 2016.

[10] C. Argyris, "Action science and organizational learning," Journal of managerial psychology, vol 10, no. 6, pp. 20-26, 1995.

[11] A. Day, S. Paquet, N. Scott and L. Hambley, "Perceived information and communication technology [ICT] demands on employee outcomes: The moderating effect of organizational ICT support," Journal of Occupational Health Psychology, vol. 17, no. 4, p. 473, 2012.

[12] K. Ninaus, S. Diehl, R. Terlutter, K. Chan and A. Huang, "Benefits and stressors–Perceived effects of ICT use on employee health and work stress: An exploratory study from Austria and Hong Kong," International journal of qualitative studies on health and well-being, vol. 10, no. 1, pp. 288-380, 2015.

[13] L. A. Huryk, "Factors influencing nurses' attitudes towards healthcare information technology," Journal of Nursing Management, vol. 18, no. 5, pp. 606-612, 2010.

[14] H. Hoen and U. Tangen, Velferdsteknologiundersøkelse (A survey of welfare technology). KS Innovasjon og utvikling Oslo. 2011.

[15] N. F. Garmann-Johnsen, T. R. Eikebrokk, M. Helmersen and K. Lindland, Agder Living Lab - Hvordan lykkes med tjenesteinnovasjon i eHelse (How to succeed with service Innovation in eHealth?), Agderforskning.no; 2018.

[16] S. K. Chokshi, D. M. Mann, "Innovating From Within: A Process Model for User-Centered Digital Development in Academic Medical Centers," JMIR human factors, vol. 5, no. 4, 2018.

[17] The British Design-Council, The Design Process: What is the Double Diamond? https://www.designcouncil.org.uk2018/ [Accessed 2019 Jan.].

[18] H. W. Chesbrough, Open innovation: The new imperative for creating and profiting from technology: Harvard Business Press; 2006.

[19] N. F. Garmann-Johnsen, M. Helmersen and T. R. Eikebrokk, "Digital Transformation in Healthcare: Enabling Employee Co-Creation through Web 2.0", AMCIS 2018 proceedings. 2018.

[20] K. Dorst, The core of 'design thinking'and its application. Design studies, vol. 32, no. 6, pp. 521-532, 2011.

[21] N. F. Garmann-Johnsen and T. R. Eikebrokk, "Critical Success Factors for Inter-Organizational Process Collaboration in Ehealth," eTELEMED 2014 proceedings. 2014.

[22] D. J. Teece, G. Pisano and A. Shuen, "Dynamic capabilities and strategic management," Strategic management journal, vol. 18, no. 7, pp. 509-533, 1997.

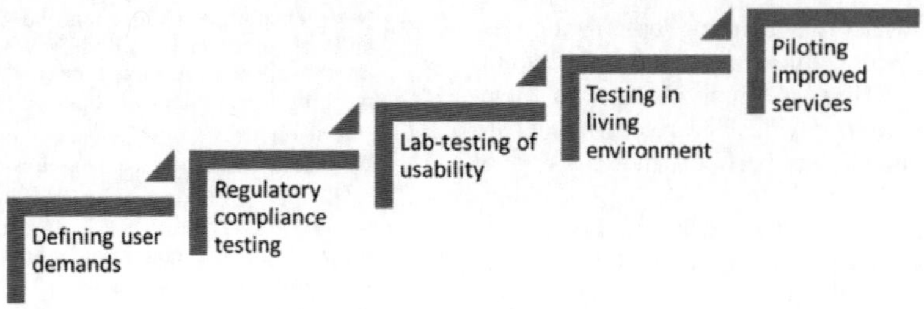

Figure 1. ALL Process model Ex Ante [15]. We find that ideation; the discovery and needs analysis and definition phase, is missing.

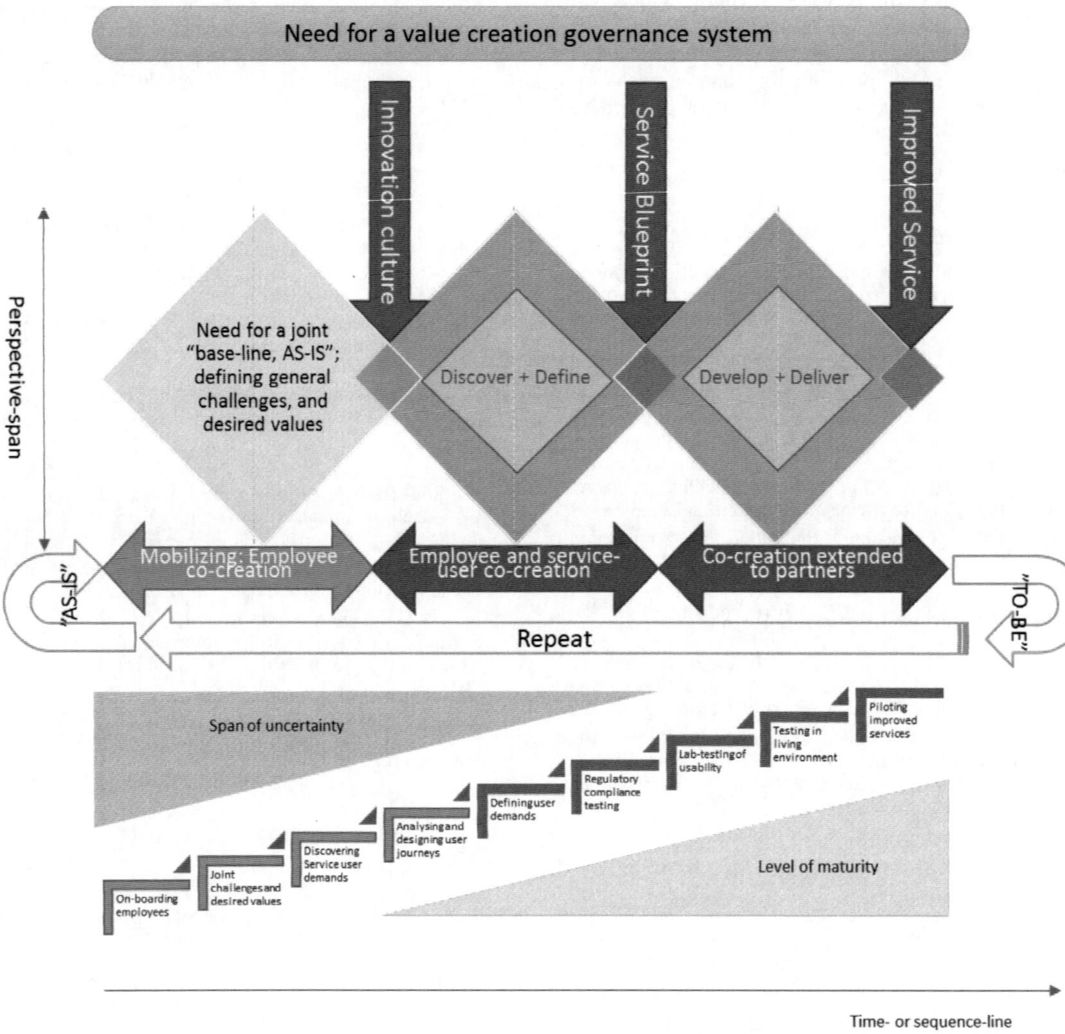

Figure 2. The "steps" model [bottom] expanded backwards, and compared with the "Double Diamond" pattern (top).

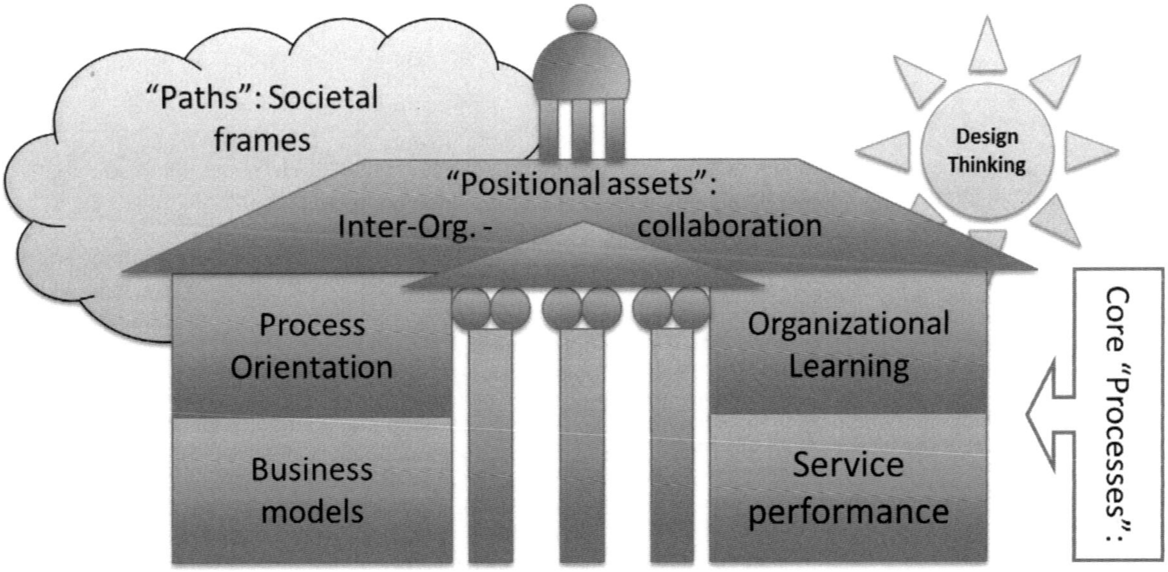

Figure 3. "Service Innovation House" - or "The Learning Organization" - A Model of Continuous Process Change and Improvement [21][22] with Design Thinking [20] as "guiding light".

Electronic Patient Communication in Norwegian Municipal Health Institutions

Halgeir Holthe
Norwegian Center for eHealth Research
University Hospital of North Norway
Tromsø, Norway
e-mail: Halgeir.holthe@ehealthresearch.no

Rune Pedersen
Norwegian Center for eHealth Research
University Hospital of North Norway
Tromsø, Norway
e-mail: Rune.pedersen@ehealthresearch.no

***Abstract*— During 2017, the Norwegian center for e-health research (NSE) conducted a national survey of the extra labor generated by incompatible Information and Communication Technology (ICT) systems in municipal health institutions. We sent out an electronic questionnaire to 283 Norwegian municipalities and addressed 5 types of health service in each community during the spring of 2017. In this paper, we describe and analyze parts of our national survey, focusing on the answers given to questions concerning direct electronic communication between the different health services and their clients. We focus on the part of the investigation concerning the municipal health services' and GPs' electronic communication with their target populations. We are primarily interested in mapping to which extent the municipalities at present use electronic means for communicating with their health service clients, and which ICT systems are applied to facilitate such communication. In our material, we are able to distinguish between 5 main municipal health services, staff work roles, and size of municipalities in terms of population. Our results suggest that around one third of the Norwegian health institutions in primary care apply electronic equipment for some of their communication with the patients. SMS for reminders has the highest score, but also appointment and prescription reservations are in common use.**

Keywords-Telemedicine; Primary care; Electronic communication.

I. INTRODUCTION

It is a national ambition in Norway to build a new common electronic communication infrastructure [1][2], which can ease the information exchange between hospitals, GPs, municipal health institutions and the public. The objective is to secure a smooth electronic dialogue between all the involved parties throughout entire patient pathways [3]. Guaranteed access to relevant and updated patient information to health professionals involved in treatment and rehabilitation is a central concern. At the patient's side of the table, access to own health information will be facilitated, at the same time as the patient is given a say in own treatment and being provided opportunities for an electronic dialogue with the professional caregivers [4]. Also, the new infrastructure is hoped to facilitate automatic extraction of data into central registers which will allow for quality surveillance, enable new research strategies, and promote health care innovations [1]. However, national ambitions are one thing, and on the ground, reality usually takes on a different complexion [5]. At present, a variety of ICT systems are in use at different levels of health care and in different institutions and geographic locations. These systems sometimes form information silos [6][7], which in turn produces extra work since the same information has to be entered into different ICT systems several times. During 2017, NSE conducted a national survey of the extra labor caused by incompatible ICT systems in municipal health institutions. We sent out an electronic questionnaire to 283 municipalities and addressed 5 types of health services in each community. During the spring of 2017, we also carried out in-depth interviews with health professionals working in Trondheim.

In this paper, we describe and analyze parts of our national survey, focusing on the answers given to questions concerning direct electronic communication between the different health services and their clients. Qualitative data from Trondheim tend to support our main conclusions from the quantitative survey. We were primarily interested in mapping to which extent Norwegian primary care services currently apply electronic tools to communicate with their clients and their families. The patients are at present provided access to own health data primarily through a central portal (helsenorge.no). However, only hospital records are accessible through helsenorge.no, while primary care health information is maintained locally. The patient centric care [8] model is still in its infancy in Norway, and our results shows that the patients' opportunities for an electronic dialogue with the health professionals are still limited.

In the introduction of this paper, we cite the relevant policy documents describing a new electronic infrastructure for primary care in Norway. In these policy documents the patients' access to own health data is emphasized. We have investigated quantitatively as well as qualitatively how health information at present is exchanged between health professionals of primary care and their clients. Our main methods as well as the materials are described in the section methods and materials.

Our main findings as described in the result section are that SMS for reminders has the highest score, but also appointment and prescription reservations are in common use. However, of the more than 1000 service institutions in Norwegian municipalities investigated in our study, only 30 % apply direct electronic communication with their clients.

II. Methods and Materials

NSE has during 2017 together with the Norwegian directorate for e-health conducted a survey of electronic communication between municipal health institutions, GPs, and their clients. The aim of the project was to study the electronic communication flow within and between municipal health institutions, and to estimate the time spent on double work generated by information silos. In this paper, we primarily focus on two objectives of the investigation:

1. To which extent do the health professionals provide updated health information to their clients by electronic means?
2. Brief mapping of ICT systems for communication with the patients currently in use in Norwegian primary care.

The investigation was divided into three phases:

1. An observational/interview phase in Trondheim to get an overview of work routines in the municipality and to learn the language by which the professionals described their own work.
2. Development and distribution of a questionnaire in accordance with the objectives of the study. The questionnaire was tested against an expert panel of five health professionals at NSE prior to distribution. We used a Questback form of around 100 questions, which was sent to key health personnel within NSE's network in each municipality. These key persons in turn distributed the questionnaire to subordinate personnel in their own community. Five municipal service areas were targeted, including nursing homes, in home care services, maternity and child services, municipal IT departments and GPs. The questions regarding direct communication between health services and clients were divided into two parts:

a) Has your institution established routines for direct electronic communication with the patients/close persons in your municipality? If so, who is involved in the electronic communication (clients, close persons or both parties)?

b) If you don't apply electronic tools for direct communication with the patients, do you have plans about implementing such services? What are the most important obstacles on the way to establishing communication with patients in your institution/municipality (lack of resources, lack of priority, legislation barriers)? The second project phase was primarily conducted during the fall of 2017.

3. Analysis and reporting of results. The work was carried out during the spring of 2018.

Trondheim is a municipality of about 220.000 inhabitants, one of the largest in Norway. We regard this fact as important, since health information handling may vary according to community size. In small municipalities (less than 5000 inhabitants), different health services are often localized within the same building. Personnel of different services often meet, and they know each other personally. In larger communities (more than 20.000 inhabitants), different services are usually located in different buildings, and personnel of different services meet less frequently face-to-face.

The transcriptions were encoded according to our checklist by several members of the staff, and the coding discussed and refined in a cyclic process during several project meetings in early 2018.

Obviously, a handful of informants cannot yield quantifiable results valid for all Norwegian municipalities. However, during our field investigation, we engaged in lengthy discussions with health personnel of all types working in our targeted institutions. All in all, we talked with around 15 people during our study, and we find it plausible that the suggestions given by them point to important aspects of the workings of electronic communication in Norwegian health care in general.

The net questionnaire of the quantitative investigation was distributed to all Norwegian municipalities (283 entities), except for municipalities in Trøndelag and Møre- and Romsdal counties. The investigation was closed on October 20, 2017, and we received 1245 answers. After cleaning of the resulting forms, a total of 1022 responses were kept for further analysis.

The compiled material about direct communication between health professionals and their clients collected for the first part of the investigation 324 responses (698 nan-values), while the second part yielded 614 (408 nan-values) and 127 (895 nan-values) responses for the last question block. In our total material 431 responses come from the larger municipalities, and 335 and 291 from medium-sized and small communities respectively (n = 1015, indicating that 7 respondents failed to fill in information about the size of their municipality). In this paper, we focus on the part of the investigation concerning the municipal health services' and GPs' electronic communication with their target populations. We are primarily interested in mapping to which extent the municipalities at present use electronic means for communicating with their health service clients, and which ICT systems are applied to facilitate such communication. In our material, we are able to distinguish between 5 main municipal health services, staff work roles, and size of municipalities in terms of population. A summary of the incoming responses showed that municipalities of different sizes are well covered, and that all health service types are represented in the materials (Table 1).

The Number of responses regarding use of electronic tools for direct communication with the clients are more frequent from large and medium size communities as compared with the small ones. The number of responses from small municipalities is only half of what we got from large and medium sized municipalities in our material (table 2). This of course has consequences when we compare number of answers to the specific questions in our Questback [10] form.

The second part of our question concerning direct communication between the municipal health services and their patients regarding who the partner in the communication with the professionals is, gave no responses. Question about plans for further implementation, and which barriers hamper such implementations, yielded 614 and 127 answers respectively. Of the 614 answers, 46 are positive responses (1-values). Hence, our material is missing or sparse for two of the questions for the rest of the material the responses are distributed as shown in table 3. Roman number I is the numbers from the first question block of the investigation, while II and III are the numbers given in the second block.

Table 1 TOTAL NUMBER OF RESPONSES IN ABSOLUTE FIGURES WITH REFERENCE TO SERVICE

Health services	Large communities	Medium-sized communities	Small communities	Total
In home service	107	72	58	237
Maternity and child service	89	69	54	212
Nursing homes	85	56	60	201
GP	73	57	47	177
IT department	55	60	21	136
Administrative	14	17	9	40
Emergency room	8	4	0	12
Total	431	335	249	1015

TABLE 2 NUMBER OF RESPONSES REGARDING USE OF ELECTRONIC TOOLS WITH RESPECT TO COMMUNITY SIZE

Electronic tools	Large communities	Medium sized communities	Small communities
SMS	144	111	69
Online booking of appointments	144	112	68
Online prescription order	144	111	68

TABLE 3 NUMBER OF RESPONSES IN TOTAL NUMBERS WITH RESPECT TO HEALTH SERVICE CATEGORY.

Service name	I	II	III
GP	152	22	4
Maternity and child service	118	88	12
In home services	30	201	36
Nursing homes	15	180	42
IT department	9	123	33
Total	324	614	127

The results were exported from Questback to excel in a full text version, but also a numeric excel report was generated. We primarily used Python 3.6 with NumPy and Pandas [11] for the numeric analysis. Python was installed by applying Anaconda 5.0.1 from Anaconda Inc. [12], on an arch Linux desktop computer, and the software read the original excel files out of the box. Python's scientific stack [13][14] represents an easy applicable and free tool for the analysis we are carrying out here. We provided access to the data for the research team via Jupyter hub. A preliminary description of the overall material is provided on the basis of the xlsx-files by Eli Kristiansen (E. Kristiansen 2018, oral communication, 5th January, NSE).

III. RESULTS

Our qualitative data suggest that direct communication with the patients by electronic means is still in its infancy in Norway. 31 per cent of our respondents to our net-form said they were currently applying electronic routines for direct communication with their clients. Of the tools in use, SMS has the highest score, followed by appointment reservation and prescription reservation on the net (table 4).

The centralized electronic services, such as helsepost.no and consultations via helsenorge.no are currently little used. Only 13% and 10% respectively of the informants saying they applied direct communication with their patients (n = 324) using these services. Looking closer at the responses explicatively saying they are applying SMS as a communication tool, the GPs and the maternity and child services use this routine most frequently (table 5). The other health services to some extent apply SMS to communicate with their clients. The figures of table 5 show that when adjusting for sample size there is no difference in the application of SMS in large versus medium-sized/ small communities.

Also, GPs apply online booking most frequently, again municipal size does not seem to influence on these work routines (table 6). The other municipal health services do not apply online booking to any extent. During our fieldwork, the GPs stated that as private enterprises GP offices often have a home site for their private enterprise up and running. Hence, online booking routines are fairly easy to implement. On the other hand, municipal health services have to work via their communities' common net pages, which might make it more difficult to implement service- specific routines for direct communication with the clients.

The numbers for online prescription order show that this routine is little used by municipal health services (table 7). Only GPs use online prescription order as a routine. The figures for the GPs' application of online prescription order is evenly distributed regarding community size, taking into account that the sample size in small communities is half of the sample sizes in large and medium sized communities.

TABLE 4 NUMBER OF RESPONDENTS (IN ABSOLUTE NUMBERS AND PERCENT) APPLYING ELECTRONIC ROUTINES FOR DIRECT COMMUNICATION WITH THEIR CLIENTS (N = 324).

Service name	Total number	Percent
SMS	266	82%
Online prescription order	122	37%
Online booking of appointments	116	35%
Other communication channels	53	16%
E-consultation	43	13%
Patientpost.no	34	10%

TABLE 5 USE OF SMS COMMUNICATION IN MUNICIPAL HEALTH SERVICES WITH RESPECT TO COMMUNITY SIZE

Health Services	Large community	Medium sized community	Small community	Total
GP	49 (42%)	44 (37%)	25 (21%)	118 (100%)
Maternity and Child services	51 (45%)	42 (37%)	21 (18%)	114 (100%)
In home Services	13 (59%)	7 (32%)	2(9%)	22 (100%)
Nursing home	4 (67%)	1 (16,67%)	1 (16,67%)	6 (100%)
IT department	3 (50%)	2 (33,33%)	1 (16,67%)	6 (100%)

TABLE 6 USE OF ONLINE BOOKING OF APPOINTMENTS IN MUNICIPAL HEALTH SERVICES WITH RESPECT TO COMMUNITY SIZE

Health Services	Large commun ity	Medium sized community	Small community	Total
GP	46 (43%)	38 (35%)	24 (22%)	108 (100%)
Maternity and Child services	1 (25%)	2 (50%)	1(25%)	4 (100%)
In home Services	1 (33%)	1(33%)	1(33%)	3 (100%)
Nursing home	1(100%)	0 (0%)	0 (0%)	1 (100%)

TABLE 7 USE OF ONLINE PRESCRIPTION ORDER IN MUNICIPAL HEALTH SERVICES WITH RESPECT TO COMMUNITY SIZE

Health Services	Large commun ity	Medium sized community	Small community	Total
GP	46 (43%)	38 (35%)	24 (22%)	108 (100%)
Maternity and Child services	1 (25%)	2 (50%)	1(25%)	4 (100%)
In home Services	1 (33%)	1(33%)	1(33%)	3 (100%)
Nursing home	1(100%)	0 (0%)	0 (0%)	1 (100%)

The general practitioners are the pivot of Norwegian primary health care, and to some extent they use electronic tools for direct communication with their patients. Working as emergency room or nurse home staff, direct communication with external clients is handled by other personnel. The same is true when the doctor work in maternity and child services. The in-home care services stated during our field investigation that one of their primary goals when rebuilding their ICT systems is to stablish direct contact with the patients' EPR when being out of office.

This way, much of the documentation required in healthcare might be down before returning to the office, i.e., the documenting tasks may be performed in the clients' home. The in-home care services, to any extent, do not communicate directly with their clients by electronic means, even if this was regarded as a constructive opportunity when rescheduling appointments. Nurse home personnel told us during qualitative interviews they primarily use telephone for direct communication with the families of their clients. Direct electronic communication with the public is hardly used, and in our quantitative material, the nurse home constitutes 19.8 % of the answers.

Maternity and child services personnel told us during qualitative interviews that in general, they used their own log systems, often in written form. In their daily routines, they need to collect written consent forms with signatures from children, as well as from adults. Today, these tasks are performed through regular mail, and hence is time- and labor consuming. An electronic system for document exchange with the clients would be more efficient than the current procedures, according to the staff.

We finally asked our informants about possible reasons for not applying direct electronic communication with their clients. Among the suggested reasons in the questionnaire was 1) lack of priority among the leaders, 2) lack of resources/competence 3) lack of technical equipment. We got 127 answers (response rate 12%) of which 895 were nan-values. Of the answers pointing to specific reasons for not applying electronic tools for direct communication with the patients 33 per cent pointed at lack of priority and 37 % pointed to lack of economic resources.

IV. DISCUSSION

The nursing homes are marked by several professions working together to provide care to their clients. Our informants told us that much of the communication with external colleagues is carried out by telephone, and the same also goes for communication with their clients' families and close persons. Some of our informants told us about communication bottlenecks due to incompatible electronic systems in different community health care institutions, and this potentially generates a lot of double work when care is to be coordinated with external parties. The nursing homes did not communicate with their clients' families electronically and usually applied telephone for this task. Our quantitative results show that the nursing homes have a lower score for all of the communication routines investigated regardless of community size. A lack of priority among municipal leaders and lack of resources seem to be the reasons for this state of affairs. In the literature, electronic communication with patients is suggested to secure continuity in the care provided [15][16].

The in-home care services' personnel focused on the lack of integration between mobile and stationary ICT systems as a main barrier to efficient digital communication at the workplace. Updated medication lists were of central concern, and our informants did not to any extent use electronic communication with their clients in their daily work. The focus is still on enhanced information exchange with professional partners, and this has to be resolved first, our informants told us. Like in the nursing homes, the in-home care services have a low score for all of the communication routines investigated, regardless of community size [17].

The Norwegian maternity and child services serve nearly all children and mothers in the communities. They are involved in general health checkups for all of their clients, vaccination programs and proactive medicine. The stations also collaborate closely with the schools. Hence, the staff reach patient categories which other health personnel may infrequently meet. The maternity and child services seem to apply means of electronic communication slightly more than nursing homes and in-home services [18]. The collaboration with municipal school may be the reason for high score on SMS. A considerable portion of the communication with other professionals and the clients is performed by written documents sent by post. Our informants told us that casework for schoolchildren is slowed down if written consent is required from the parents on behalf of their underaged children.

The GPs are the pivot of Norwegian health care, and any restriction of communication at this level has consequences for the entire patient pathway. GPs communicate extensively with patients, primary care institutions and hospitals [19][20]. According to our informants, much of the communication with other professionals in municipalities and in hospitals is conducted by telephone. In particular, the exchange of medication lists between GPs and the municipal home care services were considered as bottlenecks in the information exchange. Some of our informants claimed that restrictions in communication across municipality borders, and when patients consult different GPs constitute potential blocking of patient information exchange. Documents frequently have to be scanned, which results in data that are not electronically searchable. Some of the GPs requested a better national coordination of electronic health record systems. The GPs conducted the bulk of direct electronic patient communication which we observed in this investigation. Especially SMS reminders and online booking services were in frequent use, and more so than in the rest of primary care. This may be due to a smoother process of decision making within small work organizations.

ACKNOWLEDGMENT

We will like to thank our informants for taking part in our interviews without compensation. Also, we thank the personnel of Helseplattformen for providing valuable information. In particular, we thank Trine Hansen for scheduling the fieldwork in Trondheim during the spring of 2017.

REFERENCES

[1] Norwegian Ministry of health and care services, One citizen-One Health Record. Whitepaper 9,Oslo, 2012-2013.

[2] The Norwegian Directorate of Health and The Norwegian Directorate of eHealth, Investigation of One citizen- One HealthRecord;(Norwegian), https://ehelse.no/Documents/En%innbygger%20%en%20journal/sammendrag%Utredning%av%20%C3%89n%20innbygger%20%20%C3%A9n%20jiurnal_1.0.pdf 2018.10.02.

[3] Ministry of Health and Care Services, "The basic structure of the health service," (Norwegian), https://www.regjeringen.no/no/tema/helse-og-omsorg/sykehus/vurderes/grunnstrukturen-i-helsetjenesten/id227440/ 2018.09.15

[4] H. Slagsvold, and L. Guttormsen, "Message distribution in the municipal health service," Status report. (Norwegian), 2018.10.02 https://nhn.no/media/1160/2013-statusrapport-meldingsutbredelse-i-kommunehelsetjenesten.pdf

[5] L. W. Peute, J. Aarts, P. J. M. Bakker, and M. W. M. Jaspers, "Anatomy of a failure: A sociotechnical evaluation of a laboratory physician order entry system implementation," International Journal of Medical Informatics, vol. 79, no. 4, pp. e58-e70, 2010.

[6] L. C. Miller, B. B. Jones, R. S. Graves, and M. C. Sievert, "Merging silos: collaborating for information literacy," Journal of continuing education in nursing, vol. 41, no. 6, pp. 267-72, 2010.

[7] A. R. Miller, and C. Tucker, "Health information exchange, system size and information silos," Journal of Health Economics, vol. 33, pp. 28-42, 2014.

[8] E. H. Waagner, et al., "Improving Chronic Illness Care: Translating Evidence Into Action," Health Affairs, vol. 20, no. 6, pp. 64-78, 2001

[9] M.Bambling, et al., "Challenges faced by general practitioners and allied mental health services in providing mental health services in rural Queensland, " The Australian journal of rural health, vol.15, no. 2, pp. 126-30, 2007.

[10] Questback. https://www.questback.com/ 2018.10.02.

[11] W. McKinney, Python for Data Analysis: Data Wrangling with Pandas, NumPy, and IPython," O'Reilly Media, 2017.

[12] Anaconda Inc. https://www.anaconda.com/ 2018.10.02.

[13] F. Perez, B. E. Granger, J. D. Hunter, Python: "An Ecosystem for Scientific Computing," Computing in Science & Engineering, vol.13, no. 2, pp. 13-21, 2011.

[14] K. J. Millman, and M. Aivazis, "Python for Scientists and Engineers. Computing in Science & Engineering," vol. 13, no. 2, pp. 9-12, 2011.

[15] E. Balas, et al., "Electronic communication with patients: Evaluation of distance medicine technology," JAMA, vol. 278, no. 2, pp. 152-9, 1997

[16] A. A. Vogelsmeier, J.R.B. Halbesleben, J.R. Scott-Cawiezell, "Technology Implementation and Workarounds in the Nursing Home," Journal of the American Medical Informatics Association, vol.15, no. 1, pp. 114-9, 2008.

[17] B. Lindberg, C. Nilsson, D. Zotterman, S. Söderberg, L. Skär, "Using Information and Communication Technology in Home Care for Communication between Patients, Family Members, and Healthcare Professionals: A Systematic Review," International Journal of Telemedicine and Applications, vol. 31, 2013

[18] A. Austin, et al., "Approaches to improve the quality of maternal and newborn health care: an overview of the evidence," Reproductive Health, vol. 11, no. 2, 2014.

[19] S. Kripalani, F. LeFevre, C. O. Phillips , M. V. Williams, P. Basaviah, and D. W. Baker, "Deficits in communication and information transfer between hospital-based and primary care physicians: Implications for patient safety and continuity of care," JAMA, vol. 297, no. 8, pp. 831-41, 2007.

[20] "Primary Care Physician Shortages Could Be Eliminated Through Use Of Teams, Nonphysicians, And Electronic Communication," Health Affairs, vol.32, no.1, pp. 11-9, 2013.

Design of a Telestroke System to Optimize Healthcare Delivery for Cerebrovascular Diseases in Colombia

Brenda Ropero, Nicolás Useche
Department of Diagnostic Imaging
University Hospital Fundación Santa Fe de Bogotá
Bogotá, Colombia
e-mail: bv.ropero47@uniandes.edu.co;
jnuseche@gmail.com

Manuel F. Granja
Lyerly Neurosurgery
Baptist Neurological Institute
Jacksonville, Florida, USA
e-mail: mfgranja@gmail.com

Antonio Salazar, Juan Pérez
Laboratory of Telemedicine
University of Los Andes
Bogotá, Colombia
e-mail: ant-sala@uniandes.edu.co;
jc.perez13@uniandes.edu.co

Hernán Bayona, Carlos Martínez
Primary Stroke Center, Neurology Department
University Hospital Fundación Santa Fe de Bogotá
Bogotá, Colombia
e-mail: hernanbayonao@gmail.com;
caferma13@gmail.com

Abstract—**The purpose of this manuscript is to present the design of a software tool that supports clinical decision-making and the early transfer of patients with suspicion of cerebrovascular diseases in Colombia, We designed several clinical algorithms that comply with the latest American Heart Association and American Stroke Association (AHA/ASA) guidelines for the clinical care of a group of cerebrovascular diseases by defining multiple clinical outcomes in three different healthcare settings. Algorithms were reviewed and approved by a group of stroke experts, including a vascular neurologist, a general neurologist, and an interventional neuroradiologist. Patient data, time of symptom onset, and neurological and radiological severity scores were integrated into a comprehensive clinical workflow to increase the diagnostic sensitivity and specificity to select candidates for acute reperfusion therapies. Absolute and relative contraindications for intravenous thrombolysis and mechanical thrombectomy were incorporated to predict the need for an immediate transfer to a specialized stroke unit. These multiple variables contained in the algorithms were entered into a collaborative platform connecting three different healthcare settings with varying degrees of expertise and technological resources for stroke care. A web-based decision aid was obtained for real-time clinical decision-making. This software builds the pillars of a telestroke public-private network for the Emergency Stroke System in Colombia, guiding the clinical identification of a stroke, scoring the magnitude of the neurological deficit, and mainly suggesting whether a determined case may benefit from acute reperfusion therapies.**

Keywords-stroke; telestroke; algorithms; software; collaborative platform.

I. Introduction

Colombia is located in the northern part of South America, with an estimated population of 45.5 million by 2018 [1]. According to the Colombian National Health Observatory, by 2014, strokes were the third leading cause of death after coronary heart disease and interpersonal violence [2]. In the first half of 2018, there were 7,429 deaths attributed to stroke, and approximately 250,000 people in Colombia live with disabilities associated with cerebrovascular diseases [1]. The healthcare system delivery in Colombia is concentrated in urban areas, and access to specialized treatments for acute stroke is distributed inequitably [3]. Hence, patients located in remote areas sometimes lose the opportunity to receive proper treatment in an adequate timeframe, and some patients in urban areas are sent to healthcare facilities with no stroke handling capabilities, thus increasing the probability of permanent sequelae.

In our experience, several efforts have been made to improve the transfer of patients presenting with acute ischemic strokes from rural areas to a certified Primary Stroke Center (PSC) in Bogotá, the capital of Colombia. These experiences have demonstrated that, without a collaborative platform that assists healthcare providers in the decision-making and transfer processes, emergency services became overloaded with patients who were not real candidates, and patients with a potential benefit for acute interventions arrived late, resulting in poorer outcomes.

A system that supports real-time clinical decision-making while evaluating a patient with suspected cerebrovascular disease is not well established in Colombia, particularly for patients with the suspicion of an acute stroke with large vessel occlusion, who may be potential candidates for endovascular therapy in which the recanalization of the occluded cerebral artery results in penumbral salvage if accomplished early [4].

A collaborative platform integrating multiple clinical variables, clinical scales and scores and outcome predictors seems to be a feasible solution to increase health coverage in a country with healthcare budget constraints [5]. Considering the constantly emerging improvements in cerebrovascular disease protocols and evidence-based practices, we have developed a web-based software named Telestroke-RU (Emergency Network, which is the English translation of the Spanish "*Red de Urgencias*"), based on the latest diagnostic and therapeutic recommendations by the American Heart Association and American Stroke Association (AHA/ASA) [6]. The purpose of this software is to guide the clinical decision-making processes step-by-step when non-expert clinicians have doubts in patient management, thus serving as a communication tool with stroke experts in the context of patients presenting with a broad spectrum of cerebrovascular diseases such as ischemic stroke, hemorrhagic stroke, and transient ischemic attack (TIA). The software is available on the website https://telestroke.uniandes.academy.

This paper is organized as follows: Section II defines the three healthcare settings and describes the algorithm design process. Section III describes related work in the healthcare sector and introduces the software development. Section IV describes the results of the web-based collaborative platform. Finally, Section V presents our conclusions, future work and acknowledgements.

II. HEALTHCARE SETTINGS AND ALGORITHM DESIGNS

The software algorithms were designed according to the human and technological resources available at healthcare facilities in which a patient with a cerebrovascular disease may arrive.

A. Healthcare settings

We define three clinical settings as follows:

Primary Healthcare Setting (PHS): a basic level of healthcare delivery, where diagnostic tools are limited to the anamnesis and physical exam performed by a primary care physician and where basic blood tests are available. The purpose of this setting is to make an accurate diagnosis and to facilitate timely transfers, as shown in Figure 1. The clinical workflow designed for this setting has 10 different pathways resulting in different final possible outcomes, including priority or urgent transfer to an Intermediate or an Advanced Healthcare Setting, as shown in Figure 2.

Intermediate Healthcare Setting (IHS): intermediate level of healthcare delivery capable of performing computed tomography scans (CT scans), computed tomography angiography (CTA), and intravenous thrombolysis. Although IHS provides neurologists and radiologists, they are not available 24 hours/7 days a week to make critical decisions. The purpose of this setting is to make an accurate diagnosis of a stroke with a large vessel occlusion and optimize the early transfer of patients who are candidates for endovascular therapy. The clinical workflow designed for this setting has 39 different pathways resulting in different possible outcomes, including urgent transfer to an Advanced Healthcare Setting (AHS), referral to the Intensive Care Unit (ICU) after the administration of intravenous thrombolysis, neurosurgery referral, hospitalization or ambulatory care. Although the complete IHS algorithm is not shown due to its extensiveness, Figure 3 shows an example of one module of ischemic stroke requiring intravenous thrombolysis.

Advanced healthcare setting (AHS): an advanced level of healthcare delivery. AHS includes specialized human and technological resources available 24 hours/7 days a week for the healthcare delivery of a patient presenting with a cerebrovascular disease. These specialized human and technological resources include vascular neurologists, neuroradiologists, the ability to perform CT scans, CTAs, magnetic resonance imaging (MRI), magnetic resonance angiography (MRA), and the capacity for intravenous thrombolysis and endovascular therapy, as well as stroke ICUs. This setting can receive transfers from PHS and IHS. The clinical workflow for this setting has 1,162 different pathways, and because interfacility transfers are not necessary, it results in the following final possible outcomes: ICU referral after administration of intravenous thrombolysis or endovascular therapy, neurosurgery referral, hospitalization, or ambulatory care. Although the complete IHS algorithm is not shown due to its extensiveness, Figure 4 shows an example of one module for ischemic stroke with large vessel occlusion requiring endovascular therapy.

B. Algorithm designs

Six people were involved in the design of the telestroke system as described below. The algorithms were reviewed in weekly sessions by a group of experts working in a Joint Commission International (JCI)-certified PSC and included a vascular neurologist, a general neurologist and a neuroradiologist. Preliminary drafts of the algorithms were made with further discussions and edits for approximately two months focusing on the best and most proper way to manage patients in each setting. The group of experts finally approved the three algorithms in consensus following the latest evidence-based 2018 AHA/ASA guideline recommendations [6] and adapting them to the latest version of the national stroke guidelines of Colombia [7]. The performance of the final version of the algorithms was tested using simulated cases and dissecting the workflows into modules by a group that consisted of a physician and two engineers who were involved in developing the software. For the ease of the collection of data of clinical variables, neurological state and radiological findings, and considering the degree of expertise and technological resources for stroke holistic care of the three healthcare settings, the clinical algorithms were organized into common basic modules, as shown in Table I. The scales and predictors used in these modules, which allow assessing the severity of the event, the neurological state of the patient, and other factors which allow determining the treatment, either in situ, or the referral to a higher level service are shown in Table II. Each clinical workflow guides the diagnosis of a spectrum of cerebrovascular diseases, including ischemic stroke, hemorrhagic stroke, and TIA, allowing us to rule out conditions presenting similarly to a stroke, denoted as stroke mimics.

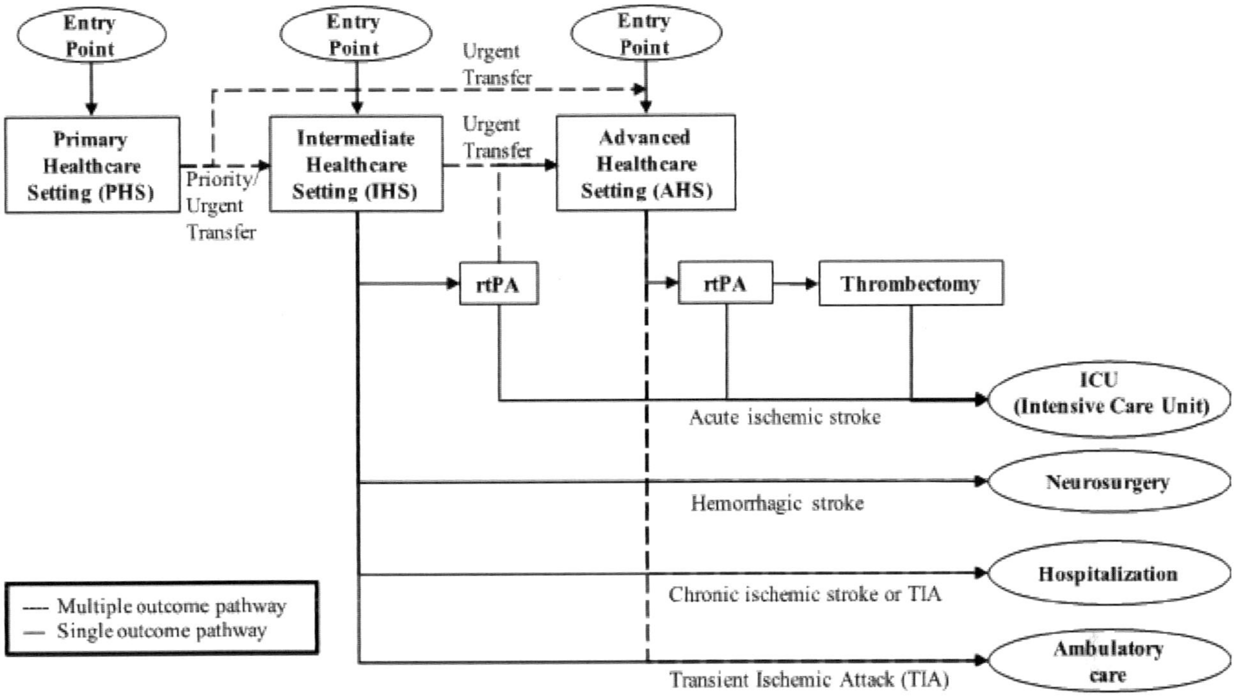

Figure 1. Interaction between the three healthcare settings and final possible diagnosis and referrals. rtPA= Recombinant tissue plasminogen activator.

III. Related Work and Software Development

This section describes previous work in the topic and the development details of our software.

A. Related Work

In our country, there are no systems in the health sector that support decisions for stroke care. Worldwide, the few existing systems are designed in different practical front-ends in the form of applications available for smartphones, tablets and web-based tools [8]-[11], or software integrated into medical records [12]. These systems rely on measurements of clinical scales or radiological scores but rarely rely on a combination of both. These are also designed mostly for acute ischemic stroke care, excluding other conditions within the spectrum of cerebrovascular diseases. Some are used for acute care, while others are used for follow-up during hospitalization or for outpatient care.

Furthermore, most existing systems use the same algorithm independent of the different levels of expertise and the technological resources of health facilities within a country. Since in real clinical scenarios the eligibility criteria for reperfusion therapies are based on a combination of clinical and radiological variables, a combination of these criteria is suggested to improve the efficiency in management and quality of care. Nevertheless, many studies implementing this type of system show a reduction of morbidity, mortality, inpatient length of stay as well as better functional outcomes [13].

B. Software development

The final version of the algorithms was arranged into a software presented in a simple and friendly interface for both administrative and healthcare provider users in the form of a collaborative platform. Access to the software was possible with the previous activation of an account by creating a username and a password. Privileges for interaction with a determined module of the software were assigned depending on the user expertise and capacity to make special contributions for the different scales and scores. Considering that in Colombia, a patient can interact with administrative staff, paramedics, triage staff, nurses, general physicians, emergency physicians, neurologists and neuroradiologists, user privileges were given depending on the role of the professional. A relational model was implemented with the following final diagnosis outcomes: ischemic stroke within the therapeutic window (acute or subacute compromising middle cerebral artery (MCA), anterior cerebral artery (ACA) and/or posterior cerebral artery (PCA) territories), chronic ischemic stroke, hemorrhagic stroke, TIA and stroke mimics.

Software code was made using PHP (Hypertext Preprocessor) and JavaScript languages. Data were stored using the MySQL 5.1.40 database (Oracle Corporation, Redwood City, CA, USA), which was created and administered by MySQL Administrator 1.1.9. The software is executed in different web browsers, such as Mozilla Firefox, Google Chrome, Safari and Microsoft Edge, so that it can be used on a computer, a tablet or a smartphone. The application is available at https://telestroke.uniandes.academy.

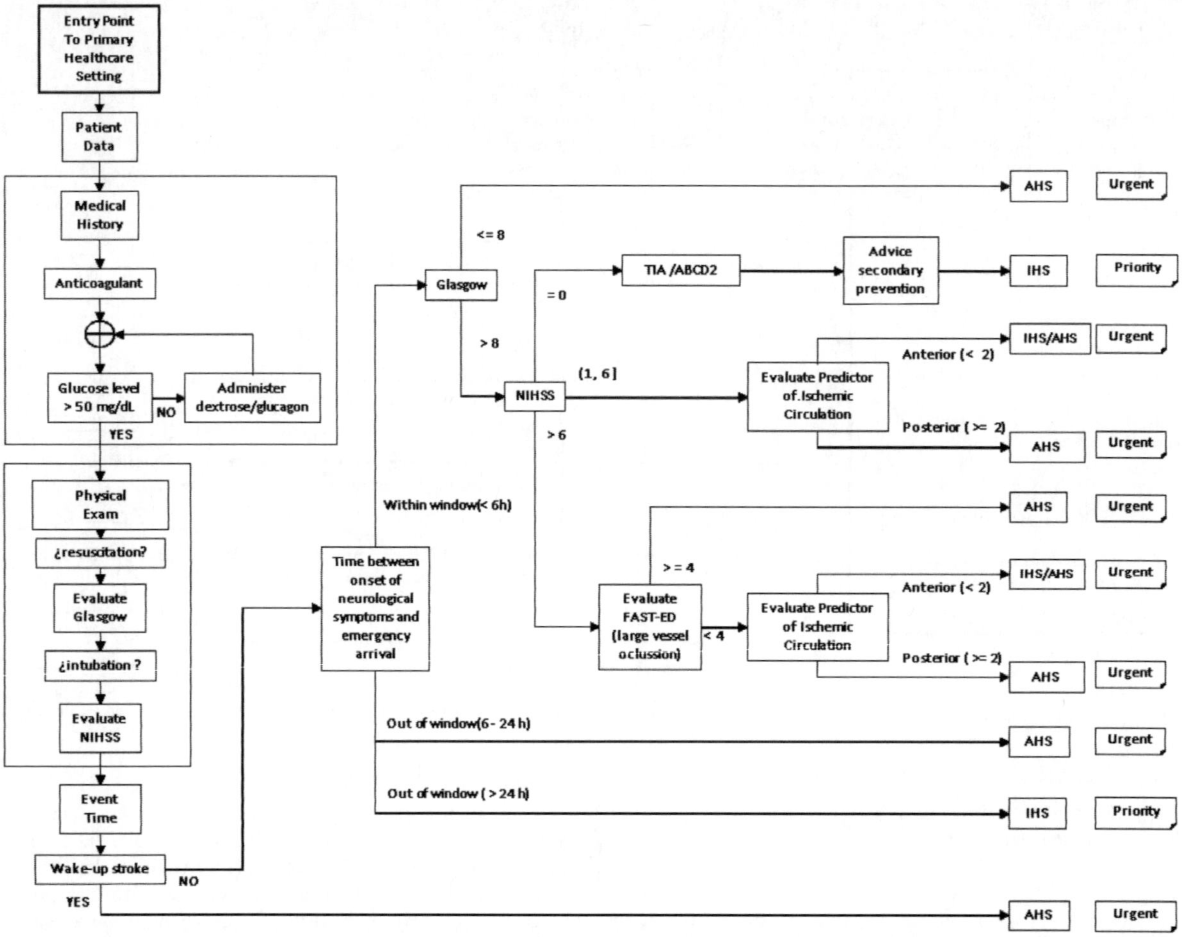

Figure 2. Detailed clinical workflow for the Primary Healthcare Setting. Algorithm results in 10 different pathways and final possible outcomes.

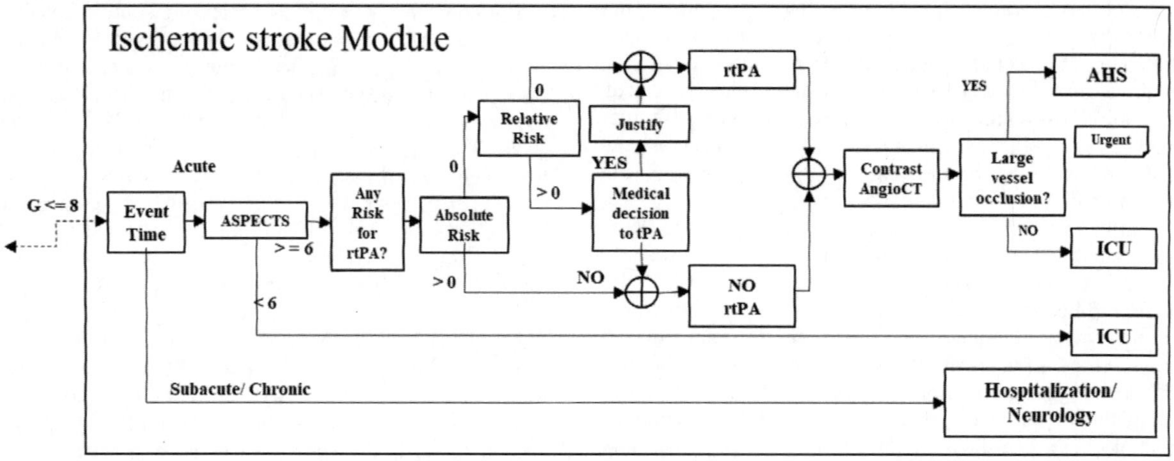

Figure 3. Ischemic stroke module showing one arm of the Intermediate Healthcare Setting algorithm. Due to its extensiveness resulting in 39 different pathways, the complete workflow is not shown. This is a case of a patient with ischemic stroke requiring intravenous thrombolysis. rtPA= Recombinant tissue plasminogen activator. --- = Intermediate pathways.

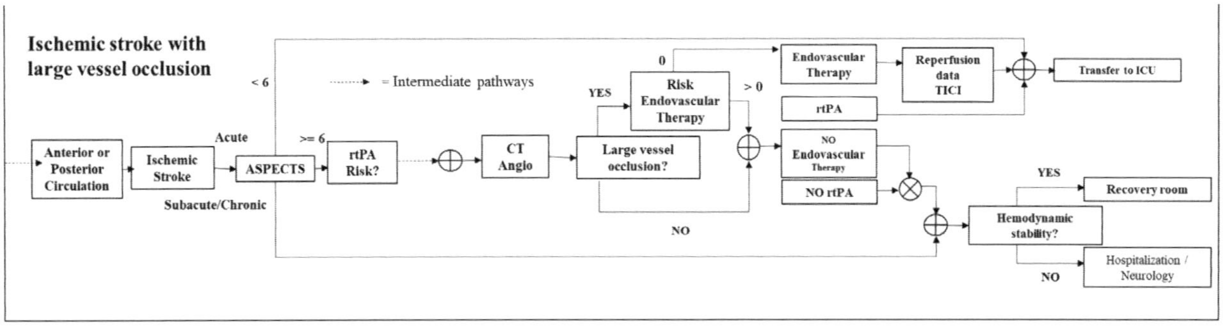

Figure 4. Ischemic stroke Module showing one arm of the Advanced Healthcare Setting. Due to its extensiveness resulting in 1,162 pathways, the complete workflow is not shown. This is a case of a patient with ischemic stroke requiring endovascular therapy. rtPA= Recombinant tissue plasminogen activator.

Tests on all the units of the workflows were run to ensure that the implementation of each unit correctly modeled the intended behavior of the diagram shown in Figure 1.

For each healthcare setting, a clinical workflow was established. Figure 2 shows an example of the algorithm implemented for the PHS, which is the most straightforward setting. The performance of the workflows for each setting was tested using simulated cases to guarantee that each unit worked.

Over 1,211 tests were run to reproduce all possible pathways among the three settings. Additional tests were run considering the frequent error of users: trying to select multiple options in single-option menus, changing a selection at the beginning of the workflow when at the end of the workflow, mistakenly selecting an option that is not possible in the current state, etc.

These tests, while not capable of modeling all possible errors, were diverse enough to detect flaws in the system so that the appropriate corrections could be made.

TABLE I. BASIC SOFTWARE MODULES

Module #	Content of modules	
	Description	*Data*
Module 1	Identification data	Name, age, sex, identification number
Module 2	Time of onset of symptoms	Day, month, year and hour
Module 3	Neurological deficit, past relevant illnesses	e.g., Left hemiparesis, global aphasia. Past Diabetes Mellitus
Module 4	Use of anticoagulant medications	e.g., Use of oral warfarin for atrial fibrillation.
Module 5	Physical exam	Heart Rate, Blood Pressure, Respiratory Rate, Weight, Laterality, rest of physical exam.
Module 6	Imaging	Compromised Cerebral territory, infarct volume, etc.
Module 7	Laboratory	Complete Blood Count, creatinine, clotting times, etc.
Module 8	Interventional procedures	Door to needle time, door to inguinal puncture, etc.

After corrections, the implementation accurately models the algorithms of the three settings. To facilitate the user experience, the interface was designed to be explicit in each question to avoid ambiguities. Furthermore, a simple manual was designed to show, step-by-step, the actions that must be followed to carry out common procedures so that the user can learn the overall mechanism of the system.

IV. RESULTS

A web-based decision aid was obtained for real-time clinical decision making using a computer, a tablet or a smartphone in Colombia. The software organizes the information contained in the algorithms into eight different basic modules as shown in Table I. A total of 11 clinical scales and radiological scores were integrated into its corresponding module to increase the sensitivity and specificity of each possible diagnosis. The privileges for the different professionals participating in stroke code responses to fill the data are shown in Table II. To ensure the safe selection of candidates for reperfusion therapies, relative and absolute contraindications for both intravenous thrombolysis and mechanical thrombectomy were also integrated into the software in different modules [14].

Three phases were planned for the validation of the performance of the Telestroke system. The first phase was initiated by one physician and two engineers who simulated a number of different cases based on a 5-year stroke database from 2014 to 2018. They collected approximately 610 past real cerebrovascular disease cases from our institution containing most of the variables shown in Tables I and II. The second phase is in progress at this moment and is intended for the training and validation of the software with neurology residents involved in daily stroke code responses. Once modifications in the software from the second phase are made, the third phase will be performed in real clinical scenarios in a network between pilot hospitals.

Depending on the patient point of entry at the Emergency System, the software guides clinical decision-making and supports early interfacility transfer for patients with a high suspicion of cerebrovascular diseases. Administrative and healthcare users can simultaneously feed the modules through directed steps entering patient data, thus beginning to solve the clinical case that models a specific pathway.

The first phase of the software validation was made by simulating diverse clinical cases. For example, a pathway simulating the diagnosis of acute stroke within the therapeutic window yields the following decision processes:

If a healthcare provider working in a PHS suspects a patient is presenting an acute stroke, the software guides the diagnosis using the Glasgow Coma Scale (GCS) [15] and the NIHSS scale [16] and supports the decision of interfacility transfer for candidates to reperfusion therapies using the time of symptom onset, the Field Assessment Stroke Triage for Emergency Destination (FAST-ED) score [17], and the relative and absolute contraindications for reperfusion therapies.

The combination of these results inside the system will determine the severity of the neurological deficit and, in this case, the decision to transfer the patient to an IHS or an AHS based on the presence of a clinically large vessel occlusion. In this case, the main purpose is to accurately diagnose a stroke and clinically evaluate the compromised cerebral territory and large vessel occlusion. The complete workflow for the PHS considering the various diagnoses and therapeutic decisions is shown in Figure 2.

TABLE II. CLINICAL SCALES AND RADIOLOGICAL SCORES USED IN THE BASIC MODULES

Module #	**Content**		
	Scale/Score	***Description***	***Responsible Professionals***
Module 3	ABCD^2 score [18]	Predictor of stroke risk after TIA [19]	General physician, emergency physician or neurologist
Module 5	Cincinnati scale [20]	Evaluation of potential stroke before physician evaluation	Pre-hospital staff, nurses, general physician
	Glasgow Coma Scale [15]	Level of consciousness	General physician or emergency physician
	NIHSS [16]	Quantify impairment caused by stroke	General physician, emergency physician, neurologist
	Posterior circulation predictor [21]	Score predicting posterior circulation involvement	Neurologist
	FAST-ED [17]	Detection of large vessel occlusion	General physician, emergency physician, neurologist
	WFNS score [22]	Grades the severity of subarachnoid hemorrhage	Neurologist or neurosurgeon
Module 6	ASPECTS [23], FISHER [24] and ICH [25]	Radiological scores for ischemic and hemorrhagic stroke	Radiologist, neurologist and/or neurosurgeon
Module 7	TICI score [26]	Thrombolysis reperfusion	Interventionist

If a healthcare provider working in an IHS suspects a patient is presenting an acute stroke, the software guides the diagnosis using the Glasgow Coma Scale (GCS) [15] and the NIHSS scale [16] and supports the decision of interfacility transfer for candidates to endovascular therapy using the time of symptom onset, the Alberta Stroke Program Computed Tomography (ASPECTS) score [23] for compromised anterior and/or posterior cerebral circulations, and the relative and absolute contraindications for endovascular therapy. The combination of these results inside the system will determine the severity of the neurological deficit and, in this case, the decision to transfer the patient to an AHS based on the presence of a radiological large vessel occlusion. If the criteria for reperfusion therapies are met, the software recommends administering intravenous thrombolysis and initiating a prompt transfer to an AHS to perform endovascular therapy and the corresponding referral to a specialized stroke ICU. Due to its extensiveness, the complete workflows for the IHS and AHS are not shown because it is not possible to appreciate the details.

In addition to the workflows for each healthcare setting, this software serves as a reference network tool. Hence, the information on the state of the patient and the results of blood tests, imaging evaluations, and procedures are available online for all health services involved in patient care.

In summary, 10, 39 and 1,162 different pathways resulting in different final possible outcomes were obtained for the Primary, Intermediate and Advanced Healthcare Settings, respectively.

V. CONCLUSION AND FUTURE WORK

The software presented here is an early step in a long-term process to strengthen the stroke emergency network between hospitals with different healthcare levels in urban and rural regions in Colombia. This novel tool can guide healthcare provider decision-making through a collaborative platform, using up-to-date clinical workflows and coordinating the consultation of stroke experts, not only for acute stroke cases but also for a broad spectrum of cerebrovascular diseases. The main goal is to identify acute phase ischemic strokes within the therapeutic window that are susceptible to acute reperfusion interventions with the intention to decrease the high number of people with resulting stroke disabilities. The collaborative platform also contributes to minimizing errors and promoting safer, quicker and high-quality clinical care. This tool is currently being implemented in simulated clinical cases with neurology residents in our institution with the objective to calibrate the software and make further updates and improvements in the interface before it is implemented in real scenarios, so the results of this phase are not shown yet. Its impact will be evaluated in a third validation phase inside a network of pilot hospitals. Once all phases are concluded, the real performance of the software will be determined in terms of comparing its impact on the timing of task completion for time-consuming processes before and after the implementation of the software, door to needle times, quality of care, connectivity, clinical outcomes at discharge, 30 days and 90 days after the event using the modified Rankin scale [27] and its performance as a learning tool platform to simulate clinical cases for healthcare providers in training participating in stroke code responses.

Considering the underreporting of stroke cases and the inconsistency between studies showing different prevalence and incidence measures across the country [28], other possible uses of this software include the collection of epidemiological data that are not currently centralized for future analyses and studies concerning cerebrovascular diseases. The implementation of this tool in an AHS may serve as a self-assessment instrument for measuring the fulfillment and adherence to the latest standards of care inside dedicated stroke units, thus contributing to providing the best care available and improving outcomes. A challenge along the way is to continually update the algorithms that feed the software, considering the ever-changing evidence-based medicine. This can be resolved with periodical reviews between stroke experts and software developers. This online collaborative platform is currently in the second phase of the aforementioned three phase-process of validation, and future work needs to be done to incorporate the real-time transmission of CT scans and the geolocation of the nearest advanced healthcare facilities based on the patient point of entry. Some possible limitations in the implementation of this software that can arise in the future might include a limited internet connection in remote areas and issues related to the articulation with ambulance transportation systems.

To our knowledge, there are no similar software tools in Colombia incorporating specific algorithms for different healthcare levels that consider the varying degree of resources and infrastructure for approaching patients with a suspicion of cerebrovascular disease regardless of the site of case occurrence. Its usability for other countries, especially for Spanish-speaking countries, is one we have not explored and depends on the completion of the three validation phases inside our country. However, we suspect it can become a tool for accomplishing accreditation requirements for stroke centers desiring a certification by an international organization once the definition of an AHS is fulfilled. The extrapolation of this system to other countries requires extensive resource investment, especially for non-Spanish speaking countries, considering the differences in healthcare system articulation and health policies. This software begins to build the pillars of a public-private stroke network initiative that guides the clinical identification of a stroke, the magnitude of the neurological deficit and, particularly, whether a determined case can benefit from an acute intervention, narrowing the times of interfacility transfers.

ACKNOWLEDGMENTS

We thank our institutions and the National Department of Science, Technology and Innovation of Colombia for funding this study (Grant 1204-744-55680).

REFERENCES

[1] DANE, "Censo Nacional 2018. Departamento Administrativo Nacional de Estadísticas," 2018.

[2] C. Castañeda-Orjuela, P. Chaparro-Narváez, and F. De la Hoz-Restrepo, "El Observatorio Nacional de Salud y la gestión del conocimiento en salud del país: Superintendencia Nacional de Salud," in *Superintendencia Nacional de Salud*, ed, 2014.

[3] F. Ruiz Gomez, T. Zapata Jaramillo, and L. Garavito Beltran, "Colombian health care system: results on equity for five health dimensions, 2003-2008," *Rev Panam Salud Publica,* vol. 33, pp. 107-15, 6 p preceding 107, Feb 2013.

[4] J. M. Wardlaw, V. Murray, E. Berge, and G. J. del Zoppo, "Thrombolysis for acute ischaemic stroke," *Cochrane Database Syst Rev,* p. Cd000213, Jul 29 2014.

[5] J. L. Amaya-Lara, "Catastrophic expenditure due to out-of-pocket health payments and its determinants in Colombian households," *Int J Equity Health,* vol. 15, p. 182, Nov 10 2016.

[6] J. Powers William, A. Rabinstein Alejandro, T. Ackerson, M. Adeoye Opeolu, C. Bambakidis Nicholas, K. Becker, *et al.*, "2018 Guidelines for the Early Management of Patients With Acute Ischemic Stroke: A Guideline for Healthcare Professionals From the American Heart Association/American Stroke Association," *Stroke,* vol. 49, pp. e46-e99, 2018/03/01 2018.

[7] MinSalud. (2015). *Clinical Practice Guideline for diagnosis, treatment and rehabilitation of Acute Ischemic Stroke in population over the age of 18*.

[8] G. Nogueira Raul, S. Silva Gisele, O. Lima Fabricio, Y.-C. Yeh, C. Fleming, D. Branco, *et al.*, "The FAST-ED App: A Smartphone Platform for the Field Triage of Patients With Stroke," *Stroke,* vol. 48, pp. 1278-1284, 2017/05/01 2017.

[9] B. Y. Andrew, C. M. Stack, J. P. Yang, and J. A. Dodds, "mStroke: "Mobile Stroke"-Improving Acute Stroke Care with Smartphone Technology," *J Stroke Cerebrovasc Dis,* vol. 26, pp. 1449-1456, Jul 2017.

[10] R. L. Dickson, D. Sumathipala, and J. Reeves, "Stop Stroke(c) Acute Care Coordination Medical Application: A Brief Report on Postimplementation Performance at a Primary Stroke Center," *J Stroke Cerebrovasc Dis,* vol. 25, pp. 1275-1279, May 2016.

[11] C. Herweh, P. A. Ringleb, G. Rauch, S. Gerry, L. Behrens, M. Mohlenbruch, *et al.*, "Performance of e-ASPECTS software in comparison to that of stroke physicians on assessing CT scans of acute ischemic stroke patients," *Int J Stroke,* vol. 11, pp. 438-45, Jun 2016.

[12] L. Katzan Irene, Y. Fan, M. Speck, J. Morton, L. Fromwiller, J. Urchek, *et al.*, "Electronic Stroke CarePath," *Circulation: Cardiovascular Quality and Outcomes,* vol. 8, pp. S179-S189, 2015/10/01 2015.

[13] J. Minnerup, H. Wersching, A. Teuber, J. Wellmann, J. Eyding, R. Weber, *et al.*, "Outcome After Thrombectomy and Intravenous Thrombolysis in Patients With Acute Ischemic Stroke: A Prospective Observational Study," *Stroke,* vol. 47, pp. 1584-92, Jun 2016.

[14] P. A. G. Sandercock and S. Ricci, "Controversies in Thrombolysis," *Curr Neurol Neurosci Rep,* vol. 17, p. 60, Aug 2017.

[15] G. Teasdale, A. Maas, F. Lecky, G. Manley, N. Stocchetti, and G. Murray, "The Glasgow Coma Scale at 40 years: standing the test of time," *Lancet Neurol,* vol. 13, pp. 844-54, Aug 2014.

[16] L. K. Kwah and J. Diong, "National Institutes of Health Stroke Scale (NIHSS)," *J Physiother,* vol. 60, p. 61, Mar 2014.

[17] O. Lima Fabricio, S. Silva Gisele, L. Furie Karen, R. Frankel Michael, H. Lev Michael, C. S. Camargo Érica, *et al.*, "Field Assessment Stroke Triage for Emergency Destination," *Stroke,* vol. 47, pp. 1997-2002, 2016/08/01 2016.

[18] K. Ghandehari, F. Ahmadi, S. Ebrahimzadeh, K. Shariatinezhad, and K. Ghandehari, "Assessment of ABCD(2) scale in patients with transient ischaemic attack or stroke," *Neurol Neurochir Pol,* vol. 46, pp. 421-7, Sep-Oct 2012.

[19] S. A. Josephson, S. Sidney, T. N. Pham, A. L. Bernstein, and S. C. Johnston, "Higher ABCD2 score predicts patients most likely to have true transient ischemic attack," *Stroke,* vol. 39, pp. 3096-8, Nov 2008.

[20] R. U. Kothari, A. Pancioli, T. Liu, T. Brott, and J. Broderick, "Cincinnati Prehospital Stroke Scale: reproducibility and validity," *Ann Emerg Med,* vol. 33, pp. 373-8, Apr 1999.

[21] S. Runchey and S. McGee, "Does this patient have a hemorrhagic stroke?: clinical findings distinguishing hemorrhagic stroke from ischemic stroke," *Jama,* vol. 303, pp. 2280-6, Jun 9 2010.

[22] W.-S. Cho, J. E. Kim, S. Q. Park, J. K. Ko, D.-W. Kim, J. C. Park, *et al.*, "Korean Clinical Practice Guidelines for Aneurysmal Subarachnoid Hemorrhage," *Journal of Korean Neurosurgical Society,* vol. 61, pp. 127-166, 2018.

[23] M. Mokin, T. Primiani Christopher, H. Siddiqui Adnan, and S. Turk Aquilla, "ASPECTS (Alberta Stroke Program Early CT Score) Measurement Using Hounsfield Unit Values When Selecting Patients for Stroke Thrombectomy," *Stroke,* vol. 48, pp. 1574-1579, 2017/06/01 2017.

[24] C. M. Fisher, J. P. Kistler, and J. M. Davis, "Relation of cerebral vasospasm to subarachnoid hemorrhage visualized by computerized tomographic scanning," *Neurosurgery,* vol. 6, pp. 1-9, Jan 1980.

[25] J. C. Hemphill, 3rd, D. C. Bonovich, L. Besmertis, G. T. Manley, and S. C. Johnston, "The ICH score: a simple, reliable grading scale for intracerebral hemorrhage," *Stroke,* vol. 32, pp. 891-7, Apr 2001.

[26] A. J. Yoo, C. Z. Simonsen, S. Prabhakaran, Z. A. Chaudhry, M. A. Issa, J. E. Fugate, *et al.*, "Refining angiographic biomarkers of revascularization: improving outcome prediction after intra-arterial therapy," *Stroke,* vol. 44, pp. 2509-2512, 2013.

[27] J. T. Wilson, A. Hareendran, M. Grant, T. Baird, U. G. Schulz, K. W. Muir, *et al.*, "Improving the assessment of outcomes in stroke: use of a structured interview to assign grades on the modified Rankin Scale," *Stroke,* vol. 33, pp. 2243-6, Sep 2002.

[28] F. A. Silva, J. G. Zarruk, C. Quintero, W. Arenas, C. Rueda-Clausen, S. Silva, *et al.*, "Enfermedad cerebrovascular en Colombia. ," *Revista Colombiana de Cardiología* vol. 13, pp. 85-89, 2006.

Computer Vision-based System for Impaired Human Vision Compensation

Povilas Daniušis[1], Andrius Budrionis[1,2], Audrius Indriulionis[1,3], and Darius Plikynas[1]

[1]Department of Business Technologies and Entrepreneurship
Gediminas Technical University
Vilnius, Lithuania
Email: {povilas.daniusis, darius.plikynas}@vgtu.lt

[2]Norwegian Centre for E-health Research
University Hospital of North Norway
Tromsø, Norway
Email: andrius.budrionis@ehealthresearch.no

[3]Laboratory of Quaternary Research
Nature Research Centre
Vilnius, Lithuania
Email: audrius.indriulionis@gmail.com

Abstract—**This paper presents a high-level architecture of a computer vision-based system for partial compensation of lost or impaired human vision. It combines standard smartphone device, external deep learning-based image processing infrastructure and audio/tactile user interface. The proposed architecture is based on input from user-centered design process, involving end-users into system development. The paper discusses user needs and expectations for electronic travelling aids for the blind and highlights limitations of the existing solutions. The suggested architecture may be used as a basis for developing computer vision-based tools for visually impaired individuals.**

Keywords–computer vision; deep learning; mobile application; aid for blind and visually impaired; audio feedback; tactile feedback; impaired human vision compensation; user-centered design.

I. INTRODUCTION

More than 250 million people worldwide have moderate to severe vision impairment, while ($\approx$ 36 million are blind) [1]. During the past decades significant effort was devoted to develop computer vision and other sensor-based aids [2]-[5] for helping the blind and visually impaired users to perceive the surrounding world better. However, computer vision is a rapidly evolving field; the proposed systems become outdated in a relatively short time and often lack accuracy and reliability in real-world conditions in comparison to the state-of-the-art technologies. In this article, we describe a system, which employs modern computer vision techniques for compensation of lost or impaired vision function in humans. Many of the previously proposed electronic aids for the blind count on highly specialized hardware, for instance smart glasses [5], Microsoft Kinnect sensors [6], helmet-mounted photo sensors and cameras [7]. Such systems often lack convenience and bring additional complexity into daily tasks, which visually impaired people currently manage by the help of a white cane. This project puts major emphasis on developing a tool, which seamlessly integrates with the hardware visually impaired persons already have on-hand and are used to (a smartphone). This paper describes a high level architecture of the technology aid for visually impaired people utilizing the latest achievements in the field of computer vision.

This paper is structured as follows. Section II describes the methodological foundations of this research. Section III focuses on the architecture of the suggested human vision compensation system. Finally, Section IV summarizes and discusses the results of the conducted analysis.

II. METHOD

Following the principles of the participatory design [8][9] we included end-users into system design process. Two visually impaired persons participated in discussion and focus group meetings together with the project team to identify the key properties of the solution. During the six discussion sessions, major attention was paid to:

1) Functionality and key features of the solution addressing real-life scenarios of visually impaired persons;
2) Technical feasibility, connecting user needs to the latest development in computer vision and assistive technologies;
3) Interface between the visually impaired person and the assistive technology;
4) Appearance, usability and potential costs of the proposed product.

Participatory design process included testing of assistive technologies for the blind and visually impaired people using widely available mobile phone apps like Seeing AI, BlindWays, Be My Eyes, Aipoly Vision, TapTapSee, etc. Semi-structured tests were performed by visually impaired participants and researchers from the project group. Strengths and weaknesses of the existing solutions are reflected in the design of the proposed system.

To have an overview on research progress on assistive technologies for the blind, a systematic literature review in three major research databases (Medline, IEEE xplore and ACM DL) was conducted covering publication period of 10 years. It revealed increasing research interest in the field (84 related publications were identified, while 17 were characterized as highly relevant for this project) and provided better understanding of the state-of-the-art in applying computer vision technologies to address the needs of the blind. The review was extended to relevant patents and patent applications worldwide. The search was performed in the following databases: EspaceNet, GooglePatents, USPTO, PatentScope, FreePatentsOnline. It revealed several technological trends in the field of assistive technologies for the blind. Based on the aforementioned reviews, our insights are presented in the Results section.

III. RESULTS

Participatory design sessions with visually impaired individuals and analysis of the aforementioned data sources provided substantial input for the design of the proposed system (Figure 1). The process made it clear that understanding of the user needs was limited in the project team. Based on the results of the participatory design sessions, scientific publications and patent analysis, we present the outline of the system for impaired human vision compensation.

The key part of the proposed system is a smartphone device, connecting User Interface (UI) and data processing components into a well-functioning ecosystem. Smartphone was selected due to its high availability and capability to perform required computations or forward visual information for processing to the external server via the Internet.

UI consists of a wearable forehead belt, integrating RGB and depth cameras, Inertial Measurement Unit (IMU), tactile feedback device and bone conduction headphones. Although more expensive than standard ones, bone conduction headphones do not block the ears and allow the user to perceive audio information from the surrounding environment and the system at the same time, which is essential for the visually impaired people.

However, some types of guidance information may be more efficiently perceived through tactile feedback device than audio channel. An example would be indicating the direction of a detected object with a vibromotor on the corresponding area on the forehead belt, instead of providing audio description. More research is needed to clarify how various types of guidance information should be represented to the user in the most convenient and efficient way. For instance, detected objects could be presented to the user in a configurable order, based on the importance of particular object classes to the user or other personalized criteria. USB and Wifi data links can be used for bidirectional data transmission between UI and smartphone component.

Although modern high-end smartphones are powerful computational devices, many computer vision algorithms require hardware of significantly higher processing capabilities (e.g., a server equipped with a high-end Graphics Processing Unit (GPU)). Therefore, an external server is used for live video data processing. In our design we envision a computational server hosting a set of computer vision algorithms implemented and exposed via web services: faster Regions with Convolutional Neural Network features (faster R-CNN) object detector [10], trained to detect important objects, Convolutional Neural Network (CNN) and Recurrent Neural Network-based (RNN) scene descriptor [11], which provides textual annotation of a given RGB image, place recognition [12], face detection and recognition [13], obstacle detection [14], action recognition [15], among others. After video analysis is completed on the server side, results are transmitted back to

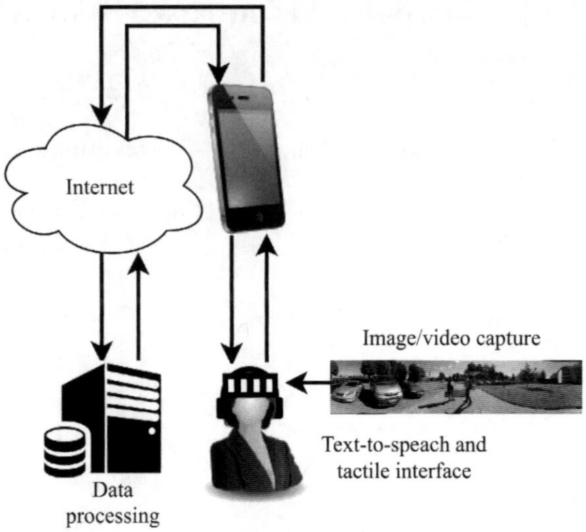

Figure 1. Schematics of vision compensation device.

the smartphone component and represented to the user through the aforementioned UI.

The user interacts with the system through gestures, voice commands or specifically designed user interface on the smartphone. User input is required for selecting operation mode of the system, i.e., switching between Optical Character Recognition (OCR), currency recognition, object detection, navigation and other available functions.

Social networking functionality was highlighted as a potentially important feature allowing other users/volunteers contribute to up-to-date and high quality guidance (for instance, temporary road works) with exact location and description. Navigating around such obstacles can be challenging for computer vision-based tools and may decrease user's trust in the tool. Integration of additional information from other users, governmental organizations and volunteers could be of high value for the visually impaired people.

Discussions on the key features of the proposed solution revealed a gap between the functionality of the existing tools, which are often focused on advanced navigation and scene description features, and actual user needs. For example, the end-users highlighted that a simplistic solution for navigating is lacking. Major focus was put on an easy-to-use and minimalistic tool, which can be trusted. Reliable object detection, direction and approximate distance to an obstacle were highlighted as the main requirements. To our surprise, only 5-8 distinctive object classes (for instance, buss stop, pedestrian crossing, doors, stairs, etc.) were of major importance while navigating. Additional information, such as type of a passing car, blossoming flowers in the park or a person riding a bicycle was considered as overwhelming and distractive.

Computer vision technology is under a rapid development and during the last years made a major breakthrough in terms of performance and efficiency. Deep learning neural networks, which are de facto standard in modern image/video processing in many real-world problems allow to achieve accuracies, similar to that of human decision (e.g., face recognition [13], object detection [10], among others). These models are now well supported by software libraries (e.g., [16]) and dedicated processing hardware is built-in even in relatively low-power computational devices, such as smartphones. These devices are also equipped with 4G Internet connectivity and sufficient computational power to perform partial or, in certain cases, even full visual data processing locally. The maturity of the aforementioned technologies suggest that our proposed computational aid for visually impaired persons may be highly feasible.

The cornerstone of this project is the interface between the visually impaired person and the assistive technology. We are proposing a combination of audio and tactile feedback, which may improve the interaction between the assistive technology and the user. Similar interfaces were suggested by research communities earlier [4][5], however, existing knowledge on user preferences and evaluation of various options of tactile feedback (types of actuators, placement on the body, frequencies, strength, etc.) is limited.

Appearance, usability and cost of the end product are of major importance to the users. The proposed assistive technology should supplement the tool that visually impaired persons used for decades - the white cane. Using a white cane requires little to no training, has very low costs and is relatively reliable. It provides information on the surrounding objects 1-2 meters in front of the user. While the cane can be used to detect nearby obstacles, electronic travel aid could be beneficial for longer range (2-10m) route planning and object detection. However, it should integrate seamlessly into the existing navigation practices of visually impaired people.

IV. Discussion

This paper outlined a high-level architecture of a computer vision-based system, which may help to partially compensate impaired or lost human vision. The main advantages of therein suggested system are: ability to use efficient (but still computationally intensive) modern computer vision algorithms via the Internet connection and present the output of the video processing to the user through a combined audio/tactile interface.

Similar functionality has been previously addressed by research communities [2]-[5] and industry [17][18]. The main difference between the aforementioned commercial solutions and the proposed system is the ability to utilize external resources for computationally intensive image and video processing tasks. While the availability of the computational power could be seen as the main advantage, it comes with high cost - dependency on a well-functioning mobile broadband. The development of the high-speed mobile networks (4G and 5G in the nearest future) is likely to make this limitation obsolete, especially in densely populated areas containing many hazards for the blind. Moreover, increasing computational power of smartphones may also allow to perform more advanced image and video processing locally. Although not always technically feasible, local processing is especially important in low connectivity areas to ensure at least partial functionality of the system. Combined audio/tactile interface may also be more convenient for the users, allowing to provide feedback in a more efficient way than using tactile or audio interfaces separately.

Both open source (e.g., [16]) and state-of-the-art commercial computer vision software libraries (e.g., [19]) may be applied implementing the suggested architecture. Moreover, a detailed specification of the suggested hardware components, following open hardware approach (providing an open repository containing a detailed list of electronic components, schematics, and 3D models of mechanical parts) could provide a solid basis for semi-standard reference platform for the researchers in the field.

It is important to emphasize that the suggested system does not aim to replace the main travelling aid of visually impaired people - the white cane. Instead, this computer vision tool aims to enhance and push the perception of the surrounding environment boundary from 1-2 meters (achieved by using the white cane) to 2-10 meters. It may improve route planning and identification of objects of interest (for instance, doors, stairs, elevators, bus stops, etc.). This functionality corresponds with the main requirement for electronic travelling aids highlighted by the end-users - direction and distance estimation to a selected object. The proposed hardware architecture can be used as a basis for various computer vision-based software modules, aiming to assist visually impaired users in daily activities (e.g., outdoor/indoor navigation, object/face recognition, obstacle detection, etc.).

The proposed architecture is based on several assumptions, which may be characterized as limitations of the system. For instance, technical feasibility of the solution is dependent on the availability of high bandwidth Internet connection ensuring access to high-power data processing components. Insufficient bandwidth may result in latency, which may not be tolerated by the users.

Economical feasibility of the proposed solution may also be questioned. Advanced technology (high-end smartphone, depth camera, tactile feedback device, bone conduction headphones) is needed to ensure reliable functioning of the system. A combination of such components may be perceived as costly by the end-users. More research is needed to demonstrate the cost-benefit analysis of the system in real-world scenarios.

Acknowledgment

This research is/was funded by the European Regional Development Fund according to the supported activity Research Projects Implemented by World-class Researcher Groups under Measure No. 01.2.2-LMT-K-718.

References

[1] R. R. A. Bourne et al., "Magnitude, temporal trends, and projections of the global prevalence of blindness and distance and near vision impairment: a systematic review and meta-analysis," Lancet Global Health, Vol. 5(9), pp. 888-897, 2017.

[2] S. Caraiman et al., "A. Computer Vision for the Visually Impaired: the Sound of Vision System," IEEE International Conference on Computer Vision Workshops, pp. 1480-1489, 2017.

[3] A. Csapó, G. Wersényi, H. Nagy, and T. A. Stockman, "A survey of assistive technologies and applications for blind users on mobile platforms: a review and foundation for research," Journal on Multimodal User Interfaces, Vol. 9, issue 4, pp. 275-286, 2015.

[4] M. Poggi and S. Mattoccia, "A wearable mobility aid for the visually impaired based on embedded 3d vision and deep learning," Proceedings of IEEE Symposium on Compututers and Communication, pp. 208-213, 2016.

[5] P. A. Zientara at al., "Third Eye: A shopping assistant for the visually impaired," Computer, Vol. 50, Issue 2, pp. 16-24, 2017.

[6] M. Owayjan, A. Hayek, H. Nassrallah, and M. Eldor, "Smart Assistive Navigation System for Blind and Visually Impaired Individuals," 2015 International Conference on Advances in Biomedical Engineering (ICABME), pp. 162-165, 2015.

[7] L. D. Dunai, I. L. Lengua, I. Tortajada, and F. B. Simon, "Obstacle detectors for visually impaired people," 2014 International Conference on Optimization of Electrical and Electronic Equipment (OPTIM), pp. 809-816, 2014.

[8] F. Kensing, J. Simonsen, and K. Bødker, "MUST: A Method for Participatory Design," Human Computer Interaction, Vol. 13, Issue 2, pp. 167-198, Jun. 1998.

[9] J. M. Carroll and M. B. Rosson, "Participatory design in community informatics," Design Studies, Vol. 28, Issue 3, pp. 243-261, May 2007.

[10] Sh. Ren, K. He, R. Girshick, and J. Sun, "Faster R-CNN: Towards real-time object detection with region proposal networks," Advances in Neural Information Processing Systems, pp. 91-99, 2015.

[11] Ch. Liu, J. Mao, F. Sha, and A. Yuille, "Attention Correctness in Neural Image Captioning," Proceedings of the Thirty-First AAAI Conference on Artificial Intelligence, pp. 4176-4182, 2017.

[12] E. Ohn-Bar, K. Kitani, and Ch. Asakawa, "Personalized Dynamics Models for Adaptive Assistive Navigation Interfaces," arXiv:1804.04118, [cs.LG], 2018.

[13] B. Amos, B. Ludwiczuk, and M. Satyanarayanan, "Openface: A general-purpose face recognition library with mobile applications," Technical report, CMU-CS-16-118, CMU School of Computer Science, 2016.

[14] I. Laina, C. Rupprecht, V. Belagiannis, F. Tombari, and N. Navab, "Deeper depth prediction with fully convolutional residual networks," Fourth International Conference on 3D Vision (3DV), pp. 239-248, 2016.

[15] G. Varol, I. Laptev, and C. Schmid, "Long-Term Temporal Convolutions for Action Recognition," IEEE Transactions on Pattern Analysis and Machine Intelligence. Vol. 40, Issue 6, pp. 1510-1517, 2016.

[16] M. Abadi et al., "TensorFlow: Large-Scale Machine Learning on Heterogeneous Distributed Systems," arXiv:1603.04467, 2016.

[17] Orcam. MyEye 2.0. [Online]. Available from https://www.orcam.com [Retrieved: January, 2019].

[18] Horus. Horus system. [Online]. Available from http://www.horus.tech [Retrieved: January, 2019].

[19] Neurotechnology, VeriLook SDK. [Online]. Available from https://www.neurotechnology.com/verilook.html [Retrieved: January, 2019].

Establishing Baseline in the Status of E-health Research in Norway

Andrius Budrionis, Karianne Lind, Inger Marie Holm, Omid Saadatfard, Rune Pedersen
Norwegian Centre for E-health Research
University Hospital of North Norway
Tromsø, Norway
e-mail: Andrius.Budrionis@ehealthresearch.no

Abstract—**E-health is a rapidly evolving research field. In Norway, it is governed by the National E-health Strategy, defining six focus areas to be addressed by both academics and industry. The strategy is relatively new and much research in the field has been performed before it was developed. Literature search and machine learning classification methods were used to map scientific publications into the focus areas of the National E-health Strategy. Results showed that all strategic areas were represented in scientific publications; one focus area, new ways to provide healthcare, attracted the most attention from research communities. This paper presents a method and baseline of e-health research activities in Norway alongside the initial results of applying this method to the existing body of literature.**

Keywords-e-health strategy; publications; review; machine learning; classification; natural language processing.

I. Introduction

Telemedicine and e-health are priority research areas in Norway. Established academic communities have a long history of research and development in the field with success stories of functional telemedicine services used in clinical practice as early as 1991 [1]. Focus on e-health was strengthened in 2016 by establishing a dedicated agency under the Ministry of Health and Care Services, The Norwegian Directorate of E-health, with the goal of establishing and managing standards and national e-health solutions that contribute to high quality and effective health services.

Norwegian government and Norwegian Directorate of E-health are planning a shared national Electronic Health Record (EHR) solution for the entire healthcare sector, which is in line with the recommendations made in the policy document One Citizen – One Health Record [2][3] published by the Ministry of Health and Care Services in Norway. This policy document outlines an important strategic direction for the healthcare sector and recommends close collaboration between a number of stakeholders in the Norwegian e-health field.

The National E-health Strategy [4] and action plan 2017-2022 [5] describes the proposed strategic direction for the goal of a digitized and integrated healthcare system that provides a simpler, better and more comprehensive experience for the service recipients. The strategy is formed around six focus areas: 1) digitization of work processes, 2) better continuity of care, 3) better use of health data, 4) new ways to provide healthcare, 5) common foundation for digital services and 6) national e-health management and increased implementation. The first four are considered functional areas with direct value for healthcare services. The last two are considered foundations that are required for the first four to be realized.

Presenting Norwegian e-health research, categorized into these focus areas is helpful for policy makers, such as the Norwegian Directorate of E-health, to promote research in higher priority areas and coordinate activities in the national e-health field. Further, e-health research institutes and e-health organizations can use this information to search for research partners and to build collaboration networks. Currently, such information is not available. The objective of this study is to address this need by creating a classification of e-health research in Norway based on the six focus areas of the National E-health Strategy. This paper presents a method and baseline measures, scoping the state of e-health research activities in Norway based on scientific publications.

The reminder of this paper is organized as follows. Section II provides a summary of methods used to produce results, which are presented in section III. Section IV discusses the key findings and limitations of this work, while section V concludes the paper.

II. Method

This project could be divided into three phases: data collection, preprocessing and analysis. The reminder of this section summarizes the methods used in every phase of the project.

A. Data collection

To collect data on production of scientific publications within e-health, three major research databases (Scopus, Web of Science (WoS) and PubMed) were queried. Publications dated 01.01.2007 - 01.06.2018 were included. Publication search was performed in two phases between May and December, 2017:

1. Phase 1 included keyword-based search (predefined list of keywords and relevant MeSH terms) and author-based search covering a list of well-known researchers in the field.
2. Phase 2 was based on an extended author-based search including authors from publications identified in Phase 1.

1) *Phase 1*

All three databases (Scopus, Web of Science (WoS) and PubMed) were queried in Phase 1. We searched in title, abstract and keywords (Scopus and WoS) or in title and abstract (PubMed) and coupled the queries with affiliation

"Norway". In the author-based searches, we searched for author name combined with affiliation "Norway".

Phase 1 combined three approaches:

- MeSH term-based search (MeSH-term for e-health, telehealth and mHealth is "telemedicine").
- Search based on predefined list of e-health terms, listed below.
- Search for publications authored by well-known researchers in the field:
  - Researchers from the Norwegian Centre for E-health Research (E-health Research)
  - Known researchers within the Norwegian e-health field (identified previously)

Lists of e-health terms, expected to be present in the majority of e-health publications, were put together by senior researchers at the E-health Research. These terms were also combined with more general terms e-health, ehealth, telemedicine, telehealth and telecare.

The following list of e-health terms was used in the keyword-based search (? denotes a single random character, * denotes multiple random characters):

General keywords: e-health, ehealth, telemedicine, telehealth, telecare

EHR-related keywords: Electronic* Health Care Record, electronic*health record, electronic* healthcare record, electronic* medical record, electronic* patient record, patient health record, decision support, health information system, information infrastructure, information security, integration, process support, regionali?ation, semantic interoperability, standardi?ation, terminology, usability, privacy, archetypes, user interface.

Health analytics keywords: analy*, health analy*, large dataset*, big data, predictive analy*, computational epidemiology, health intelligence, artificial intelligence, machine learning, natural language processing, text and data mining, statistical analy*.

M-health keywords: Mhealth, m-health, homecare, sensor system*, medical app*, health app*, app*, mobile, remote monitoring, medical device, usability, information security, privacy, health record, self-management, wearable*, sensor*, self-generated data, economic impact*, facilitator*, mental health, tracking, empowerment, guidelines.

2) *Phase 2*

After phase 1 and deduplication, the dataset consisted of approximately 2000 references. Duplicates were removed following a simplified version of a method described by Bramer et al 2016 [6].

A list of the first and second author was compiled from the publications identified in Phase 1. Names, which were not searched for in Phase 1, were used to query Scopus database.

After de-duplication of the entire dataset it contained 3028 references. References were exported to a comma separated CSV file, including author, title, year, abstract, keywords, URL, DOI, Reference Type and Author Address.

B. Data preprocessing

Data analysis was performed in Python 3 environment using the latest versions of Natural Language Processing Toolkit (NLTK) [7] for free-text processing and scikit-learn [8] for analysis. Data analysis focused solely on publication title, keywords and abstract fields. Preprocessing included removal of stop-words and numeric values. Words in the free-text fields were stemmed (Snowball stemmer) and processed using Term Frequency – Inverse Document Frequency (TF-IDF) vectorizer. It resulted in a numerical representation of importance of n-grams (1 or 2 words in length) in the corpus.

A random sample of publications (N = 1700) was manually labelled assigning them to one of the 6 classes originating from the Norwegian National E-health Strategy [4]. References unrelated to e-health were also marked. Manual labelling was performed by one of 5 independent reviewers, who discussed classification criteria beforehand. Publications, which could not be classified by a single reviewer due to uncertainty regarding the correct class were discussed in common meetings where consensus class was determined.

Description of the classes in the strategy was used as classification criteria. Typical projects, which fit these classes are:

1) Digitization of work processes – improvement of work processes for healthcare professionals.
2) Better continuity of care – improvement of healthcare services for patients.
3) Better use of health data – health data analytics driven projects.
4) New ways to provide healthcare – novel services in healthcare, which were not available before.
5) Common foundation for digital services – infrastructure for large scale digital services.
6) National e-health management and increased implementation – national e-health solutions.

Labelled data were randomly split into training (80%) and testing (20%) data for supervised machine learning analysis.

C. Data analysis

Unsupervised machine learning methods were applied to cluster the publications and explore the data. In depth analysis was performed using supervised machine learning algorithms. Classification of publications was performed in two steps: binary classification (e-health/not e-health) and multi-class classification of e-health publications.

III. RESULTS

Results from data collection and analysis are presented in the reminder of this section.

A. Data collection

Data collection was performed in two phases including removal of duplicates. It resulted in 3028 publications, which were included in data analysis.

B. Supervised 2-class model

To clean the dataset from irrelevant publications, four binary classifiers (Linear SVC, Naïve Bayes, Logistic Regression and K-nearest neighbor) were trained and tested using 10-fold cross-validation on the labelled data (e-health/not e-health). Logistic regression classifier demonstrated the best performance for this problem and was selected for further analyses. The performance of the 2-class classifier is presented in Table 1.

TABLE I. PERFORMANCE OF THE 2-CLASS CLASSIFIER

	Precision	Recall	f1-score
Not e-health	0.93	0.58	0.71
E-health	0.72	0.96	0.82

While interpreting the results, attention was directed towards the e-health class. The performance measures listed in Table 1 indicate, that the classifier is able to identify 96% of the relevant publications in the dataset (recall = 0.96), however, it also has a relatively high percentage of false positives in the e-health class (precision = 0.72). Regardless of the false positives rate, the classifier identifies almost all relevant e-health publications required for further analysis. The trained classifier was used to classify the unlabeled part of the dataset.

C. Unsupervised model

Data labelling stage could be perceived as biased due to overlap between the classes and human factors involved in the process. To adjust for the potential biases, unsupervised clustering of e-health publications was performed to check whether publication could be automatically assigned to a class characterized by content similarity. Principal Component Analysis (PCA) and k-means algorithm were fitted to visualize the high dimensional data in 3 dimensions (Figure 1).

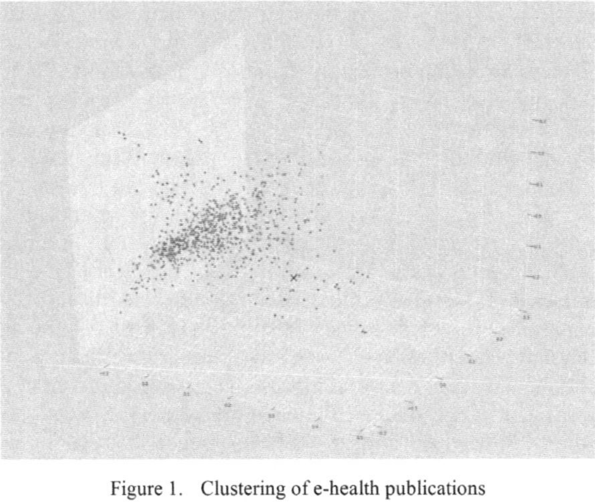

Figure 1. Clustering of e-health publications

Cluster analysis (before and after removing not e-health publications) identified no clear boundaries between the clusters and was not pursued further.

D. Supervised 6-class model

The 2-class model cleaned the dataset from the most of irrelevant publications (N = 1377). The cleaned dataset (N = 1651) showed uneven distribution of publication among classes with one class being overrepresented. The overrepresented class was down sampled in the training data

TABLE II. PERFORMANCE OF THE 6-CLASS CLASSIFIER

	Precision	Recall	f1-score
1. Digitization of work processes	0.70	0.58	0.63
2. Better continuity of care	0.61	0.62	0.62
3. Better use of health data	0.62	0.71	0.67
4. New ways to provide healthcare	0.74	0.77	0.75
5. Common foundation for digital services	0.53	0.62	0.57
6. National e-health management and increased implementation	0.66	0.64	0.65

to ensure that it is not overrepresented in the classification model. Four classifiers (Linear SVC, Naïve Bayes, Logistic Regression and K-nearest neighbor) were fitted and tested using 10-fold cross-validation, Naive Bayes model showed the best performance (Table 2). The trained classifier was used to classify the unlabeled e-health publications. Results are represented in Figure 2.

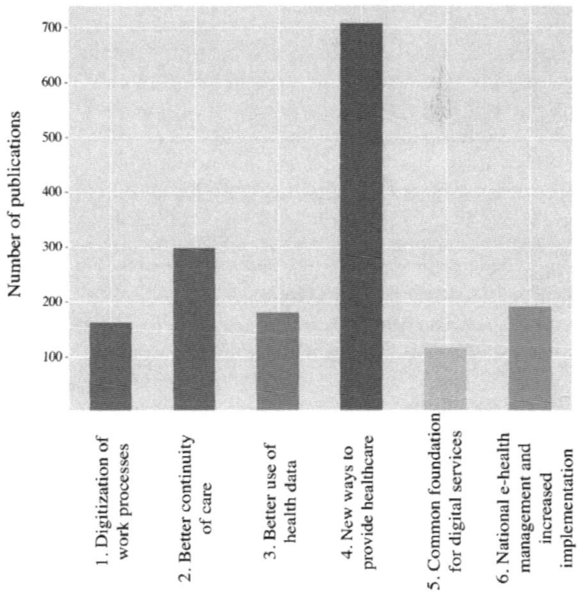

Figure 2. E-health publications classified according to the National E-health Strategy [4]

Top-10 keywords representing each class (sorted by importance):

1) Digitization of work processes: nurse, patient, use, care, hospital, electronic, record, health, work, support
2) Better continuity of care: patient, care, inform, health, communication, design, nurse, user, service, use
3) Better use of health data: data, patient, health, record, predict, inform, use, electronic, fall, medical
4) New ways to provide healthcare: diabetes, patient, health, use, social, service, care, technology, group, design
5) Common foundation for digital services: secure, health, standard, information, develop, service, data, ehealth, use, medical
6) National e-health management and increased implementation: telemedicine, health, information, implement, studies, infrastructure, technology, use, e-health, care

IV. DISCUSSION

Implications of the findings, methodological limitations and future works are discussed in this section.

A. *Interpretation of findings*

This project aimed to shed light on the status of e-health research activities in Norway with regards to the strategic documents [4][5]. The results of this project provide an overview on how the academical effort to bring the e-health field forward is reflected in scientific publications, and how these publications map to the focus areas defined in the National E-health Strategy (Figure 2). Disregarding the potential overlap between the focus areas and other methodological limitations, the project demonstrates the high pace of producing scientific results in the field. The most of research effort falls into class 4 (new ways to provide healthcare). This may not be a surprising finding, since the most of activities in e-health could be categorized as new ways to deliver healthcare.

The other publications distributed more evenly into the 5 classes. Class 5 (common foundation for digital services) had the weakest representation in scientific papers. This may be explained by the specifics of the Norwegian healthcare system. Norwegian healthcare is publicly owned and funded. Class 5 focus on the publications dealing with the infrastructure for delivering healthcare services to the citizens. Such infrastructure is partly developed from off-the-shelf components; procurement procedures are often based on other aspects than research (for instance, cost, reliability, flexibility, etc.). Space for research in this focus area is limited.

B. *Methodological limitations*

Methods to achieve the aforementioned results could be questioned. Level of uncertainty varied throughout the project, therefore, results should be interpreted in the context of the following limitations.

Data collection process was not strictly structured, therefore some publications may have been left out from further analyses. Phase 1 was focused on predefined keywords, which may not be completely representative for the entire field. Databases used in the search process do not include national publication channels, which often publish results in Norwegian.

Phase 2 used a list of the 1st and 2nd authors compiled from the publications identified in the Phase 1. All other authors were ignored. The search was performed only in Scopus database; PubMed and WoS added very few results in the previous phase. Data labelling process inherited the uncertainty originating from the National E-health Strategy. Focus areas are not defined to form mutually exclusive classes and there is a clear overlap between some of them. At the same time, publications often cover several focus areas and are difficult to assign to a single class. It was reflected in data labelling process, together with the additional uncertainty caused by human factors and background of the labelers. During the labelling process a class of e-health publications, which did not fit well into any focus areas was identified. Such publications deal with medical education, social media use for health purposes, reviews of various e-health topics and user health data storage solutions. These topics should be better addressed in the next versions of the e-health strategy.

Data analysis had to deal with the uncertainty inherited from the previous steps in the process. It may be reflected in relatively low precision and recall measures, especially in the 6-class model (Table 2). It may also be influenced by the false positives in 2-class model (precision = 0.72), which left some noise in the data filtering process. All the aforementioned aspects need to be taken into consideration when interpreting the results from this study.

C. *Alternative classification strategies*

The project started out with an idea to classify the publications according to the four research arenas at E-health Research: citizen services, patient pathways, health data and services for health professionals. The purpose was to establish a clear link between e-health research production in Norway and focus areas at E-health Research, which are logically distinct, and thus make classification easier. While the focus areas at E-health Research are well-aligned with the National E-health Strategy, they do not directly map to one another.

A decision to classify the identified publications according to the focus areas in the National E-health Strategy was a consensus reached in the project team. Regardless of the importance of measuring e-health research status towards national guidelines, focus areas defined in the strategy may not the best choice for classifying scientific production in the field. The National E-health Strategy is relatively new and much development in the field have taken place before it was made public. Focus areas are not meant to form mutually exclusive classes; research publications are often interdisciplinary and cover several focus areas. This situation causes uncertainty in classification, which might be avoided by selecting more distinctive classes. However, connection to the existing e-health strategy would be lost. One question that could be asked is whether the National E-health Strategy should be used to classify and potentially influence e-health research, or it is e-health research that should influence the e-health strategy?

D. Future work

This paper aimed to establish a baseline in the status of e-health research in Norway. Studying the development of the field is meant to be a continuous process, which is repeated periodically, preferably every third year. Results can be used as an input for the policy-makers in organizing, coordinating and allocating research funding, strengthening the weak focus areas and research institutions. Lessons learned during this iteration will contribute to a more structured and easier reproducible data collection and analysis process in future iterations. Further analyses will focus on classifying the publications according to the research institutions, mapping the focus areas of the National E-health Strategy to the interests of the most important academic actors in the country. The project aims to communicate the findings in a visual and easily understandable manner, therefore results will be represented as an interactive periodically updated map.

V. Conclusion

This paper presented a method and initial results scoping the status of e-health research in Norway based on scientific publications from the last decade. It mapped the published research results to the focus areas of the National E-health Strategy. Findings show that all focus areas are represented in the previous and ongoing research activities. Most of the publications fell into the focus area dealing with "new ways to provide healthcare services". The focus area "common foundation for digital services" had the weakest representation in the identified scientific publications.

References

[1] I. Nordrum et al., "Remote frozen section service: A telepathology project in northern Norway," *Human Pathology*, vol. 22, no. 6, pp. 514–518, Jun. 1991.

[2] Norwegian Ministry of Health and Care Services, "Meld. St. 9 (2012–2013)," *Regjeringen.no*, 30-Nov-2012. [Online]. Available: http://www.webcitation.org/73nsLIeoU. [Accessed: 09-Nov-2018].

[3] Norwegian Directorate of E-health, "Utredning av «Én innbygger – én journal»," [Report "One Citizen - One Health Record"] Dec-2015. [Online]. Available: http://www.webcitation.org/73nstSwCb. [Accessed: 09-Nov-2018].

[4] Norwegian Directorate of E-health "Nasjonal e-helsestrategi 2017-2022." [National e-health strategy 2017-2022] [Online]. Available: http://www.webcitation.org/73ntEa1LR. [Accessed: 09-Nov-2018].

[5] Norwegian Directorate of E-health, "Nasjonal handlingsplan for e-helse 2017-2022." [National action plan for e-health 2017-2022] [Online]. Available: http://www.webcitation.org/73ntLCceo. [Accessed: 09-Nov-2018].

[6] W. M. Bramer, D. Giustini, G. B. de Jonge, L. Holland, and T. Bekhuis, "De-duplication of database search results for systematic reviews in EndNote," *J Med Libr Assoc*, vol. 104, no. 3, pp. 240–243, Jul. 2016.

[7] S. Bird, E. Klein, and E. Loper, *Natural Language Processing with Python*, 1st ed. O'Reilly Media, Inc., 2009.

[8] F. Pedregosa, G. Varoquaux, A. Gramfort, V. Michel, and B. Thirion, "Scikit-learn: Machine Learning in Python," *J. Mach. Learn. Res.*, vol. 12, pp. 2825–2830, Nov. 2011.

Taking Adaptable Co-design Action: Flexible Learning Between Health Experts, End-Users and Technology Experts in the Early Stage of eHealth Design

Jan Aidemark
Department of Informatics
Linnaeus University
Växjö, Sweden
email: Jan.Aidemark@lnu.se

Linda Askenäs
Department of Informatics
Linnaeus University
Växjö, Sweden
email: Linda.askenas@lnu.se

Abstract— **Co-design, as a way of integrating users' perspectives and technical possibilities, has been subject to a multitude of interpretations. To facilitate the development of patient-empowerment services, co-design is a central approach needed to ensure high acceptance among users. In this research, we offer a flexible co-design approach by building on adaptive and continuous planning based on sets of co-design actions directed towards participants' mutual learning. The research is based on the experiences of a co-design process within the area of eHealth, specifically the fall prevention area, as a special case of self-training systems and self-management systems. The project team comprises a variety of member groups, including the elderly, technical developers, business developers, methods experts and project management. The key success factor was identified among the persons of the different groups as achieving possibilities for learning. To make learning processes requires adaptability and flexibility in interacting different planning actions in order to balance both the need for technology development and the understanding of users' needs.**

Keywords: Co-design; learning; adaptable; eHealth; fall detection.

I. INTRODUCTION

Understanding the interaction between actors in a systems development process is a multi-faceted affair, open to interpretations of the different preferences of diverse participants. Within eHealth, this is even more complicated due to the nature of the application area, i.e. dealing with sick people, on the one hand, and with professional people responsible for delivering safe and approved care, on the other. The central approach in the field of eHealth is Experience-Based Co-Design (EBCD). EBCD builds on the notion that collecting facts about patients and professionals enables the building of effective and useful support systems [1][2]. Other approaches to co-design stress, e.g., co-creation [3], the transfer of power from technology and design aspects to user perspectives. Empowering users to be active in decision-making is arguably an important step towards better information systems. It can thus be assumed that good co-design should include both facts about practice, users and professionals and the power to act. To achieve positive effects of flexible co-design like these requires, to a great extent, the creation and transfer of knowledge. The co-design process as such could therefore be understood as a learning process. Ward et al. [4] argue that there is a gap within the peer-reviewed literature on how to conduct co-design in practice. In this research, we look at how co-design can become an effective learning experience for the participants by setting it up as an adaptable and flexible method and applying a continuous process of re-planning. We target the early formative parts of the design process, which Sander and Stappers [3] call the fuzzy front end. This is done to ensure that the project receives a proper start and a development of all the parties involved by raising them to sufficient levels of relevant knowledge. Different groups within the co-design process have different needs, which necessitates a well-balanced set of co-design actions that are continuously re-evaluated and re-planned in the process. Although the research revolves around a fall prevention project and the development of technical solutions, we focus in this paper on the properties of the co-design process and reflect on how the experiences can be used to improve co-design practice. Since it has been claimed that it is the flexibility of EBCD that gives good results [5], the aim of this paper is to look closer into how to actually achieve this. The paper is structured accordingly by a quick initial view of the current state of EBCD and related co-design methods, followed by an overview of the research setting and the systems development project forming the core of the research. As a result, a generalized example is provided of how to structure co-design processes in a flexible and adaptable way on the basis of the experiences of the development project. The paper ends with a discussion of central issues related to how to structure a project for learning that helps parties of all types to be active and productive co-designers. A neutral project management with a co-design expertise background could be one path towards a better project structure where all participants can gain sufficient knowledge to be empowered to become creative co-designers. The paper is organized according to the following structure: Section 2 provides a theory background with focus on EBCD and co-design; Section 3 gives the research approach and an overview of the case; Section 4 is the analysis of co-design actions; Section 5 contains a discussion of the result, with concluding remarks and future research.

II. THEORETICAL BACKGROUND: EBCD AND CO-DESIGN PERSPECTIVES

Co-design in general and EBCD in particular currently attract strong interest, not least in the face of the current surge in eHealth applications. Hence, we give an overview of current topics, issues and debates that seem important for improving the practice of co-design for eHealth development.

The basic approach of EBCD is the design of new solutions, (e.g., ways of working, IT systems or new procedures) based on facts gathered from professionals and patients, molded through joint sessions of design, and hopefully leading to new ways of working that are acceptable to all parties. Bate and Robert [1] provide a basic EBCD cycle, including set-up, gathering experiences, and co-design in groups.

In the area of health, co-design can be seen as a framework for managing an intervention project [6][7]). Co-design functions as a way of creating change in a project research style and as such becomes a ´method for organizational development.

The practice and real effects of EBCD are studied in a review of EBCD studies. Gleeson et al. [8] note that a great many EBCD projects only deal with minor changes in work practice, which are more helpful in implementation than in finding novel solutions. They [8] also observe that surveys are a common way of collecting facts for EBCD projects.

In a larger perspective, EBCD could be understood as a spirit that is supposed to drive a development project including important values like empowerment, trust, autonomy and self-determination among both users and professionals [9]. By this approach, we are looking for more than a technique for cooperation during design.

Brocklehurst et al. [10] point to the fundamental change that EBCD could bring about, from a traditional top-down approach to developing a design that is based on the end user's situation and interests. This is supposed to be especially true when it comes to co-design with elderly people who are moving into a life of greater dependence on support and help from others.

EBCD, as noted by Matthews et al. [11], can be used not only for collecting facts from users but also for engaging them in the co-production, design and development of new services.

In addition to being an important property of the method area [5], flexibility is, in our view, important for the engagement of patients, families and staff.

The situation for both professionals and patients changes when they enter into a co-design process. They are then forced to look upon their situation from new perspectives and to step out of their traditional positions where they feel comfortable [12].

As healthcare staff and patient experiences with patient processes differ, this creates many complex problem situations. To understand and deal with these, a collaborative approach is needed to capture all areas needing improvement and to create solutions that respect all parties. [13].

Hill et al. [14] argue that theories of adult learning need to be part of co-design in order to support user learning during the project. This is a necessity, particularly in light of the different backgrounds of eHealth users, as they are patients and often find themselves in difficult situations.

Since knowledge processes are regarded as central in the research [15], Langley et al. argue that co-design should be understood as a process where knowledge is discovered, created and shared between different groups of stakeholders.

III. RESEARCH SETTING

The research approach builds on the interpretations of what takes place in an action research tradition [16]. The research setting is an innovative development project in the area of eHealth fall prevention solutions for the elderly. The context of designing fall prevention systems is a drive for helping the elderly to continue living at home. This was designed as an innovative technical project in need of a co-design process to ensure a well-functioning connection to potential users. The general approach to fall prevention was to have a data-driven learning system for changing habits and improving the understanding of personal risk behaviors. To attain these goals, the project has devised a co-design process, whose focus in this research lies on the formative first half year.

During this period, an initial technical solution was developed in cooperation between different user groups, including experts, professionals and end users. Five categories of participants were involved in this process: technical experts, end users, professional users, design experts, and project managers. The end users were still relatively healthy elderly people, who were able to live independently in apartments or houses. There were 5 participants in this category, whose age ranged between 70 and 75. From the technical side, 3 representatives from private companies (from which they also received further support) were active in the co-design in crafting the practical IT solutions. The design and management team consisted of project management for the healthcare sector and eHealth experts from universities, a total of 6 people. The health care professionals had 3 representatives, including physicians and physiotherapists. Besides, invited quests were added to some of the meetings. In total, approximately 16 persons were active at the co-design meetings. Two types of meetings were used: physical meetings in a conference room setting, and web conference, voice and video meetings. Both recordings and field notes were used to capture the proceedings. Interspersed between these were periods of field action, patient observations or home testing, as well as of technology development. These activities were prepared in planning meetings. Together, these formed a series of co-design cycles, four of which were performed during the first part of the project. The results of this paper focus on the way actions were mixed during the different cycles.

As the project constituted an action research effort, the authors were active members of the design team. Hence, active observation formed a way of capturing and understanding the nature of the co-design process. The research stance was interpretative and self-reflecting, aiming at giving a generalizable account of what worked during the project, while simultaneously learning from its mistakes and successes. The further aim was to apply a "reflective practice" approach [17][18] to better understand current practices and provide generalized ones for a knowledge- and learning-oriented way of working with co-design. To do research in the manner of "reflective practice" is a new way of creating a critical awareness of current practices and of facilitating change [19]. Schön [17] works with a view of knowledge and knowledge generation that is based on an epistemology of action. Although experts with their practical knowledge often contribute in a tacit or unspoken form, Schön sees this as a possibility of attaining intellectual rigor in scientific work. The basis of the approach, referred to as reflection in action, lies in the capacity of reflecting on the knowledge that emerges from action.

IV. Result: Co-design as adaptive learning

Based on the experiences of the case study, sets of co-design actions organized into cycles of design, which are planned and adapted continuously, can be proposed. We present this in two steps, first in sets of design actions, which are here somewhat generalized in comparison with the actual project. After this, we provide examples of how to combine these into cycles of design.

A. Co-design action sets

The actions are divided into four sets plus a planning set. Each set consists of a number of possible types of actions. The planning set deals with the formation of the co-design actions and their enactment during various meetings and stages of the project. The four action sets form the movement of the design process including the initiation of purpose and mission, the requirements collection and the testing and solution formulation, all guided by planning actions. The list is not exhaustive, but rather a set of key examples that serve as illustrations of how co-design actions can be devised. The list follows a clear and straightforward logic usually associated with a "waterfall" model. However, in implementing the practical cycles of co-design, these should be mixed and repeated, as they are needed to achieve the desired solutions to a desired state, as well as bringing the co-designers to a state of knowledge and understanding to enable them to make informed decisions about what a good design entails. The sets include the following actions:

1) Action set: Planning Actions

- Preparing the co-design project. This entails setting the general direction of the project and stating the purpose and role of co-design. It is very important to be explicit about what outcomes are expected of the co-design process. This action is also important for understanding what the biases of the project are and what taken-for-granted assumptions are held by those setting up the co-design process.

- Preparing the co-design process. This includes, e.g., the style and location of meetings, and the resources and competences required. Setting dates, outlines of meetings, as well as the principles of the planning and coordination of the process, is also included.

- Evaluations and feedback of the project progress. This means tracking the learning and knowledge development of the project participants. On the basis of an understanding of how the co-designers develop during the project, the process should be adapted to the needs and pre-conditions of the members of the team. Active evaluation and analysis form keys to a flexible and adaptive co-design process. To strengthen this feedback process, it is an advantage if end user representatives are invited to planning sessions.

- Recruitment of participating co-designers. Finding and motivating people to take part in a project is a major undertaking. The selection of participants could be decisive for the outcome of the design effort. Their abilities to contribute and remain throughout the project are important to consider to be able to structure the process and meet the goals and expectations of both the users-to-be and the project at large.

- Ethics board approval. As co-design often takes place in the context of research efforts, the ethical aspect of this project could be problematic, as the method also works with discovery and change, which entails that many of its aspects may be unknown beforehand. This makes ethical approval difficult unless the board understands the nature of co-design and agrees with its benefits.

2) Actions set: Initiating phase: facts and purpose

- Ascertaining the facts about the current health care area. In the current project, this means including lectures, e.g., by fall prevention experts and by experts advising people on how to live safer at home. As this is a mainstay of EBCD, it should establish a base line of knowledge shared by all.

- Technology orientation. Experts on current IT trends provide an overview for the design group of the technical possibilities. Providing users and health care professionals with sufficient knowledge is another cornerstone of co-design. To give both groups the right message is a pedagogical challenge. It is also a question of giving just enough information and not drowning people in tech facts, as the insight disseminated must be useful for taking part in discussions and making a personal judgement on the possibilities and usefulness of the technology.

- Project purpose. It entails, in this case, to explain the purpose and mission of fall prevention. The focus lies on current project goals, especially on the purpose of the co-design process. The questions to be addressed here concern what the project is set out to achieve, why co-design is employed as part of the project, and what are the expectations of co-design. The different participant groups need different messages here, since catering for all is decisive for the continued process.

- Story telling. The participants of this project were asked to give their own accounts of the problem area, in this case personal or other people's experiences of fall situations. This meant that all participants told their stories, including tech people, old persons (users), health care experts, project leaders and researchers. The purpose of this was to share an insight into the problems of fall prevention but also to make everyone personally engaged in the problem. This does not only involve knowing about the particular stories, but also creating a personal understanding, which is important for all, not least for the professionals.

3) Action set: Requirements – initial design

- Initial suggestions for technical solutions. Solution providers give presentations and demonstrations, including both those engaged in the project and selected external solution providers. The sessions comprise the initial testing of solutions, whereby the design group is given the opportunity to see and interact with the solutions (in dry runs) in a conference setting.

- Technology adaptation. Based on experiences, a selection is made of solutions that seem to fit and of redevelopment, when needed. The initial testing, in turn, leads to adapting solutions to current situations and needs.

- Preparing for field testing. This comprises the presentation and demonstration of solutions in a conference setting, including actions such as practical planning for field tests, the assignment of technology to people, and instructions for its operation.

- Field testing: installation and practical support. Packages of solutions are installed on site in the participants' homes. If and when needed, revisits or telephone consultancy to fix problems, capture feedback or change equipment are performed. The co-designers use the equipment at home and in everyday life, making notes of problems and experiences of use in general.

- Open dialogues. This may be regarded as a key action in a focus group style for creating a deeper understanding of problem scenarios and of how technology can support a better life situation. Small mixed groups of people from the different sides of the design groups are engaged in open conversations around topics provided by a session leader. The open and dialogue-styled format is important for allowing all participants to give their views on the topic. The purpose is not to create a consensus around certain solutions but to open up new views on problematic situations, users' needs and technological possibilities.

4) Action set: Design of the solution.

- Feedback sessions from users. Continuous feedback may consist of, e.g., conversations, telephone calls, emails, or meetings between technical staff and testing co-design users. Both physical and web-conference meetings were used to sum up the findings.

- Technical evaluation. Based on field testing, technical evaluations of how the solution performed is conducted. Technical deficiencies or problems of practical use are analyzed, and alternative solutions are developed. If changes are deemed serious enough, it might be necessary to reiterate the field testing of the solution.

- In-depth analysis of the key systems that have emerged during the process and how these have performed vis-à-vis goals and expectations. Here an expanded analysis is undertaken, connecting back to, for instance, overall project goals, more general applications, business cases or general feasibility.

- The design of a solution package for large-scale field testing. This could be seen as the final stopping point of the first part of the project and a period of transition to a second part consisting of large-scale testing.

B. Cycles of flexible and adaptive co-design

There is need for a continuous adaption of the co-design process to ensure that the different participating parties evolve on a par with each other. This means a continuous process involving the re-planning of actions taken during the design process. The project works with a general pattern of four types or phases of project activities: Planning session, physical meetings, home activities/testing, and web sessions with user feedback. Four full cycles were performed in the reported project, ending with a final physical meeting of the whole design group. It should be noted that "planning actions" are not limited only to the "planning" phase of the cycle. To illustrate this, an example of one such cycle is provided.

TABLE 1. EXAMPLE OF ONE CO-DESIGN CYCLE.

Phase	Actions
Planning, including representatives of all groups	- Feedback analysis - Agenda setting, meeting - Design group communication
Meeting - Physical meeting with all of the parties of the design process.	- Facts about fall detection - Technology presentations - Group discussion about living longer at home
Field work - the user at home acting and observing.	- Gear testing at home, - Recording habits - Critical incident reports
Web meeting, feedback and reflection meeting	- Feedback of experiences such as gear testing - Suggesting future actions

To achieve an adaptable approach, we combine actions during the different phases, based on the outcomes from the continuous planning and feedback actions. Table 2 combines the phases of a cycle in a co-design process. The key to a successful planning of the next cycle is repeated feedback action during all the phases of the cycle. The feedback is the challenge of the project and is central to involving the end-users in particular in the design process. As the planning session is central, representatives of the different parties should be present, including the users. To instill a feeling of empowerment especially in the end user is necessary for the dynamics of an innovative design process.

V. DISCUSSION AND FUTURE WORK

The challenges of co-design and the need for an adaptive process can be discussed from several angles. The main arguments revolve around the notion of biases, as they always influence the proceedings of the design process. To elevate the knowledge of those concerned with the project is, as we argue in this paper, a key aspect in avoiding the problems of skewed design and ineffective solutions. In the following, we discuss a number of issues that can help avoiding such problems in the project.

There is need for co-design methodology expertise, considering the set of people with separate skills that form the driving force of the project. Too often, designers with a background in health care are project leaders and thus become the ones defining the rationale of the project, as well as planning the co-design process. In co-design, a very complex stakeholder landscape must be managed. To pick the key driving persons from one of the dominant parties will eventually tilt the progression in that direction. However careful or active those forming a project management are in their effort to create a balance, they will be affected by their professional background. A neutral corner of project management with a methodology expert background could be one path to explore, as a means of working with these types of problems.

Learning during co-design helps participants to evolve in areas of which their knowledge is low was present at the very outset of the project. Self-understanding and the articulation of user experiences are equally important as learning new things. To evoke the everyday experience of living during design sessions might be harder. Users need to be given time to work with their own experiences to be able to retell them in the design setting.

One challenge is to clarify and explain the mission and purpose of the project as well as how to convey an understanding of its rationale. A project means different things to different groups. As initial expectations can vary in a very high degree, it can be a hard task to convey the core message to everyone. However, a unitary view of the project might not always be an advantage, but everyone must at least be given the chance of understanding (a topic further discussed in a coming section). A solid piece of advice would be never to take anything for granted, not even the most basic assumptions.

It is necessary to achieve a human-technology balance. There is a challenge in how to set up a project so as to achieve a good balance between creativity and innovation based on the needs of users and the possibilities that technical experts offer or perceive as possible. This division seems still to be a very important one, which remains very hard to handle. The experience from the current project is that it is actually harder to get the technical professionals to engage in user perspectives than for users to adopt a technological way of thinking. The human-technology balance question as posed in this section underlines the need of raising both experts and users to relevant knowledge levels in their respective areas. This includes the necessity of becoming aware of hidden biases and assumptions. To achieve this requires a more neutral project management perspective which is not entrenched in any particular corner.

An EBCD project in a research setting functions under special conditions. The differences between an organizational development situation and a research setting could strongly influence how EBCD works out. One very important aspect here could be the ethics approval, which is required in a research context. An approval from an ethics board is customary, but the understanding of EBCD and its open-ended plan of action could create difficulties to obtain this. As the process is a search for unknown solutions, it might be hard to predict all that will happen. This could be seen as a weakness in the eyes of an ethics board.

There is a hard question of how to deal with a multiple and complex goal structure. The sentiment of co-design all too often seems be a feeling of constituting a big family that will somehow get along. This might not be the case, nor even very desirable. Co-design seems to build on a consensus view of the group and the design issues. A mechanism for handling disagreements seems in general to be lacking. The experiences of this project show that technology restraints often obtain the role of putting discipline into the design group, by claiming "this cannot be done", or "this will be too expensive" and similar arguments that often stop discussions. A great many actions within a co-design process aim at getting "all on board". There could be more active ways of exploring the potentials of dissent as a source of innovation and thinking outside of the box solutions. On the methodology side, this should be explored by using tools and methods that can handle such rather delicate processes. A comparison could be made with 'critical systems thinking' schools [20] where expressions of dissent are regarded as a fact and systems design as a way of resolving them and creating better solutions for more people.

VI. Conclusions

The conclusion of this research is that a standard model of co-design in the area of eHealth, such as EBCD, must be viewed as a learning vehicle that enables the different parties to acquire the necessary knowledge to be productive in the project. The basic model of EBCD seems to rely on facts that can be acquired from participants. In this article, we argue for co-design as a formative process of creation and discovery based on experiences of both patients and professionals, especially in the early "fuzzy front-end" stage, as focused on in this paper. An adaptable approach to co-design is needed which tracks the learning progress and adds necessary co-design action to ensure the ability of all participants to grow and learn. The empowerment process includes all the different parties, like users, health care professionals and tech people. As there is a clear trap if the co-design planners and leaders have too clear a connection to any of these areas, a more definite co-design expertise should be at the helm of the process. Compared to a traditional EBCD approach [2], a more iterative and multi-track approach is proposed here, for the purpose of elevating all parties to a point where they can be active and co-responsible. There is a cross connection of fact and empowerment dimensions in co-design that warrants more research into its theory and practice. The outcome of this particular project shows the necessity for high levels of engagement and drive to find the right solutions that will work in practice both in a technological sense and in the life world of the user. To achieve this reality the knowledge and learning dimension must be at the core of the co-design process. No co-design is possible unless there are well planned processes of what could be described as "co-learning". In this study, we have proposed some outlines of what a method of co-design could look like, but further research should be performed to formulate a consistent concept of co-design by co-learning.

ACKNOWLEDGMENT

This research was supported by Celtic+ and Vinnova, founding the project FRONT-VL, Project-ID: C2016/2-5.

REFERENCES

[1] P. Bate and G. Robert, Bringing user experience to healthcare improvement, Oxford: Radcliffe, 2007.

[2] S. Donetto, V. Tsianakas and G. Robert, Using Experience-based Co-design to improve the quality of healthcare: mapping where we are now and establishing future directions, London: King's College London, 2014.

[3] E. B-N. Sander and J. P. Stappers, "Co-creation and the new landscapes of design", International Journal of Cocreation in Design and the Arts. 4(1) pp. 5-18, 2008.

[4] M. Ward, et al., "Using Co-Design to Develop a Collective Leadership Intervention for Healthcare Teams to Improve Safety Culture", International Journal of Environmental Research and Public Health, 15(6), 1182, pp. 1-17, 2018.

[5] R. Blackwell, K. Lowton, G. Robert, C. Grudzen and Grocott, P. "Using Experience-based Co-design with older patients, their families and staff to improve palliative care experiences in the Emergency Department: A reflective critique on the process and outcomes", International Journal of Nursing Studies, 68, pp. 83-94, 2017.

[6] S. Hintistan, N. Nural, D. Cilingir and A. Gursoy, "The effect of nursing telephone counselling upon chemotherapy related symptoms and quality of life among patients with lung cancer", European Journal of Oncology Nursing, 18, pp. S14, 2014.

[7] P. Outlaw, S. Tripathi, J. Baldwin, and Z. Malpus, "Using patient experiences to develop services for chronic pain", British Journal of Pain, 12(2), pp. 122-131, 2018.

[8] H. Gleeson, A. Calderon, V. Swami, J. Deighton, M. Wolpert, and J. Edbrooke-Childs, "Systematic review of approaches to using patient experience data for quality improvement in healthcare settings", BMJ Open, 6(8), E011907, pp. 1-11 2016.

[9] V. Palmer, et al., "The Participatory Zeitgeist: An explanatory theoretical model of change in an era of coproduction and codesign in healthcare improvement", Medical Humanities, 28, pp. 1-11, July, 2018.

[10] P. R. Brocklehurst, et al., "How do we incorporate patient views into the design of healthcare services for older people: a discussion paper", BMC Oral Health, 18(1), pp. 1–7, 2018.

[11] E. Matthews, M. Cowman, and S. Denieffe, "Using experience - based co - design for the development of physical activity provision in rehabilitation and recovery mental health care" Journal of Psychiatric and Mental Health Nursing, 24(7), pp. 545-552, 2017.

[12] S. Gustavsson, and T. Andersson, "Patient involvement 2.0: Experience-based co-design supported by action research", Action Research, pp. 1-23, August, 2017.

[13] S. Gustavsson, "Improvements in neonatal care; using experience-based co-design", International Journal of Health Care Quality Assurance, 27(5), pp. 427-438, 2014.

[14] D. Hill, L. Walter, J. Casas, K. DiDomenico, J. Szymczak, and A. Feudtner, "The codesign of an interdisciplinary team-based intervention regarding initiating palliative care in pediatric oncology", Supportive Care in Cancer, 26(9), pp. 3249-3256, 2018.

[15] J. Langley, D. Wolstenholme, and J. Cooke, "Collective making' as knowledge mobilisation: The contribution of participatory design in the co-creation of knowledge in healthcare", BMC Health Services Research, 18(1), pp. 1-10, 2018.

[16] C. Argyris, R. Putnam, and D. M. Smith, Action Science: Concepts, Methods, and Skills for Research and Intervention, San Francisco: Jossey-Bass, 1985.

[17] D. Schön, The reflective practitioner: How professionals think in action, Aldershot: Arena, 1995.

[18] B. Taylor, Reflective practice for nurses and midwives: a practical guide, Allen and Unwin, St Leonards, 2000.

[19] B. Taylor, B. Bulmer, L. Hill, C. Luxford, J. Mcfarlane, J. Reed, and K. Stirling, "Exploring idealism in palliative nursing care through reflective practice and action research", International Journal of Palliative Nursing, 8(7), pp. 324-30, 2002.

[20] R. L. Flood, and M. C. Jackson, Creative Problem Solving: Total Systems Intervention, Wiley, Chichester, 1991.

TABLE 2. ACTIONS AND MEETINGS MATRIX DURING A DESIGN CYCLE.

Actions type	**Cycle phase**			
	Planning	**Physical meeting**	**Field test: User at home**	**Web conference**
Planning Actions				
Initiating phase: facts and purpose				
Requirements – initial design				
Design of the initial solution				

A Personalised Lifestyle Management Framework for Decision Support System in Mental Health

Sakirulai Olufemi Isiaq
Department of Computing
Southampton Solent University
Southampton, United Kingdom
email:femi.isiaq@solent.ac.uk

Katie Hamling
Department of Computing
Southampton Solent University
Southampton, United Kingdom
email:katie.hamling@solent.ac.uk

Abstract— **According to the World Health Organisation (WHO), over 600 million people suffer a form of mental health associated symptoms and about 350 million people are affected by depression, worldwide. Although symptoms such as restlessness, isolation and lack of confidence among many others have been attributed to the causes of depression, its alleviation has proven challenging over the years. Therefore, depression has become one of the most challenging health difficulty to manage in recent times. While common depression alleviation techniques include antidepressant medications and various therapies, such as Cognitive Behaviors Therapy (CBT), this work embraces a technology-based lifestyle management approach to effectively manage depression, in terms of Diagnosis, Prevention and Alleviation (DPA). A robust iterative-Documents and Stakeholders Analysis for Development (*i-DaSAD*) methodology was adopted to develop a Framework for Lifestyle and Depression Management – *FLaDM*. This framework underpins a decision support system that revealed its suitability for the development of lifestyle management system for depression, particularly in the area of early diagnosis, prevention and alleviation.**

Keywords - eHealth; stakeholders; depression management; diagnosis; prevention; alleviation; i-DaSAD; FLaDM.

I. INTRODUCTION

In recent years, mental health conditions have increasingly become a reckoning factor. In the United States, one in five adults (44.7 million in 2016) suffers a form of mental health disorder [1] i.e., a failing of certain aspects of human health, such as intelligence, imagination and thought. Similarly, according to EPIBUL [2], one in five residents has experienced some common form of mental issues during their lifetime.

Recent records have shown between 7 and 12 per cent of the adult population lives with mental health maladies such as depression and anxiety and a significant proportion of these sufferers do not get any form of intervention. Currently, about 322,000,000 people suffer from depression and 264,000,000 people experience anxiety worldwide [3]. By 2020, depression has been predicted to be the second most impactful condition with estimated annual mental illness cost of about 6 trillion USD by 2030, thereby, doubling its cost in 2010 [4]. In developed nations, depression has not only become one of the most common health disorder but it is also responsible for a greater burden of disability than any other cause [5]. Currently, depression is responsible for more loss of years at work than any other disability or health conditions. By 2026, it is expected to cost up to 3 billion USD in England [4]-[6].

As the demand for mental health treatment becomes higher, the gap between treatment and accessibility is increasingly becoming wider, estimated between 35% and 50% [7]. Although a significant number of mental health conditions remained widely undiagnosed and untreated, the causes and alleviations techniques are not unknown. Moreso, only fewer than half of those diagnosed receive treatment due to reasons substantiated to stigma, ineffective therapies and inadequate access to mental health supports and resources amongst other reasons.

Conventionally, the diagnosis of mental health disorders such as depression involves the utilisation of tools like the DSM-5 [8] criteria by healthcare practitioners (GP or a psychiatrist) to determine the presence of depression symptoms. Subsequently, medications and/or prescriptions are recommended as appropriate. However, more than half of the world's population lives in a country with two psychiatrists per hundred thousand people [9]. In additions, other diagnostic tools such as the Beck Depression Inventory (BDI) [10] requires a level of sophistication (minimum of fifth or sixth grade) to sufficiently comprehend the tool items [11]. Besides, a clinical study conducted in 2006 about conventional treatment indicated common antidepressants produce responses in just under half of the participants and was only able to achieve 28% full remission [12]. Conversely, effective lifestyle changes and management have proven to lift people out of mental health conditions such as depression without the possibility of reoccurrence or known side effect. Since the demand for mental health treatment clearly outnumbers the available practitioners, the demand for support tools and/or potent approach of treatment remains undoubtedly imperative.

The recorded advances in technology have been influencing the pivotal role played by technology in the daily management of human activities. Perhaps, it is not improper to consider technological approaches for proactive management of human mental health states. For example, smartphone technology combines mobile communication and computation in a handheld device, facilitating mobile computing as a point of care [13]. Nowadays, mobile devices and the internet facilitate different health activities and communication online. For instance, a computer-based CBT program via the internet has proven to be clinically effective just like face to face [7]. Therefore, this approach is becoming increasingly popular due to associated benefits that include the ability to reach larger audiences with low cost among others[14]. Nevertheless, the lack of known technological standards raises potential questions about the reliability and acceptability of this approach. For example, online tests and BDI for depression face difficulties such as exaggeration or misunderstanding, which can sway diagnosis thus, constituting a negative effect. This paper proposed a research-oriented *Framework for Lifestyle and Depression Management* - FLaDM to eliminate such anomalies. The framework is based on a robust i-DaSAD methodology, which considers stakeholders and other components necessary for the development of a personalised support system for depression management. The remainder of this paper is divided into sections as follows: Section II covers related work while Section III and IV discuss i-DaSAD methodology and the framework (FLaDM) respectively. The framework was evaluated using a developed decision support system in section V, while the conclusion and future work are presented in Section VI.

II. Related Work

Easy accessibility and affordability increase the deployment of mobile devices for the management of mental health conditions. Mobile devices now play valuable roles in regular face-to-face therapy [7] making such service accessible from anywhere at any time. Evidence emerging indicates that psychotherapy delivery via video conferencing and telephone gives equivalent efficacy to face-to-face delivery [15]. Therefore, with fewer psychiatrists indicated in Section I, sufferers of mental health conditions may access psychotherapy treatments anytime from any part of the world. Over the years, significant advances have been recorded in the application of technology for health management and decision-making. For example, the traditional approach of measuring blood alcohol, nicotine and vitamin D levels will soon be substituted with other technological advances in the near future. Although yet to be widely adopted, devices such as PROOF alcohol tracking wearable [16], BACtrack alcohol tracker [17], SmartStop wearable device [18], and Chrono Band [19] are gaining prominence in carrying out these activities.

In recent times, internet-based CBT protocols are becoming more prominent for treating common mental disorders [14] and have proven impactful and useful, particularly in reaching out to a larger number of people. In addition, behavioural intervention technologies (BITs) - a technological application of behavioural and psychological intervention to address behavioural, cognitive and affective targets - are commonly used to support physical and behavioural mental health [15]. More recently, mobile applications are used to trigger momentary interventions to assist mental health candidates in difficult and stressful situations, particularly when treatment is urgent. To expressly explain the technological applications of lifestyle management in terms of Diagnosis, Prevention and Alleviation (DPA) of mental health, this work laid emphasis on depression as a case study. While various depression applications are readily available for users, many are less robust in terms of managing depression symptoms. Importantly, many depression applications are prone to the risk of misdiagnosis causing the users a potential health retrogression through incorrect health management.

A few of the highly rated and readily available applications include Depression Test [20], which presents users with diagnosis and progress tracking functionalities. While Moodtrack diary [21] and Depression screening test [22] allows activities progress tracks and diagnosis respectively, applications, such as Pacifica [23] and Relieve depression PRO [24] provide prevention and alleviation functionalities, although without diagnosis or personalisation features. To the best of our knowledge, most applications have not been able to successfully combine all the necessary aspects of lifestyle management in terms of Diagnosis, Prevention, Alleviation, Personalisation and Tracking, DPA/PT. Hence, our proposition of a Framework for Lifestyle and Depression Management - FLaDM as a standard framework for the development of a depression management decision support system. Following, Section III explains iterative-Document and Stakeholders Analysis Development, i-DaSAD, methodology adopted for the development of FLaDM framework.

III. Iterative-Document And Stakeholder Analysis Development (i-DaSAD)

To adequately replicate the process of development, study methods need to be clearly defined. Kothari [25] highlighted various methodologies including descriptive versus analytical, quantitative versus qualitative, and conceptual versus empirical among others. This work considered a mixed-method approach that combines elements of multiple methodologies from the initiation phase of defining a generic area of interest, to the final phase of artefact development. Document analysis and stakeholder analysis were carried out in conjunction with other scientific methods to develop the FlaDM artefact. The methodology combines different suitable techniques or methods in each of its five phases to derive a robust artefact. The first phase of the methodology, defining a generic area of interest, is explained as follows.

A. *Defining Generic Area of Interest (G-AoI)*

The i-DaSAD methodology commences with the identification of a generic area-of-interest. This generic area of interest is usually instigated by investigators' curiosity, research questions or matters arising that require resolutions. For instance, an area of interest can be as general as 'mental health' triggered by questions such as: why is mental health becoming increasingly prevalent? A substantial study of literature is expected at this stage to build a context around the history and state-of-the-art of the topic in question. Document analysis including publications, primary and secondary research and white papers certainly leads to the definition of a specific area-of-interest. For instance, exploring the body of knowledge about mental health can be narrowed to an area of focus such as eating disorder, anxiety or depression. An output of this phase is usually specifics of the general area of interest. Although there can be multiple specifics from a general interest, each specific are treated independently to improve clarity. For example, while mental health is a generic area of interest, its specifics include depression and anxiety among others. These specifics are examined independently in separate iterations.

B. *Evaluating Specific Area of Interest (S-AoI)*

In this phase, further document exploration is conducted on the identified specific area-of-interests. For example, if the area of interest has been narrowed down to depression, barriers, symptoms and alleviation of depression amongst other attributes are evaluated at this stage. An outcome of this phase is a comprehensive knowledge and data about the specifics. Data acquisition and statistical analysis about the specific area of focus are also carried out at this stage. For instance, statistical analysis of secondary data about depression from multiple sources is carried out at this stage. A pyramidal approach is adopted, i.e., having more general information at the bottom up. For example, if depression is the main focus, at the bottom will be statistics about depression followed by the type, method of diagnosis, prevention and alleviation and associated limitations respectively. Having comprehensive knowledge about the specific area-of-interest will help accomplish a robust analysis in the following phase (Identify Stakeholders, Problems & Solutions). Statistics about depression, diagnosis, prevention, alleviation and associated technologies were identified at this stage.

C. *Identifying Stakeholders, Problems & Solutions*

Further data analysis is performed on the outcome of the preceding phase (evaluating specific area-of-interest). A stakeholder analysis is conducted to identify all stakeholders. Problems and potential solutions about the specific area-of-interest are then derived from the previous phase of data collection. For example, in our analysis, we identified stakeholder categories to include poor people, students, career, medical practitioner among others. Following is the identification of potential solutions for the identified stakeholders' problems. For instance, one of the identified problems derived from the knowledge evaluation is stigmatisation and a corresponding solution is the increase of awareness and making the process of contacts and supports easier. Techniques such as root cause analysis are employed to identify similarities between symptoms and causes in order to define the hierarchy of importance. Findings in terms of problems and solutions will then be validated using tools such as questionnaires, surveys, interviews or focus group amongst others to improve stakeholders' involvement. In this case, participants regarded as potential stakeholders were asked to complete a 'seventeen-question' survey via an online medium. Results of the survey were used to authenticate the designed stakeholders, problems and potential solutions about depression. The validation exercise of this phase provides clarity on the level of the importance of the stakeholders, problems and potential solutions.

D. *Mapping Stakeholders, Problems and Solutions*

Knowledge about the identified stakeholders, potential problems and solutions are mapped based on the priority of importance or interest in this phase. Techniques such as priority matrix and rate and ranking scale are used to formulate stakeholders' solutions for an intended artefact. For instance, symptoms of depression include 'recurrent thoughts of suicide', 'recurrent thought of self-harm', 'affected sleep pattern', 'reduced mood' and 'affected appetite' etc. The ultimate consequence of these symptoms is death. The closest to death among these symptoms is a 'recurrent thoughts of suicide', followed by 'recurrent thought if self-harm'. A ranking and rating scale is utilised to rank and rate these symptoms in the order of impacts. Also, a priority matrix is used to determine the influence of these symptoms on the system functionalities. Further validation techniques such as focus groups and interviews are used to validate the proposed stakeholders' solutions. In this case, a survey result of 114 respondents was evaluated to validate the proposed framework components. For example, a question such as 'will you like to see an update of your mental wellbeing' validates the necessity of a 'progress track' as a measure of state in the implied framework.

E. *Developing the Artefact*

The final stage is the development of the artefact based on the needs of the stakeholders, corresponding problems and intended solutions. Key defined components are integrated at this stage to produce an artefact. After the ranking and validation of important components, the uppermost components are composed using techniques such as associative rules. To make a system a whole, self-dependent and more generalised components are placed at the uttermost layer while dependent or associated components are further embedded with their corresponding associated components. Taking that the processes within the methodology (in Figure 1) are iterative, multiple and various techniques, methods, approaches can be adopted in different phases until all requirements of each phase are satisfied. A clear benefit of the i-DaSAD methodology is its element of robustness, which allows the necessary requirements to be satisfied iteratively. See Figure 2 for FLaDM components and their corresponding layers. Section IV below explains the FLaDM and the corresponding components as derived artefact of the i-DaSAD process.

Figure 1. Schematic representation of i-DaSAD

IV. FRAMEWORK FOR LIFESTYLE AND DEPRESSION MANAGEMENT (FLADM)

Artefact realised through i-DaSAD methodology as discussed in section III is the *Framework for Lifestyle and Depression Management* (FLaDM). An iterative process of survey and analysis of the literature on mental health, depression systems and stakeholders yielded the framework, which serves as a base for the development of a depression management system. Our proposal is that FLaDM based systems will consider the core components of the framework for the development an effective depression management system in terms diagnosis, prevention, alleviation, personalisation, and progress tracking, DPA/PT.

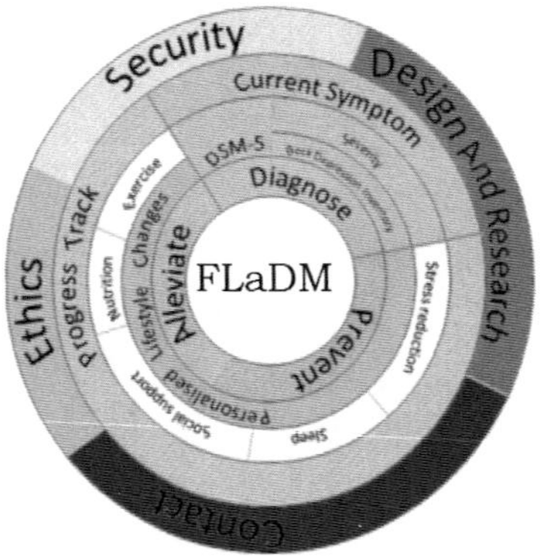

Fig 2. Framework for Lifestyle and Depression Management (FLaDM)

A. *Diagnosis (Presence & Severity)*

The diagnosis component of the framework comprises of two sub-components (primary and secondary). The primary sub-component focuses on examining the presence of depression symptoms in a user and secondary sub-component considers its severity if present. A flexible approach is embraced in order to allow developers adopt a preferred standard diagnostic tools, questionnaire or criteria. For example, we adopted the DSM-5 [8] as diagnostic tools for the presence of depression and the Beck Depression Inventory, BDI is utilised to determine its severity. Other tools such as ICD 10, PH9 [26] and GAD7 [27] may be flexibly adopted by the developer for depression and anxiety respectively. As earlier discussed in Section II (related work) that a minimum level of sophistication is required to comprehensively utilise the BDI tool, it is expected that a decision support system caters for its simplification. In this case, users with a basic smartphone operation should find it easy to consume the information meaningfully.

B. *Prevention/Alleviation*

The Prevention and Alleviation components consider the lifestyle change required for a potential or identified depressed candidate. Similar to the diagnosis, these components allow development's flexibility thereby, providing system developers with the opportunity to determine the lifestyle activities they classify suitable for depression management. As discussed in Section III, comprehensive document analysis was used to derive key lifestyle activities associated with depression. For instance, the study of 1046 women by Jacka et al., [28] reflected an odds ratio of 0.87 and 0.08 lower for depression with diets of fresh fruit and vegetables, grains and meats, than heavily processed diet respectively. This study provides the associative significance of diets with depression and its management. Analysis by Tanaka et al., [29] in the Komo-Ise study found that 1.3% fewer men had depression with exercise, and positive associations with sleeping 6-9 hours per night, small amounts of alcohol consumption as compared to an 'over' or 'under' alcohol consumption, and negative associations with poor physical health and chronic disease. Similarly, Gregory et al., [30] analyses of 1556 participants on wave-4 G1219 and G1219 twins longitudinal studies found a significant correlation between poor sleep and depression/anxiety symptoms. Also, the lack of sleep may prevent brain repair, making it exceedingly prone to various symptoms of depression [31]. Analysis of these works paves way for the consideration of lifestyle activities management such as stress, sleep, diet and exercise amongst others. While the framework permits the inclusion of other activities to be managed, these activities are considered fundamental lifestyle for our system.

C. *Personalisation & Progress Tracking*

Forasmuch as FLaDM considers depression management from the perspective of lifestyle management, a significant consideration is given to lifestyle variation. Therefore, personalisation is considered as a prerequisite component for Diagnosis, Prevention and Alleviation. This is similar to the action of a conventional practitioners while dealing with individual cases. Personalising recommendation of lifestyle changes will not only improve the effectiveness of the system but it will also help enhances the decision support process of the system. In doing so, such systems are expected to retain the capability of tracking the progress of corresponding activities to provide effective supports. For example, in this case, progress tracking is accomplished using CESD-R scale [32] to measure the users' progress in specific areas of the scale.

D. *DPA/PT Wrapper*

The DPA/PT Wrapper is the outermost layer shielding all other components of the framework. The wrapper consists of four basic features namely; design and research, contacts, ethics, and security. The features are expected to be considered for any decision support management system in mental health. Considering the state of mental health stakeholders, particularly the mentally ill candidate, it is expected that any adopted management for diagnosis, prevention or alleviation is highly intuitive, supportive, less demanding, moderately automated. Perhaps, such a system should be undoubtedly simple so as not to aggravate the state of stakeholders in any way. Therefore, the design of such systems requires utmost consideration of simplicity and usability of the system. Concepts including impacts of colours, fonts, navigation and overall comprehension of such system are expectedly thought through. That is, a significant research needs to be considered in the process of design. Also, the importance of the topic in terms of sensitivity to stakeholders necessitate the adoption of robust ethics and security architecture. For example, in considering the process of data handling within the system, issues such as user privacy, encryption and access level are thoroughly considered. In taking cognisance that a decision support system is not expected to replace practitioners, FLaDM based systems are expected to make provision for easier communication between stakeholders, i.e., the practitioner

(GP, psychiatrist, career) and patients' communications must be uncomplicated.

V. FLADM- BASED DECISION SUPPORT SYSTEM

In order to evaluate the framework, a FLaDM based decision support system was developed to measure its effectiveness and usability. A Hierarchical Task Analysis (HTA) approach is utlised for definition of system tasks. A schematic representation of the system hierarchical layout is shown in Figure 3.

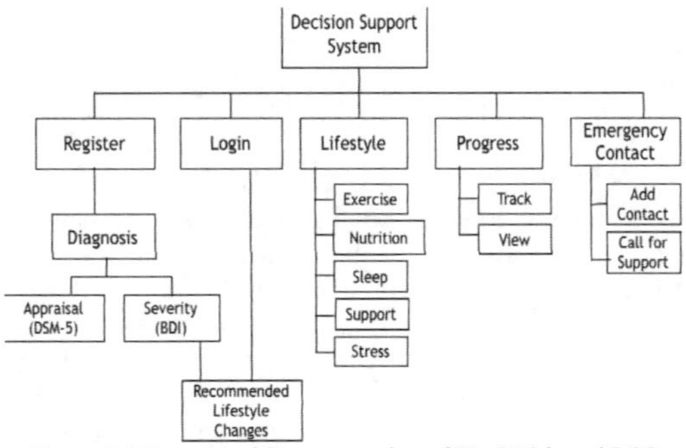

Figure 3. Hierarchical Representation of FLaDM-based DSS

Test metrics were designed to measure the effectiveness of the framework. The test activities were carried in a usability laboratory. It is noteworthy that the test was conducted to measure the suitability of the framework for system implementation. Since all usability problems can be derived from conducting the test with five participants, not so much will be derived from any more participants [33]. Effectively, five participants were recruited for the usability test of a FLaDM based system. A supplementary 'ten-question' pre-test questionnaire and 'fifteen-question' post-test questionnaire were completed by the participants to investigate the system's effectiveness, particularly in terms of diagnosis, prevention and alleviation. Examples of Pre-test and post-test questions are as shown in table I.

TABLE I. EXCERPT OF QUESTIONS FROM TEST QUESTIONNAIRE

No	Pre-test Questions	Post-test Questions
1	Have you ever been diagnosed with depression	Do you feel the application should ask for more personal information?
2	Are you at risk of suffering from depression in future?	Do you think three emergence contacts is suitable for the application?
3	What make you at risk of depression in future?	How do you find the process of diagnosis in general?
4	When were you diagnosed with depression?	Do you think the emergency contacts should include medical practitioners?
5	What are your known symptoms of depression?	Have you been correctly diagnosed using the application?

Five test metrics were designed to evaluate area of concentration including *Registration* (setting up the user), *Diagnosis* (presence and severity), *Lifestyle management and recommendations* (symptom prevention and alleviation) and *Contacts* (stakeholder communication). During the test, participants were requested to provide a self-rating perception of the tasks completed under each metric using a rating scale of 1-10 with 10 being the highest and 1 being the lowest. To ensure participants' actions correlates with the task and their rating, a 'eye-tracking software' was utilised to track participant's eye movement as seen in Figure 4. The red dot indicates the user's area of concentration at a specific time.

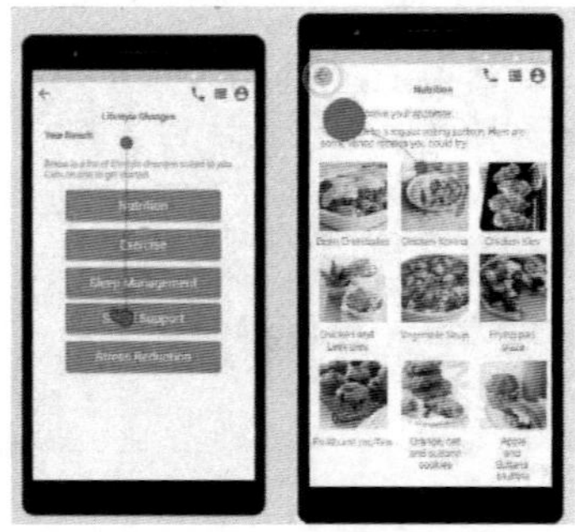

Figure 4. Eye tracking system indicating users concentration

Results of evaluations indicated appropriate diagnosis with overall users rating of 93.5% accuracy. It is interesting to note that all participants believed the lifestyle recommendations for prevention and alleviation are appropriate with 97% accuracy. Table II. represents the test metrics with a corresponding user rating of functional accuracy in terms of DPA/PT and ease of use.

TABLE II. TEST METRICS VERSUS USER RATINGS.

Metrics No	Test Metrics	Overall participants rating
1	Registration	96.5%
2	Diagnosing for Presence (DSM-5)	93.5%
3	Diagnosing Severity (BDI)	93.5%
4	Lifestyle change and progress track	97%
5	Emergency contacts and ease	82%

Although with a smaller sample size, usability problems are experienced with great variation among the test participants. Three of the participants are confirmed to suffer depression as established in the post-test question, one participant was confirmed not depressed and one was on the verge of being depressed. All participants are correctly diagnosed. Also, the test participants were of varied genders, ages and backgrounds, yet all participants were able to use the application for DPA. In addition, all participants expressed confidences in the recommended lifestyle activities to help alleviate their situations. All participants agreed that the most important feature is the emergency contact mechanism that connects them with the relevant support authority (GP and career, etc.) when needed. Most participants expressed that the emergency contact feature should be more intuitive and accessible from any part of the system.

VI CONCLUSION AND FUTURE WORK

The study conducted by Trivedi, M. H. et al. [12] indicated a normal antidepressant produces responses just under half of the participants and only achieved 28% full remission; hence, the need for other treatment options. This work adopts a lifestyle change approach for the development of a framework for lifestyle and depression management that underpins a decision support system. The approach involves extensive document and stakeholder analysis to identify the necessary stakeholders and their requirements for depression management DSS. The framework was evaluated in a usability laboratory to appraise the ease of use, intuitiveness and effectiveness. Results evidenced the suitability of the framework's adoption for the development of DSS for depression management. Further work will be carried out on expanding the framework to other categories of mental health disorder. Also, the system evaluation will be conducted with a larger sample. In additions, the adoption of machine learning and artificial intelligence techniques in the recommendation of change in lifestyle activities will be explored in the area of decision support system development.

REFERENCES

[1] National Institute of Mental Health, (NIMH). "Transforming the understanding and treatment of mental illnesses". [Online]. Available from: https://www.nimh.nih.gov/health/statistics/mental-illness.shtml. [retrieved: 2019.01.03].

[2] Samele, Chiara, Stuart Frew, and Norman Urquia, eds. Mental Health Systems in the European Union Member States, Status of Mental Health in Populations and Benefits to be Expected from Investments Into Mental Health: European Profile of Prevention and Promotion of Mental Health (EuroPoPP-MH):[main Report]. European Commision, Executive Agencz for Health and Consumers Tender, 2013.

[3] World Health Organization, "Suicide data," [Online]. Available from: www.Who.Int/Mental_health/Prevention/Suicide/Suicideprevent/En, 2015, [retrieved: December, 2018

[4] D. E. Bloom, E. T. Cafiero, E. Jané-Llopis, S. Abrahams-Gessel, L. R. Bloom and S. Fathima, "The Global Economic Burden of Noncommunicable Diseases (World Economic Forum, Geneva)," 2011.

[5] M. Lech, L. Low and K. E. Ooi, "Detection and prediction of clinical depression," in Mental Health Informatics, Springer, 2014, pp. 185-199.

[6] P. R. McCrone, S. Dhanasiri, A. Patel, M. Knapp and S. Lawton-Smith, Paying the price: the cost of mental health care in England to 2026, King's Fund, 2008, .

[7] D. Becker, "Acceptance of mobile mental health treatment applications," Procedia Computer Science, vol. 98, pp. 220-227, 2016.

[8] American Psychiatric Association, "Diagnostic and Statistical Manual of Mental Disorders," vol. 2018, "". October 1, 2018.

[9] K. Smith, "Mental health: a world of depression," Nature News, vol. 515, pp. 180, 2014.

[10] Psych Congress, "Beck Depression Inventory-II (BDI-II)," vol. 2018, "". 2018.

[11] G. Groth-Marnat, Handbook of psychological assessment, John Wiley & Sons, 2009, .

[12] M. H. Trivedi, A.J. Rush, S. R. Wisniewski, A. A. Nierenberg, D. Warden, L. Ritz, G. Norquist, R. H. Howland, B. Lebowitz and P. J. McGrath, "Evaluation of outcomes with citalopram for depression using measurement-based care in STAR* D: implications for clinical practice," Am.J.Psychiatry, vol. 163, pp. 28-40, 2006.

[13] R. Alnanih, O. Ormandjieva and T. Radhakrishnan, "Context-based user stereotype model for mobile user interfaces in health care applications," Procedia Computer Science, vol. 19, pp. 1020-1027, 2013.

[14] S. Rosenbaum, J. M. Newby, Z. Steel, G. Andrews and P. B. Ward, "Online physical activity interventions for mental disorders: a systematic review," Internet Interventions, vol. 2, pp. 214-220, 2015.

[15] D. C. Mohr, M. N. Burns, S. M. Schueller, G. Clarke and M. Klinkman, "Behavioral intervention technologies: evidence review and recommendations for future research in mental health," Gen.Hosp.Psychiatry, vol. 35, pp. 332-338, 2013.

[16] PROOF(TM) Weables, "Alcohol Tracking Wearable," [Online]. Available from: www.proofwearable.com/.

[17] BACtrack, 2018 [Online]. Available from: www.bactrack.com

[18] Digital Health Age, "SmartStop wearable to help smokers quit," 2016, [Online]. Available from: digitalhealthage.com

[19] Vandrico Inc, "Chrono Solution, 2018" [Online]. Available from: www.vandrico.com. [retrieved: 2018.12.10].

[20] Psych Central, "Depression Test," vol. 2018, "". 2018.

[21] PsyberGuide, "MoodTrack Diary:Mood Tracker," vol. 2018, "". 2018.

[22] BC Partners for Mental Health and Addictions Information, "Online Screenings," vol. 2018, "". 2018.

[23] Pacifica, "Pacifica is more than a meditation app," vol. 2018, .

[24] L. Surf City Apps, "Relieve Depression Pro," vol. 2018, "". 2018.

[25] C. R. Kothari, Research methodology: Methods and techniques, New Age International, 2004, .

[26] Caremark, "Patient Health Questionnaire—PHQ-9," www.Psycho-Oncology.Info/PHQ9_depression.Pdf, "August ". 2003.

[27] Psychiatric Times, "GAD-7," vol. 2018, "Apr 26, ". 2013.

[28] F. N. Jacka, J. A. Pasco, A. Mykletun, L. J. Williams, A. M. Hodge, S. L. O'reilly, G. C. Nicholson, M. A. Kotowicz and M. Berk, "Association of Western and traditional diets with depression and anxiety in women," Am.J.Psychiatry, vol. 167, pp. 305-311, 2010.

[29] H. Tanaka, Y. Sasazawa, S. Suzuki, M. Nakazawa and H. Koyama, "Health status and lifestyle factors as predictors of depression in middle-aged and elderly Japanese adults: a seven-year follow-up of the Komo-Ise cohort study," BMC Psychiatry, vol. 11, pp. 20, 2011.

[30] A. M. Gregory, F. V. Rijsdijk, T. C. Eley, D. J. Buysse, M. N. Schneider, M. Parsons and N. L. Barclay, "A longitudinal twin and sibling study of associations between insomnia and depression symptoms in young adults," Sleep, vol. 39, pp. 1985-1992, 2016.

[31] Pallab Ghosh, "Why do we sleep?" vol. 2018, "May, 2015.

[32] Center for Epidemiologic Studies Depression Scale, CESD-R, "Depression", "". 2018.

[33] R. Molich, K. Bachmann, J. Biesterfeldt and W. Quesenbery, "Five users will find 85% of the usability problems," in and other myths about usability testing. Proceedings of the Usability Professional's Association-International Conference, 2010.

Predicting If Older Adults Perform Cognitive Tasks Using Body Joint Movements From RGB-D Videos

Jenna Ryan*, Sarah Inzerillo*, Jordan Helmick†, Ali Boolani‡, Natasaha Kholgade Banerjee* and Sean Banerjee*
*Department of Computer Science, Clarkson University, Potsdam, NY, USA
†MedExpress, Morgantown, WV, USA
‡Department of Physical Therapy, Clarkson University, Potsdam, NY, USA
Email: {ryanjc, inzeris, aboolani, nbanerje, sbanerje}@clarkson.edu, jordan.helmick@medexpress.com

Abstract*—We present an approach that performs automated detection of whether an older adult has performed cognitive tasks such as form filling or problem solving using RGB-D video data of older adults collected using the Microsoft Kinect v2 sensor. Our approach uses the variances of 25 joint points on the 3D skeleton obtained from the Kinect for training random forest classifiers to detect if cognitive tasks are performed, based on deviations in postural sway induced by cognitive tasks. We validate our approach using a dataset of 10 subjects performing the test on standing with eyes closed in the Berg Balance Scale (BBS) series of diagnostic tests before and after cognitive tasks. Using leave-one-subject-out cross-validation, we obtain an average detection accuracy of 69.5%, with accuracies of 60% and 79% at detecting that the test on standing with eyes closed was performed prior to and after cognitive tasks respectively. Our approach can be incorporated into intelligent health care systems to detect whether older adults have performed cognitively demanding activities that may induce stress or fatigue, and allow early intervention well before the occurrence of adverse events such as falls.

Keywords–RGB-D; Kinect; cognitive; fatigue; random forest

I. Introduction

The spread of low-cost sensors and computing in the consumer space has enabled the rise of approaches to perform automated health assessment in older adults. RGB-D sensors such as the Microsoft Kinect have been used for detection of falls [1], [2] and frailty [3], and encouragement of exercise [4]. Adverse events such as falls in older adults place high financial pressure on the health care system. The cost of fatal and non-fatal falls in the US was $19.2 billion in 2000 [5] and $23.3 billion in 2008 [6]. Instead of detecting adverse events after they have occurred, financial burden on the health care system can be significantly reduced by monitoring older adult health for signs of fatigue, which has been found to be correlated with fall risk in older adults [7]. In this work, we use random forests on RGB-D data from the Microsoft Kinect v2 sensor to perform automated detection of whether an older adult has performed cognitive tasks using variation in sway of standing posture before and after cognitive tasks have been performed. We define cognitive tasks in this work as activities involving reasoning and decision making, that involve the use of working memory [8]. The cognitive tasks used in our work include problem solving and questionnaire filling. Cognitive tasks have been found to be correlated to increase in postural sway [9], [10] due to weakening of sensory systems related to balance in older adults [11]. Increased sway has been found to be related to rise in fatigue [12]. Using off-the-shelf sensors such as the Kinect to detect the influence of cognitive tasks on balance in older adults can enable intervention in the event of stress induced by cognitive activities in daily living.

The primary challenge of our work is that cognitive tasks show a subtle change in postural sway. Traditionally, physical therapists detect deviation in postural sway by having subjects perform balance tests such as tests under the Berg Balance Scale (BBS) [13]. The BBS consists of 13 tests performed in various postures with eyes open and one test performed standing with eyes closed. The physical therapist provides a numerical score between 0 and 56 on assessing all 14 tests, with a lower value indicating lesser balance ability. However, due to variations between scores of multiple physical therapists, several studies [14], [15] have been conducted to determine the reliability of the BBS score in detecting a change in postural sway. The minimum detectable changes at an absolute reliability of 95% has been cited as 4 for scores between 45 and 56 by Donoghue et al. [14] and between 2.8 to 6.6 for eleven studies reviewed by Downs et al. [15] with BBS scores above 20. In the dataset used in our work where 10 subjects were evaluated doing BBS tests prior to and after cognitive tasks, the BBS score after cognitive tasks was found to be reduced by 1.6 points on average with scores ranging from 49 to 56. Since the reduction is lower than the thresholds discussed in Donoghue et al. [14] and Downs et al. [15], the BBS score alone cannot be used to detect movement changes induced by cognitive tasks.

While there exist approaches to detect changes in gait and posture due to physical fatigue [16]–[18], these approaches use body-mounted sensors that hinder natural motion and may introduce noise that overpowers the subtle influence of cognitive tasks. Additionally, these approaches use sensors placed on a narrow range of locations, e.g., on the foot [16], [17] or on the lower back [18], which may prove insufficient to detect the influence of cognitive tasks on full body postural sway. Our work addresses the subtle influence of cognitive tasks over sway in the head, shoulders, arms, hands, spine, legs, and feet by using variances in the positions of 25 3D joints from the Kinect, which being a non-contact sensor allows natural unweighted motion. We use the 25 body joint variances as input to the random forest classifier, which provides a binary output on whether the person performed cognitive tasks. We enhance the contribution of sense of body joint location, i.e., proprioception, toward balance by using the BBS test on standing with eyes closed to eliminate the influence of visual cues, which have been found to have the largest contribution toward balance amongst the visual, proprioceptive, and vestibular balance mechanisms [19].

We use a linear mixed model with joint variances from the 10 subjects analyzed in this work to demonstrate that the 25 joint variances have a statistically significant effect in distinguishing the data prior to and after cognitive tasks.

We use a similar linear mixed model to demonstrate that in a control dataset where the same 10 subjects perform two rounds of the BBS tests on a different day with a break, i.e., without cognitive tasks between the rounds, the 25 joint variances do not have a statistically significant contribution in distinguishing between the pre- and post-break conditions. Using a leave-one-out cross-validation approach for training each random forest using 9 subjects and testing with the left out subject, we obtain an overall average accuracy of 69.5%, with average accuracies of 60% and 79.0% in detecting that the test on standing with eyes closed was performed prior to or after performing cognitive tasks respectively. We also obtain an average accuracy of 44.5% on the control dataset in distinguishing between pre- and post-break condition, which being close to random chance indicates that postural sway remains consistent in the absence of cognitive tasks.

The remainder of the paper is organized as follows. In Section II, we discuss contact based approaches for detecting fatigue, non-contact based approaches for detecting older adult health, and fatigue detection in driving using non-contact based sensors. In Section III, we describe the data collection procedure during the experimental and control days for our study. In Section IV, we describe our approach for detecting deviations in balance using joint variance analysis. In Section V, we describe the linear mixture model that we designed to determine that joint variances are appropriate predictors of whether older adults have performed cognitive tasks. In Section VI, we provide details of the random forest classifier we trained to determine if an older adult had performed cognitive tasks. In Section VII, we discuss the performance of our classifier and provide insights on where misclassifications occur. We conclude the paper in Section VIII and provide potential areas of future work.

II. Related Work

To the best of our knowledge, there exists no work on detecting whether a subject performs cognitive tasks from routine motions such as walking or standing. As cognitive tasks are known to impact postural sway [9], [10] which is related to fatigue [11], we discuss approaches to detect physical fatigue using contact-based and non-contact sensors, and fatigue during the cognitive task of driving.

Contact-Based Sensors for Detection of Fatigue. Traditional physiological approaches for automated fatigue detection have focused on the use of surface electromyography (sEMG) to measure muscle fatigue during physical exertion; a review of these approaches may be found in Al-Mulla et al. [20]. While sEMG electrodes do not puncture the skin, they require skin contact and use a significant amount of hardware, thereby being cumbersome for fatigue detection in everyday environments. There also exist approaches to use body-mounted sensors, such as pressure sensors at six points on the foot to estimate fatigue during walking [17] or running [16] gait as a function of maximum pressure. These gait-based approaches have multiple cues from asymmetries and variations in rotation or translation of the upper body, hips, legs, and feet during walking or running gait. Our work on detecting changes in standing body posture handles the challenge of working with a smaller set of cues restricted to body sway.

In the area of detection of posture balance, Wall et al. [21] provide a balance prosthesis that measures head tilt using inertial sensors mounted on the shoulders and trunk, and displays the tilt to the user for user-driven balance re-adjustment. Their approach does not perform detection of differences in sway. Shahzad et al. [18] estimate balance impairment for fall risk by predicting the score of the BBS for older adults using acceleration data obtained from a triaxial accelerometer positioned at the lower back. Their approach shows a separation between the mean BBS scores of non-fallers at 53.3 ± 2.9 and fallers at 45.5 ± 7.2. The mean separation is greater than the thresholds discussed in Donoghue et al. [14] and Downs et al. [15]. This enables Shahzad et al. to use linear least squares and LASSO for regression of the BBS score. In our work, the average separation of 1.6 between BBS scores before and after cognitive tasks is much smaller than the Donoghue et al. or Downs et al. thresholds for change detection, preventing the use of the BBS score as a predictor of cognitive tasks. Our approach overcomes this issue by using the skeleton obtained with the Kinect sensor to detect whether cognitive tasks are performed from the effect they have in the increasing the sway of 25 body joints.

Physical fatigue or aging handled by the above approaches show gross changes in movement which can be picked up by body-mounted sensors. However, body-mounted sensors hinder natural user motions, and due to their added weight may prevent separation of data prior to and after cognitive tasks that have a subtle influence on motion. Our approach of using non-contact sensors retains natural motions, enabling us to detect the weak influence of cognitive tasks.

Non-contact RGB-D Sensors to Assess Older Adult Health. Despite the popularity of cameras and RGB-D sensors such as the Kinect in the consumer space, approaches to perform automated detection of anomalous health patterns in older adults using non-contact at-a-distance sensors have been limited. There exist approaches to perform detection of falls by tracking the vertical state of a subject [1] or the acceleration and distance of the center of mass [2]. However, these approaches enable intervention only after a fall has occurred. In contrast, our work evaluates postural instabilities to enable early intervention well before a high-risk incident such as a fall arises. Gianara et al. [3] use the Kinect to perform frailty assessment on subjects performing the Timed Up and Go (TUG) [22] test. They determine that gait parameters such as walking time and speed, distance covered, and swing correlate well with the Tillburg Frailty Indicator (TFI) score, while postural parameters such as torso inclination do not show a strong correlation with the TFI score. Unlike our work, they do not perform prediction of frailty in a novel individual.

Fatigue Detection During Driving. The body of work that resembles ours in using non-contact sensors to detect fatigue due to cognitive tasks is work on driving fatigue detection, surveyed extensively in Wang et al. [23]. Here, the cognitive tasks involved include paying attention to roads, signs, lanes, pedestrians, intersections, and other drivers. These approaches use a camera to analyze the user's face, whereas our work analyzes the body and assumes that the face may not be visible, an issue that can occur when the user looks away from the sensor, has their back toward the sensor, or is at a distance where the resolution is insufficient for face analysis.

Additionally, these approaches detect gross observable changes such as eyelid open versus closed [24]–[30], head tilt due to nodding versus upright head [24], [28], eye-gaze narrowed versus wide [24], and yawning versus not yawning [24], [28], [30], [31]. Our work detects the presence of subtle changes in body sway when cognitive tasks such as questionnaire filling and problem solving are performed by elderly individuals in whom body posture prior to induced fatigue shows low stability due to age [32].

III. Data Collection Procedure

Our dataset consists of 10 older adults with 6 female and 4 male subjects. The subjects in our dataset range in age from 57 to 68 years with a mean age of 63.50 ± 4.28 years. To reduce the effect of confounding factors, we ensure that all subjects have not been diagnosed with any neurological conditions, have no pre-existing balance disorders or recent orthopedic surgery, have visual acuity, and do not require assistive devices to stand or walk. We collect data from the subjects on two separate days—an experimental day when the subjects perform cognitive tasks and a control day when the subjects do not perform cognitive tasks and instead have a rest break. The time in between the two collection days ranges from 1 to 9 days, with an average of 2.4 ± 2.5 days, and with the order of experimental and control day randomized across subjects.

On both the control and experimental day we measure the subject's vitals, i.e. height, weight, body composition, and radial pulse heart rate/blood pressure, and administer three questionnaires on current feelings of fatigue, mood and motivation. We then administer three standard diagnostic tests to measure static and dynamic balance. The tests include 14 assessments under the BBS [13], the Timed Up and Go (TUG) [22] test, and the 30 Second Chair Stand test [33]. On the control day, we give the subject a 1 hour break. On the experimental day, we have the subject perform cognitive tasks by asking them to fill out questionnaires on demographics, medical history, physical activity, sleep, grit, hope, satisfaction with life, interest, self management, and food frequency, and to perform problem solving activities such as Serial Subtract 3, Serial Subtract 7, Continuous Performance Test, Rapid Visual Information Processing Test, Trails B Test, and 8 minute Tapping Test [34]–[36]. The subject performs cognitive tasks on a tablet. After the break on the control day or cognitive tasks on the experimental day, we ask the subject to repeat the 30 Second Chair Stand, TUG, and BBS tests. We record the subject's vitals, and ask the subject to re-fill the three questionnaires on current feelings of fatigue, mood and motivation. The control day activities range from 64.7 to 71.6 minutes, with an average of 68.5 ± 2.0 minutes. The experimental day activities range from 93.6 and 120.4 minutes, with an average of 107.5 ± 8.5 minutes.

We developed a custom C# based application using the Kinect SDK to collect RGB-D video, face, and joint data for each subject. The RGB-D video data consists of both color and depth frames captured at 30 frames per second. The face data consists of 1347 3D face points for each frame of the video. The joint data consists of 25 3D joint points along with corresponding x and y coordinate points in color and depth space for each frame of the video. In Figure 1, we show color images and corresponding 3D skeletons for 4 different subjects. The right side of Figure 1 shows the 3D Kinect skeleton from a frontal view with labels for joints along the medial axis and on the left side of the subject.

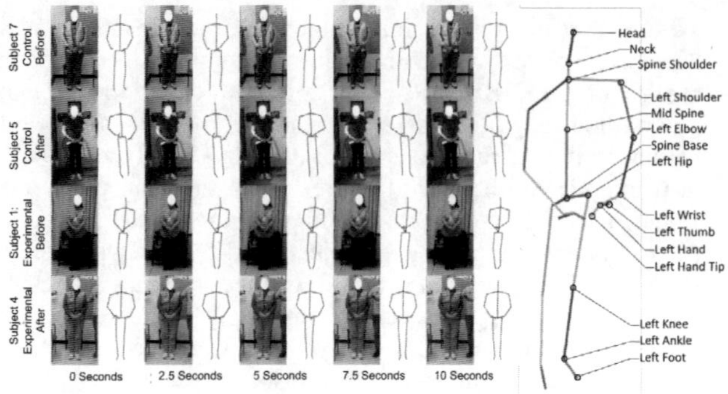

Figure 1. Left: Color and skeleton data from four subjects for various captures. Right: Kinect skeleton labeled with medial axis and left side joints.

IV. Representing Balance Deviation using Joint Variances

To eliminate the influence of visual stimuli which show the largest contribution to balance [19], we use RGB-D videos representing the 'Standing with Eyes Closed' test from the BBS assessments for each subject. The 'Standing with Eyes Closed' test involves a subject standing upright with eyes closed for 10 seconds. Due to the short period of the test, it provides the added advantage of avoiding noise in the data due to motions such as straightening clothes, speaking to the experimenter for clarification, or accidental gesturing. The longer duration eyes open tests contain such motions, which while natural in everyday settings, overpower changes in motion due to the subtle impact of cognitive tasks. Our recognition of deviation in balance is based on the premise that increased movement in the balance tests is related to an increase in fatigue [12]. To evaluate amount of movement, we provide an algorithm in MATLAB that estimates the net variance in each body joint by computing the sum of the variances in the X, Y, and Z dimensions from the 3D joint data collected by the Kinect. The variances in the spread of 3D joint points represents the deviation of the subject from a mean standing position.

Figure 2 shows 3D plots for the Kinect joints data obtained for a sample subject performing the 'Standing with Eyes Closed' test on the control day prior to and after the break, and on the experimental day prior to and after the performance of cognitive tasks. Each plot shows the 3D skeleton from the first frame of the data, together with clusters of 3D points at each joint representing the locations of the joints throughout the capture. As each cluster indicates, deviations in balance inherent to the human body introduce variations in the locations of the joints. The cluster of points from the left shoulder is magnified at the bottom of each plot. Note that on the experimental day after cognitive tasks, the shoulder points trace a more spread out trajectory in comparison to the experimental day prior to cognitive tasks and the control day after the break. The joint variance captures this spread. Also, while there is a high spread of points on the control day prior to the break, the bulk of the points are concentrated

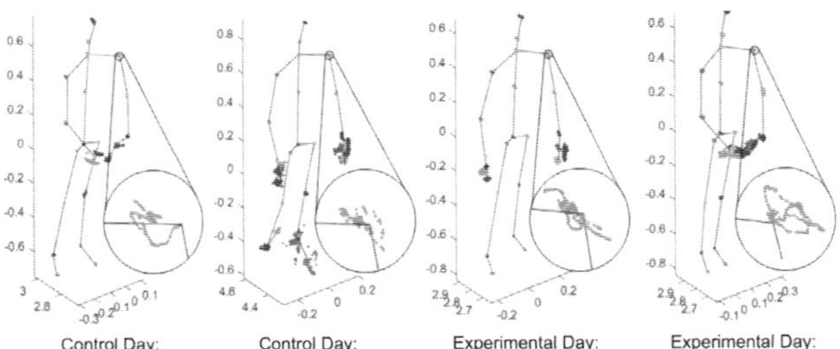

Figure 2. Positions of 3D joints for control and experimental day before and after. Each inset shows the left shoulder points zoomed in.

around the shoulder, and the trailing points are largely outliers. The joint variance in this case is weighted toward the higher concentration of points near the mean.

V. Hypothesis Testing on Diagnostic Tests and Joint Variances

The three diagnostic tests—30 Second Chair Stand, TUG and BBS—provide a holistic understanding of changes in static and dynamic balance. To determine if cognitive tasks have an observable impact on the diagnostic test scores, we perform a two-sided paired Wilcoxon Sign Rank Test to test the following sets of hypotheses on both the control and experimental day:

Null: Difference between [diagnostic test] scores before and after is 0.
Alt.: Difference between [diagnostic test] scores before and after is not 0.

Here [diagnostic test] refers to the 30 Second Chair Stand test, the TUG test, and the BBS tests. Since the three diagnostic tests are performed sequentially, and a subject's performance may be impacted by a prior test we apply a Bonferroni correction and reject the null hypothesis if $p < 0.05/3$, i.e., if $p < 0.016$. On both the control and experimental day we fail to reject the null hypothesis for all three balance tests, indicating that cognitive tasks do not create a statistically significant difference in the diagnostic test scores. For the 30 Second Chair Stand test, we obtain a p-value of 0.76 for the control day, and 0.34 for the experimental day. For the TUG test we obtain a p-value of 0.68 for the control day, and 0.37 for the experimental day. For the BBS, we obtain a p-value of 0.11 for the control day, and 0.06 for the experimental day. By failing to reject all three null hypotheses on the diagnostic tests, we show that cognitive tasks do not show a gross change in static and dynamic balance tests, indicating that the test scores are poor predictors of whether cognitive tasks are performed.

The 25 joint variances computed in Section IV provide a fine-grained understanding of how cognitive tasks affect movement. We construct a linear mixed model on both the experimental and control days to determine whether cognitive tasks impact the joint variances in the 'Standing with Eyes Closed' test before and after cognitive tasks are performed. Using a log-linear transformation, in the experimental day we obtain a confidence interval bound of $[30.80\%, 98.59\%]$ for the percent difference between the joint variances before and after cognitive tasks. The percentages indicate a statistically significant increase of 30.80% to 98.59% in joint variance after cognitive tasks. On the control day, we obtain a confidence interval bound of $[-34.31\%, 3.76\%]$ for the difference between the joint variances before and after the rest break. In this case, since 0 lies within the confidence interval, we show no statistically significant difference in joint variance before and after the rest break. The statistically significant percent increase in joint variances after cognitive tasks on the experimental day substantiates the need for fine-grained joint-based analysis.

TABLE I. Accuracy of predicting whether cognitive tasks have been performed on the experimental day and on the control day (Pred. = Predicted).

	Experimental Day		Control Day	
	Pred. Before	Pred. After	Pred. Before	Pred. After
Before	**60.0%**	40.0%	**42.0%**	58.0%
After	21.0%	**79.0%**	53.0%	**47.0%**

VI. Predicting If Cognitive Tasks Are Performed Using Random Forests

As determined by the linear mixed model, joint variances on the experimental day have strong discriminative power to distinguish whether the 'Standing with Eyes Closed' test was performed before or after cognitive tasks. We use the joint variances as input features in a random forest classifier, with the output being 0 prior to cognitive tasks and 1 after cognitive tasks. We use the `treebagger` function in MATLAB to train a random forest with 500 decision trees and $\sqrt{n}$ predictors for each decision split as recommended in Breiman [37], where n is the number of features, i.e., $n = 25$. We use a leave-one-out cross-validation approach for testing our model on data from the experimental day, where in each fold, 9 of the 10 subjects are used for training, while the left-out subject in that fold is used for testing. The joint variances from the data captured before and after cognitive tasks are removed from the training data for the left-out subject, enabling our approach to perform prediction of cognitively induced fatigue without prior knowledge of the subject. The randomness in formation of decision of trees induces slight differences in the classification at each run of the random forest algorithm. To account for the randomness, we aggregate the results of the random forest classification for each leave-one-out fold over 10 re-runs of the random forest algorithm. To determine if the random forest classifier substantiates the control hypothesis in Section V by detecting close to chance, we similarly use leave-one-out cross-validation to train and test the random forest classifier on data from the control day, where the joint variances are calculated prior to and after the rest break.

VII. Results

We show overall results of classification using random forests in Table I. Results of classifying whether the subject performed the 'Standing with Eyes Closed' test before and after cognitive tasks on the experimental day are shown on left. We obtain an overall classification accuracy of 69.5% on the experimental day, where the average and standard deviation are computed over 10 runs of the random forest classifier. We obtain a classification accuracy of 60.0% for the 'Standing with Eyes Closed' test being performed before cognitive tasks,

TABLE II. PER-SUBJECT EXPERIMENTAL DAY CLASSIFICATION BEFORE AND AFTER COGNITIVE TASKS.

Subject	1	2	3	4	5	6	7	8	9	10
Before	100%	0%	100%	100%	100%	0%	0%	0%	100%	100%
After	100%	80%	100%	100%	100%	100%	20%	100%	0%	90%

and an accuracy of 79.0% for the test being performed after cognitive tasks. The false positive rate is 40.0%, while the false negative rate is 21.0%. The lower false negative rate using the standard random forest algorithm with $\sqrt{n}$ split predictors proves advantageous to our work, as it is essential to accurately detect stress induced by cognitively demanding tasks for intervention. Higher false positives, while inconvenient, do not adversely affect the health of the monitored individual. The right side of Table I shows results of classifying whether the subject performed the test prior to or after the break on the control day. The overall classification accuracy is 44.5% which being close to chance at 50% substantiates the hypothesis test in Section V that in the absence of cognitive tasks, discernible changes in movement may not be produced. Table II shows per-subject classifications on the experimental day averaged over all classifier runs. The classifier correctly classifies subjects 1, 3, 4, and 5; 6 and 8 after cognitive tasks; and 10 before cognitive tasks. It generally shows correct classification for 2 and 10 after cognitive tasks.

Incorrect classifications for data prior to cognitive tasks occur due to clothing adjustment for subject 2 as seen in Figure 3(a), and influence of hand motions toward the center and experimenter hand support for subject 7 as seen at the top of Figure 3(b), all of which cause the joint points to show large deviations from the mean posture. Incorrect classification for data after cognitive tasks may occur due to directed motions corresponding to a downward shrug observed at the bottom of Figure 3(b), and due to a steady movement of the shoulders along the camera axis as observed for the skeleton of the upper body for subject 6 on the left of Figure 3(c) which is larger than the motion after cognitive tasks on the right. The motions in both subjects may resemble the hand motions of subject 2 adjusting clothing prior to cognitive tasks. These steady, largely uni-directional, movements may not be related to balance which tends to show movements with spread along multiple dimensions as in the shoulder points in Figure 2.

As shown by Figure 3(d), subject 9 shows dense clusters of points both prior to and after cognitive tasks in the upper body, causing the data obtained after cognitive tasks to be classified as being prior to cognitive tasks. One reason for the misclassification may be that the subject had a higher amount of physical and mental energy in comparison to the average physical and mental energy across all subjects on the experimental day. According to the summarization of the physical and mental energy for each subject obtained from self-reported responses to the three questionnaires for fatigue, mood, and motivation discussed in Section III, the average physical and mental energy of all 10 subjects prior to cognitive tasks was 184.7 ± 55.0 and 193.8 ± 64.9 respectively, while the average physical and mental energy after cognitive tasks was 155.1 ± 71.1 and 150.2 ± 81.0 respectively. Subject 9 showed higher values of physical and mental fatigue prior to cognitive tasks at 266 and 263, and after cognitive tasks at 259 and 243.

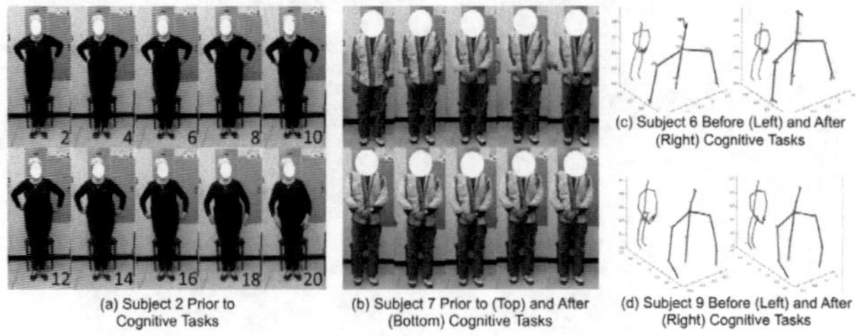

Figure 3. Frames demonstrating reasons for incorrect classification.

VIII. DISCUSSION

In this paper, we have presented an approach to predict if cognitive tasks have been performed by an older adult by analyzing deviations in the standing posture of subjects using the Kinect. Even though the Kinect has been discontinued, our approach for movement analysis can be performed using third party skeleton tracking [38] with other depth sensors such as the Asus Xtion Pro or Intel RealSense. Our approach provides an average accuracy of 69.5%, with 79% accuracy in detecting that cognitive tasks have been performed. By using low-cost non-contact sensors, our approach enables detection of the subtle effect of cognitive tasks on motion, and enables implementation of monitoring technologies in homes and health care facilities without body-mounted equipment. In this work, we focus our analysis on subjects performing the 'Stand with Eyes Closed' test on the BBS scale to eliminate the effect of visual cues. In future work, we will analyze eye gaze patterns from video data to account for the contribution of the visual system in re-adjusting for balance after cognitive tasks are performed in the open eyes tests. To account for potential subject-dependency on balance as indicated by subject 9 in Section VII, we will build subject-specific models for detection of the effect of cognitive tasks by using data on the control day to learn regular subject behavior. Such a system may be propagated to everyday environments for older adult health monitoring by using occasional input from health-care providers to update movement models. We will also analyze the performance of classifiers such as linear regression, logistic regression, and support vector machines to determine optimal classifiers prior to propagation in the field.

While our approach on joint variances enables detection of balance based on increase in deviation from upright posture, extraneous motions such as clothing adjustments, hand gestures, and body motions while speaking can overpower the subtle effect of cognitive tasks. However, such motions are natural in everyday interactions, and must be accounted for in a health monitoring system deployed in average user spaces. The signature of extraneous motions unrelated to balance may show a higher directional dependence, e.g., downward movement of a cluster of points belonging to the hand during clothing adjustment, whereas joint motions related to balance may have a higher degree of isotropy. In future work, we will perform principal components analysis to use the magnitude of movement along various directions at each joint in predicting the performance of cognitive tasks based on balance ability. We will also perform an extended data collection on movement patterns of subjects in everyday environments.

ACKNOWLEDGEMENTS

This work was partially supported by the National Science Foundation (NSF) grant #1730183.

REFERENCES

[1] E. E. Stone and M. Skubic, "Fall detection in homes of older adults using the microsoft kinect." IEEE J. Biomedical and Health Informatics, vol. 19, no. 1, 2015, pp. 290–301.

[2] T. Xu, Y. Zhou, and Z. Ma, "Athocare: An intelligent elder care at home system," in International Conference on Digital Human Modeling and Applications in Health, Safety, Ergonomics and Risk Management. Springer, 2016, pp. 298–305.

[3] E. Gianaria, M. Grangetto, M. Roppolo, A. Mulasso, and E. Rabaglietti, "Kinect-based gait analysis for automatic frailty syndrome assessment," in International Conference on Image Processing. IEEE, 2016, pp. 1314–1318.

[4] S. Ganesan and L. Anthony, "Using the kinect to encourage older adults to exercise: a prototype," in CHI'12 Extended Abstracts on Human Factors in Computing Systems. ACM, 2012, pp. 2297–2302.

[5] J. A. Stevens, P. S. Corso, E. A. Finkelstein, and T. R. Miller, "The costs of fatal and non-fatal falls among older adults," Injury prevention, vol. 12, no. 5, 2006, pp. 290–295.

[6] J. Davis, M. Robertson, M. Ashe, T. Liu-Ambrose, K. Khan, and C. Marra, "International comparison of cost of falls in older adults living in the community: a systematic review," Osteoporosis international, vol. 21, no. 8, 2010, pp. 1295–1306.

[7] S. Morrison, S. R. Colberg, H. K. Parson, S. Neumann, R. Handel, E. J. Vinik, J. Paulson, and A. I. Vinik, "Walking-induced fatigue leads to increased falls risk in older adults," Journal of the American Medical Directors Association, vol. 17, no. 5, 2016, pp. 402–409.

[8] A. Diamond, "Executive functions," Annual review of psychology, vol. 64, 2013, pp. 135–168.

[9] C. Smolders, M. Doumas, and R. T. Krampe, "Posture and cognition interfere in later adulthood even without concurrent response production," Human Movement Science, vol. 29, no. 5, 2010, pp. 809–819.

[10] U. Granacher, S. A. Bridenbaugh, T. Muehlbauer, A. Wehrle, and R. W. Kressig, "Age-related effects on postural control under multi-task conditions," Gerontology, vol. 57, no. 3, 2011, pp. 247–255.

[11] E. V. Sullivan, J. Rose, T. Rohlfing, and A. Pfefferbaum, "Postural sway reduction in aging men and women: relation to brain structure, cognitive status, and stabilizing factors," Neurobiology of aging, vol. 30, no. 5, 2009, pp. 793–807.

[12] A. Nardone, J. Tarantola, A. Giordano, and M. Schieppati, "Fatigue effects on body balance," Electroencephalography and Clinical Neurophysiology/Electromyography and Motor Control, vol. 105, no. 4, 1997, pp. 309–320.

[13] K. Berg, S. Wood-Dauphine, J. Williams, and D. Gayton, "Measuring balance in the elderly: preliminary development of an instrument," Physiotherapy Canada, vol. 41, no. 6, 1989, pp. 304–311.

[14] D. Donoghue and E. K. Stokes, "How much change is true change? the minimum detectable change of the berg balance scale in elderly people," Journal of Rehabilitation Medicine, vol. 41, no. 5, 2009, pp. 343–346.

[15] S. Downs, J. Marquez, and P. Chiarelli, "The berg balance scale has high intra-and inter-rater reliability but absolute reliability varies across the scale: a systematic review," Journal of physiotherapy, vol. 59, no. 2, 2013, pp. 93–99.

[16] K. Yonekawa, T. Yonezawa, J. Nakazawa, and H. Tokuda, "Fash: Detecting tiredness of walking people using pressure sensors," in Mobile and Ubiquitous Systems: Networking & Services, MobiQuitous, 2009. MobiQuitous' 09. 6th Annual International. IEEE, 2009, pp. 1–6.

[17] I. Tanaka, S. Terada, R. Tsuchiya, D. Hanawa, and K. Oguchi, "Fatigue estimation using foot pressure sensors," in International Conference on Telecommunications and Signal Processing. IEEE, 2013, pp. 600–602.

[18] A. Shahzad, S. Ko, S. Lee, J.-A. Lee, and K. Kim, "Quantitative assessment of balance impairment for fall-risk estimation using wearable triaxial accelerometer," IEEE Sensors Journal, vol. 17, no. 20, 2017, pp. 6743–6751.

[19] E. E. Hansson, A. Beckman, and A. Håkansson, "Effect of vision, proprioception, and the position of the vestibular organ on postural sway," Acta oto-laryngologica, vol. 130, no. 12, 2010, pp. 1358–1363.

[20] M. R. Al-Mulla, F. Sepulveda, and M. Colley, "A review of non-invasive techniques to detect and predict localised muscle fatigue," Sensors, vol. 11, no. 4, 2011, pp. 3545–3594.

[21] C. Wall, M. S. Weinberg, P. B. Schmidt, and D. E. Krebs, "Balance prosthesis based on micromechanical sensors using vibrotactile feedback of tilt," IEEE Transactions on Biomedical Engineering, vol. 48, no. 10, 2001, pp. 1153–1161.

[22] D. Podsiadlo and S. Richardson, "The timed up & go: a test of basic functional mobility for frail elderly persons," Journal of the American geriatrics Society, vol. 39, no. 2, 1991, pp. 142–148.

[23] Q. Wang, J. Yang, M. Ren, and Y. Zheng, "Driver fatigue detection: a survey," in Intelligent Control and Automation, 2006. WCICA 2006. The Sixth World Congress on, vol. 2. IEEE, 2006, pp. 8587–8591.

[24] Q. Ji, Z. Zhu, and P. Lan, "Real-time nonintrusive monitoring and prediction of driver fatigue," IEEE transactions on vehicular technology, vol. 53, no. 4, 2004, pp. 1052–1068.

[25] W. Dong and X. Wu, "Fatigue detection based on the distance of eyelid," in Proceedings of the IEEE International Workshop on VLSI Design and Video Technology. IEEE, 2005, pp. 365–368.

[26] M. S. Devi and P. R. Bajaj, "Driver fatigue detection based on eye tracking," in International Conference on Emerging Trends in Engineering and Technology. IEEE, 2008, pp. 649–652.

[27] R. C. Coetzer and G. P. Hancke, "Eye detection for a real-time vehicle driver fatigue monitoring system," in Intelligent Vehicles Symposium. IEEE, 2011, pp. 66–71.

[28] R. Jiménez, F. Prieto, and V. H. Grisales, "Detection of the tiredness level of drivers using machine vision techniques," in Electronics, Robotics and Automotive Mechanics Conference. IEEE, 2011, pp. 97–102.

[29] J. He, S. Roberson, B. Fields, J. Peng, S. Cielocha, and J. Coltea, "Fatigue detection using smartphones," Journal of Ergonomics, vol. 3, no. 03, 2013, pp. 1–7.

[30] T. Azim, M. A. Jaffar, and A. M. Mirza, "Fully automated real time fatigue detection of drivers through fuzzy expert systems," Applied Soft Computing, vol. 18, 2014, pp. 25–38.

[31] Y. Cao and B.-L. Lu, "Real-time head detection with kinect for driving fatigue detection," in International Conference on Neural Information Processing. Springer, 2013, pp. 600–607.

[32] D. L. Sturnieks, R. St George, and S. R. Lord, "Balance disorders in the elderly," Neurophysiologie Clinique/Clinical Neurophysiology, vol. 38, no. 6, 2008, pp. 467–478.

[33] C. J. Jones, R. E. Rikli, and W. C. Beam, "A 30-s chair-stand test as a measure of lower body strength in community-residing older adults," Research quarterly for exercise and sport, vol. 70, no. 2, 1999, pp. 113–119.

[34] A. B. Scholey, S. J. French, P. J. Morris, D. O. Kennedy, A. L. Milne, and C. F. Haskell, "Consumption of cocoa flavanols results in acute improvements in mood and cognitive performance during sustained mental effort," Journal of Psychopharmacology, vol. 24, no. 10, 2010, pp. 1505–1514.

[35] T. N. Tombaugh, "Trail making test a and b: normative data stratified by age and education," Archives of clinical neuropsychology, vol. 19, no. 2, 2004, pp. 203–214.

[36] H. Nagasaki, H. Itoh, K. Hashizume, T. Furuna, H. Maruyama, and T. Kinugasa, "Walking patterns and finger rhythm of older adults," Perceptual and motor skills, vol. 82, no. 2, 1996, pp. 435–447.

[37] L. Breiman, "Random forests," Machine learning, vol. 45, no. 1, 2001, pp. 5–32.

[38] "Nuitrack full body skeletal tracking software," https://nuitrack.com/.

Machine Learning Methods for Detection of Epileptic Seizures with Long-Term Wearable Devices

Salima Houta*, Pinar Bisgin*, and Pascal Dulich†,
*Digitalization in Healthcare, Fraunhofer Institute for Software and Systems Engineering, Dortmund, Germany,
Email: salima.houta@isst.fraunhofer.de, pinar.bisgin@isst.fraunhofer.de
†Department of Computer Science, Technical University of Dortmund, Germany
Email: pascal.dulich@udo.edu

Abstract—**The detection of epileptic seizures plays a major role in patient safety and therapy. Although several research projects on mobile seizure detection have already been conducted, there are still no approaches that are able to reliably detect different seizure types in the home environment. The challenge lies in the variety of symptoms of certain seizure types. The present research describes the project EPItect, which aims to detect epileptic seizures with the help of an In-Ear sensor and to set up a networking infrastructure to exchange medical data between relevant actors. We contribute a machine learning framework for the detection of epileptic seizures and exemplify the application using the example of detection of Generalized Tonic-Clonic Seizures using acceleration data from the In-Ear sensor.**

Keywords–Epilepsy; Seizures; EPItect; SUDEP; Automated seizure detection; Wearables; IHE; HL7; Accelerometer; Classification; k-NN.

I. INTRODUCTION

Epilepsies are among the most common neurological diseases worldwide. Depending on the degree of severity, affected persons can live a life with great restrictions on their autonomy. Characteristic symptoms are recurring epileptic seizures, which can be very stressful for the affected persons, relatives and carers due to the unpredictability of the time at which seizures occur, as well as the impairment of consciousness and the loss of control over different body functions. Among other things, the mortality of people with epilepsy is increased by a factor of 2-3 due to severe epileptic seizures (e.g., failure of the respiratory center) and seizure consequences (e.g., accidents, suffocation) [1][2]. The early detection of seizures can possibly help to take of appropriate safety measures for the person concerned and to reduce sudden unexpected death in epilepsy (SUDEP). By using technical solutions for better supervision (e.g., video cameras, pulse oximeters) or rooming-in of relatives the SUDEP incidence in an epilepsy center shows a decreasing trend between 1981 and 2016 [3]. In addition to such early detection, an accurate recording of the seizures also helps in the individual planning of the therapy. In order to reduce seizure frequency or, at best, to achieve complete seizure control, a central component of medical treatment is the suppression of seizures by medication. Proper documentation of epileptic seizures by patients or relatives plays an important role in coordinating therapy. The documentation can be done on paper or web-based seizure calendars (e.g., EPI-Vista®) [4]. However, previous studies show that approximately 50% of seizures are not documented and approximately two-thirds of patients provide incorrect data [5][6]. The main reasons for the faulty seizure documentation are, for example, the disturbed perception of one's own seizures, amnesia for seizure or later forgetting of the seizure that has taken place. The seizure documentation by relatives or caregivers is also prone to failure as relatives do not notice symptom-poor epileptic seizures [7].

In this paper we propose and validate a ML Framework using the example of acceleration data of long-term wearable devices. The paper is organized as follows: Section 2 presets a brief literature review about epileptic seizure detection. Section 3 describes the project EPItect including the components of the technical solution. Section 4 describes briefly the data set of the In-Ear sensor employed in our research. This section presets information regarding the methods used in this study. You can find, also, information related to the performance evaluation criteria employed. Section 5 provides the assessment procedures used and the experimental results obtained. Finally, Section 6 describes the conclusion derived from the study and some thoughts with regard to future work.

II. RELATED WORK

Recently, various models have been proposed for the detection of epileptic seizures. Continuous electroencephalographic (EEG) monitoring is the current gold-standard for seizure diagnosis. Algorithms for the automatic detection of EEG-based seizures have been developed in various research projects [8][9]. However, EEG monitoring is expensive, time consuming and needs professional installation and observation. For the detection of epileptic seizures in the home environment mobile sensors are needed. The knowledge about the symptomatology of epileptic seizures can help to select the fittest seizure detection device for each seizure type.

Ulate-Campos et al. [10] literature study showed the effectiveness of various seizure detection devices for certain seizure types. Generalized Tonic-Clonic Seizures (GTCS) produces bilateral, convulsive tonic contraction followed by generalized clonic muscle contractions. They also manifest with a loss of consciousness and marked autonomic distribances. The main findings for GTCS are movements and physiological signals like heart or respiratory rate [10]. By using accelerometry or electromyography, generalized tonic-clonic seizures can usually be easily identified as epileptic seizures based on their characteristic movement patterns and differentiated from everyday movements. To detect motor phenomena, accelerometers (wrist-worn or body-mounted instruments that measure changes in speed or acceleration) and electromyography devices (for example, mounted on the chest, measure electrical muscle activity) are used [11][12].

The following sections show some study results using various medical devices and sensors. Sensitivity for the detection of clonic, tonic, hypermotoric and generalized tonic-clonic seizures was between $80-90\%$ in several studies [13][14][15]. In [16], Lin et al. developed a small headband for epilepsy patients. The headband is connected to a smartphone and records EEG signals to detect seizures in real-time. Once a seizure is detected, the headband will trigger the apps on the smartphone to locate the patient. Sixteen features indexes (entropy and the powers of 15 frequency bands from 0Hz and 15Hz) were input into the classifier to verify the seizures occurrences. They used the linear classifier called Linear Discriminant Analysis (LDA). The LDA is often used to reduce the dimension by separating the data into different classes and minimizing the data distribution of the same class in the feature space [16]. In another study, Arends et al. [17] employed epileptic nocturnal seizure detection by combining heart rate and movement. The sensor system is a bracelet that was fixed around the upper arm on the side where the seizures were known to start. The algorithm determined the heart rate. Simultaneously, a signal quality index was calculated for each heart rate value. If the signal quality index is $> 80\%$, seizure detection starts. Otherwise, the accelerometer is used to detect seizures. They obtained a signal quality of 94 % for the heart rate and up to 100% for accelerometry. This study demonstrated that it was possible to reliably detect major motor seizures using a combination of heart rate and accelerometer. In [18], an automatic seizure detection algorithm based on EEG, EMG and ECG with an overall detection sensitivity of 86% was developed. The average sensitivity of the developed algorithm depends on the seizure type an the diagnosis. The best result was achieved for the detection of focal seizures evolving to bilateral tonic-clonic. The described work involves small numbers of patients in the evaluation and is predominantly carried out in an experimental environment. A home environment was used in [17]. However, a limitation of seizure detection on nocturnal seizures was made.

The main contribution of this work is the Machine Learning Framework (ML-Framework) for detecting epileptic seizures in experimental and real term environments. Using the example of the acceleration data of the In-Ear sensor, the ML-Framework was initially applied unimodally. Current research is testing the application of multimodal approaches for seizure detection.

III. EPITECT

The focus of the project EPItect is to develop a non-invasive sensor system, which reliably detects those bio signals that enable automated detection of epileptic seizures. The sensor is placed in the external auditory canal (similar to a classic hearing aid). The data are made available to selected persons via mobile devices. In this way, the personal environment can also be included if necessary. This specially developed in-the-ear sensor technology and a networking infrastructure based on (inter-)national communication standards (e.g., Integrating the Healthcare Enterprise, Elektronische Fallakte 2.0, HL7 Fast Healthcare Interoperability Resources) are the basis for several IT applications, which are also integrated into the existing medical-nursing processes. In addition, the signals of the In-Ear sensor and the recorded data such as context information about seizures can give scientist much more reliable data to make better diagnosis, because the frequency and severity of seizures can be recorded better. The anonymization and cross-patient aggregation of the data also enables clinical research, for example regarding the drug that reduces the seizures most effectively or different context parameters, which trigger epileptic seizures.

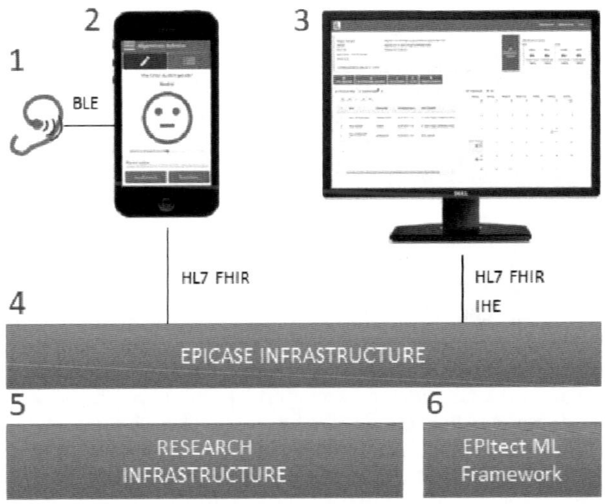

Figure 1. EPItect architecture.

The components of the technological solution are shown in Figure 1: the In-Ear sensor (EPISENS), the mobile application (myEPI), the portal (EPICASE Portal) and the networking infrastructure (EPICASE Infrastructure). EPISENS (1) includes sensors to optimize seizure detection and seizure counting. It sends vital data, raw data and alarm events via Bluetooth Low Energy to the myEPI App. myEPI App (2) is a mobile companion for the patient. The app includes an alarm module. Upon receipt of alarm events, selected persons (e.g., parents or partners of an affected person) should be informed. The patient can use a simple action on the smartphone to confirm the seizure event or classify it as a false alarm. This information is used in the next step to optimize the specificity of the algorithms developed. In the app, the patient also has the opportunity to collect additional data (contextual information on seizure events, mood, medication administration, side effects). He can selectively release data for doctors or relatives. The data is transmitted securely via the EPICASE infrastructure (4) and can be viewed by relevant actors via the EPICASE portal (3). The EPICASE portal is a case based communication portal for patients as well for professional and informal caregivers. It enables exchange of treatment-relevant data (e.g.,. medication order, medication administration, seizure documentation, diagnosis). The EPICASE infrastructure connects the IT applications. It is based on international standards and fully complies with data protection and data security requirements. The project EPItect also provides a research infrastructure (5) for pseudonymization, data capturing and integrating and storage of case based generated data. The integrated data is the basis for our machine learning framework (6).

The consortium of the project EPItect coordinated by the epileptologists of the University Hospital Bonn consists of five institutions and two associated partners in Germany: Department of Epileptology at the University Hospital Bonn, Fraunhofer Institute for Software and Systems Technology ISST, Department of Neuropediatrics of the University Kiel (UKSH), the North German Epilepsy Center in Schwentinental-Raisdorf,

Cosinuss GmbH Munich, the University for Healthcare Professions in Bochum, and the Epilepsy Bundes-Elternverband e.V. [National Epilepsy Parents Network] in Wuppertal.

IV. METHODS

This section first introduces the sensor platform and clinical study. Afterwards an overview of our ML Framework will be given. Then the framework will be applied to the example of acceleration data using the k-NN classification method.

A. Sensor Platform

The sensor system is based on a sensor concept called "earconnect" which was developed by cosinuss°. The sensor elements are integrated into a silicone screen, which is placed in the ear canal similar to a sports headphone. Behind the pinna sit microcontroller, power supply and the radio connection. There, the analog signals are digitized and then extracted by means of various filters and algorithms, the required vital data. A wearable sensor platform is designed in the EPItect project, which contains 3D accelerometer, PPG, and vital signs (heartrate, temperature). Acceleration data is suitable in order to detect tonic-clonic seizures that are characterized by severe motor symptoms [19]. The entire platform is fixed in the ear at the test subject (see Figure 2).

B. Study

For the evaluation of the technologies, clinical studies are carried out at the participating specialist clinics. The University Hospital Bonn and the Department of Neuropediatrics of the University Kiel have initiated a study with several patients. In the first phase, 170 patients have been recruited to test the biosensors. For these patients, EEG, ECG and In-Ear sensor data were collected over an average period of four days. On the basis of the EEG data, physicians have recorded seizures occurring (period, type of seizure). More than 490 seizures were recorded by January 2019. The data are used to identify relevant biosignals and biosignal patterns and to develop algorithms. Subsequently, the algorithms are validated with test data. This work focuses on tonic-clonic seizures.

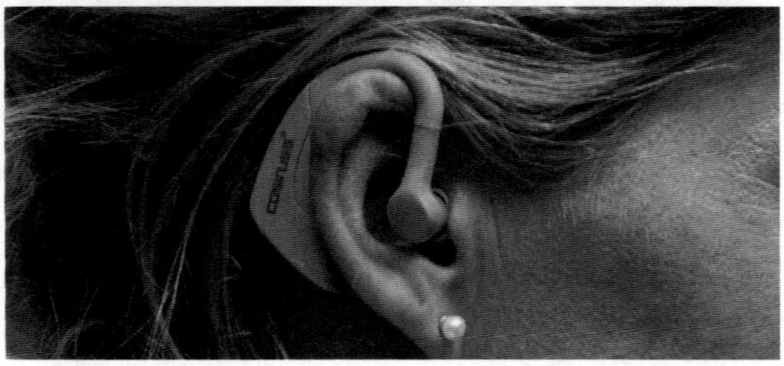

Figure 2. The In-Ear sensor ©cosinuss°.

C. ML-Framework

We would argue that the use of a structured experimental approach to the problem of seizure detection is useful to obtain the best possible results with all given data sets. In this section, an intelligent architecture for seizure detection based on the CRISP-DM [20] is presented. Figure 3 gives an overview in the steps of our EPItect ML-Framework. The ML-framework covers both the Experimental Environment and the Real Term Environment.

1) Experimental Environment: The Experimental Environment is used to develop and optimize seizure detection models using controlled conditions for the duration of the study. The main steps are: domain and data understanding, data preprocessing, feature extraction, model selection and evaluation. Domain understanding includes understanding the problem and the goal of the modelling. This means for example the understanding of the symptoms of an epileptic seizure. The understanding is obtained by literature review and by involving neurological experts and also affected persons in the project. A main step is to understand the data, which plays a major role. A detailed analysis of the data and the understanding of the goal would help to avoid later problems. The project integrates a variety of data sources: ECG, EEG, PPG, vital data (heart rate, temperature), seizure labeling and classification, patient meta data (gender, age). The data preparation includes tasks like selecting, cleaning, integrating and formatting data. The data integration and formatting are critical tasks for multimodal approaches. The feature extraction depends on the data and the objective of the model. A feature set can be selected by experts opinion or feature selection algorithms. We test several different feature selection approaches considering their selected classification models. The next step is the modelling process. To select models for the experimentation, a literature review and the identification of similar research activities that have previously been successful are required. Each selected model can be trained on according features and feature sets. A continual evaluation and adjusting of features and models will identify the best model. It is important to preserve the order of the training data for the seizure detection problem, so that upcoming classification are based on previous results. From this we get a new classification model. To evaluate the trained model, one would classify match results into seizure and non-seizure and then determine sensitivity, specificity, positive and negative predictive value. The implementation of a seizure detection system must be valid. The representation is performed using the confusion matrix to estimate the performance of learning algorithms and the generated classifiers. The confusion matrix records the correct and misclassifed features for each class. A comprehensive rating can be obtained from the Receiver-Operating-Characteristic curve. The ROC curve can be used to find the best possible value of a parameter [21] and to assess the trade-off between sensitivity and specificity [22]. The ROC analysis allows the rating of the classifier performance to be independent and complete rather than just accuracy.

2) Real Term Environment: Once a trained model is accepted, it is made available for use (Real Term Environment). For our project, this means deploying the models to mobile devices and sensors. Further training of the models, for example to take into account individual circumstances of the patient, is possible if the mobile devices and sensors provide good computing power and sufficient memory. Unlike the Experimental Environment, the verification of the model is done by the patients' labeling of alarm events which are triggered by the trained model. In order for the trained model to be accepted by the user, it is important that there is no high amount of false alarms. In a survey of patients and care environment (n = 305), we have found that on average a maximum of 2/10 false alarms are accepted. The activities in everyday life (for example: sports, activities that trigger emotional states such as excitement) have a great influence on the signals and possibly on the applicability of the trained model. Further training of the model is therefore an important task and will be addressed in a second study.

D. Seizure Classification

For data analysis, Weka toolkit was used. The Weka toolkit is a machine learning toolbox with many existing algorithms. It also allows to evaluate the algorithms for a particular dataset using cross validation. From the accelerometer data was the three vector component obtained as measurement variables.

1) Data Preprocessing: All observation data that does not involve tonic-clonic seizures are filtered.

2) Feature Extraction: Features were extracted from the raw accelerometer data using a window size of 128 and 50% overlapping between consecutive windows. At a sampling frequency of 50Hz, each window represents 3 seconds. For each measurement variable, the following quantities were derived in time and frequency space: (1) arithmetic mean [23], (2) median [24], (3) variance [23], (4) maximum, minimum and range [23] and (5) skewness and curtosis [25].

3) Classification: For our intended goal of embedded system classification we focused on classifiers that could be implemented in computationally efficient manner. There are a large set of classifiers. Our choice was to classify the data with the k-NN algorithm. The k-NN is an instance-based classifier based on majority voting of its neighbors: The k-NN algorithm finds a close group of k objects in the training dataset with the target object in the training data and predicts the class of closest objects to the target object [26].

4) Evaluation Metrics: The results can be evaluate by sensitivity (SEN), specificity (SPE), positive (PPV) and negative predictive value (NPV):

$$\text{SEN} = \frac{\text{TP}}{\text{TP} + \text{FN}} \qquad \text{SPE} = \frac{\text{TN}}{\text{TN} + \text{FP}} \tag{1}$$

$$\text{PPV} = \frac{\text{TP}}{\text{TP} + \text{FP}} \qquad \text{NPV} = \frac{\text{TN}}{\text{TN} + \text{FN}} \tag{2}$$

where TP, TN, FP, and FN denote true positives, true negatives, false positives, false negatives, respectively [27].

5) Computation time and memory requirements: Computational time and memory requirements are essential for the decision if an algorithm is suitable for a particular purpose. The software cannot work efficiently if they are too high.

V. Results

The experimental results of the k-NN algorithm with different k values of $\{1, 2, 3, 5, 10\}$ are summarized in Table 1. With $k = 1$ with the full feature set a sensitivity 65.1% is obtained. It has been shown that the k-NN algorithm has the potential to

TABLE I. k-NN ALGORITHM RESULTS.

k	*sens [%]*
1	65.1
2	52.4
3	60.6
5	54.4
10	45.9

recognize tonic-clonic seizures. While some seizures are well classified, distinguishing different types of seizures is more challenging and leads to lower classification performance. With the reduced sensor set, classification performance worsens as expected. The number of detected seizures are quite low. For a better classification performance it is required to combine different sensors to get a higher classification rate.

Specific memory and runtime tests must be performed to determine if k-NN is possible on the EPItect sensor.

VI. Conclusion and Future Work

With the use of acceleration data we detect tonic-clonic seizures with a sensitivity of 65.1% for $k = 1$ and 60.6% ($k = 3$). Following improvements will increase the classification performance: With a reduced feature set, classification performance worsens as expected. Using more features can improve the seizure detection. For this reason, many more features should be generated. From the large amount of features, feature selection methods can be used to reduce the optimal number of features to avoid the "curse of dimensionality" [28]. Die length of the clonic-tonic seizures vary from one to another. The fixed window size can be the reason for the misclassification. Therefore, the window size should be adjusted dynamically.

The optimization of the developed model is continued using the following steps feature extraction, model selection, classification and evaluation of the ML Frameworks to obtain a better trained model.

A. Combination of bio signals

In summary, the studies to date, despite their rather good sensitivity, do not yet show sufficient specificity for the correct recognition of non-epileptic events in order to make meaningful use of these methods for automated seizure counting. In addition, seizures with dominant motor phenomena were predominantly investigated. Given the variety of seizure symptoms, multimodal synchronous measurement and analysis of various body signals (e.g., simultaneous measurement of heart rate, skin resistance, and acceleration of limb movements) appears to be most promising to achieve high sensitivity and specificity in seizure detection.

Our current work addresses the implementation of multimodal approaches using the ML Framework. An important task is the combination of different data sources (e.g., ECG and PPG) for the development of models. We use PPG and ECG data to determine the pulse transit time and evaluate whether the pulse transit time has an impact on the model for the detection of epileptic seizures. The PTT is related to the blood pressure. It describes the time that a pulse wave needs to arrive [29]. The time of the heart contraction and the beginning of the pulse waves are needed to determine the blood pressure. In future work, we will analyze the PTT-data to identify epileptic seizures.

B. Clinical study under everyday conditions

In a subsequent clinical study (from January 2019), the technological solutions will be tested for everyday suitability. For this, 30 affected children and adolescents and 30 affected adults each receive an ear sensor and the mobile companion solution for one week each. The informal carers of the patient (e.g., parents), as well as the professional nurses received access to the EPICASE portal and can participate in the data exchange process. The study addresses the following subgoals.

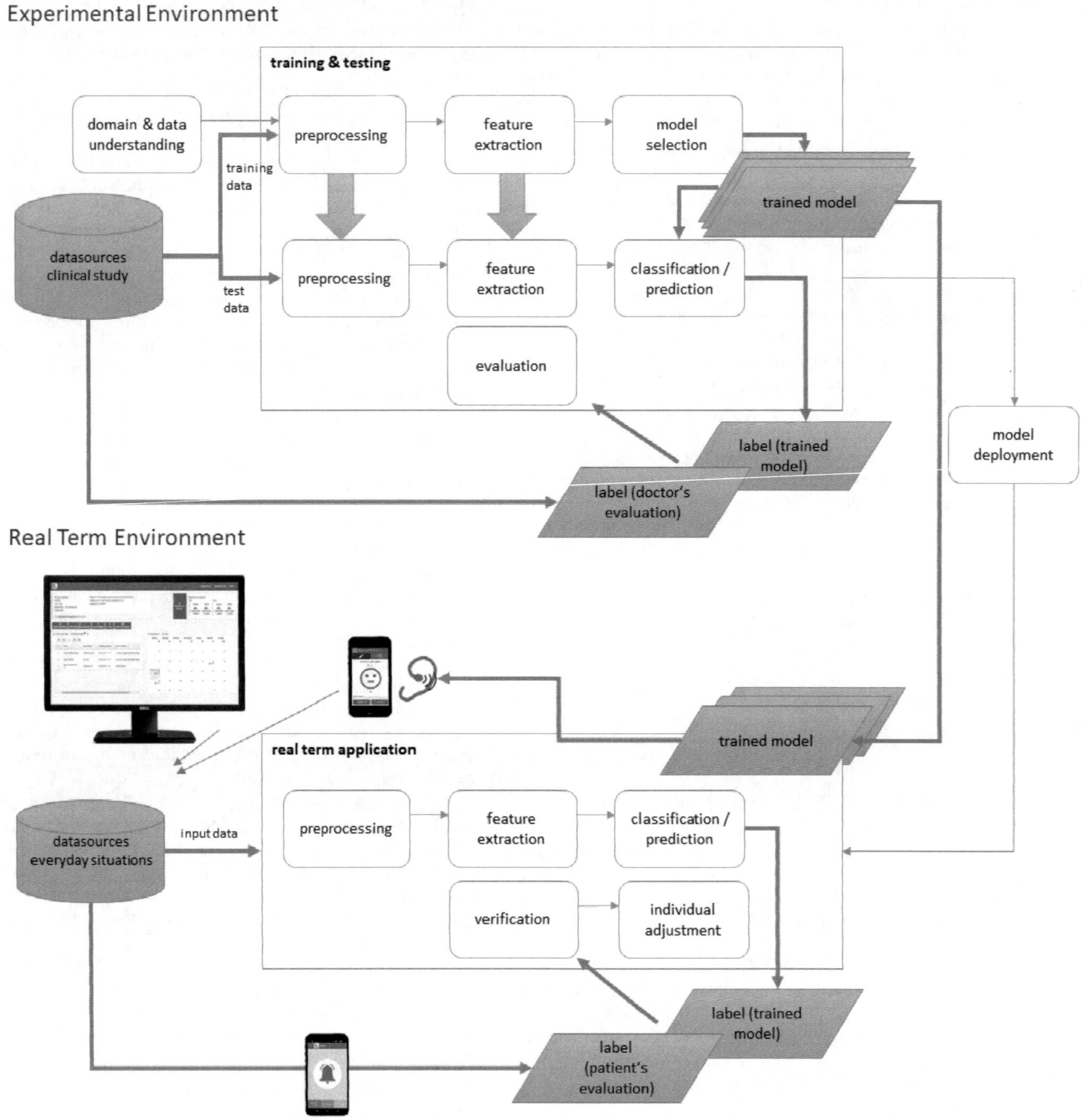

Figure 3. The ML-Framework.

1) Validation of the algorithms in everyday situations: It is to be examined whether the developed algorithms are applicable to everyday situations. Since the data was collected during the first phase while lying down or sitting, it is to be expected that the algorithm must be adjusted iteratively.

2) Effects of the technologies: The aim is to examine the impact of all technologies (sensor, app, portal) on automated and manual seizure records, quality of life and care processes.

ACKNOWLEDGMENT

The authors acknowledgement the financial support by the Federal Ministry of Education and Research of Germany in the framework of EPItect (project number 16SV7482).

REFERENCES

[1] L. Forsgren, E. Beghi, A. Oun, and M. Sillanpää, "The epidemiology of epilepsy in europe–a systematic review," European Journal of neurology, vol. 12, no. 4, 2005, pp. 245–253.

[2] M. Thom, "Neuropathology of epilepsy: epilepsy-related deaths and sudep," Diagnostic Histopathology, vol. 25, no. 1, 2018, pp. 23–33.

[3] R. Schulz, C. G. Bien, and T. W. May, "Decreasing sudep incidence in a tertiary epilepsy center between 1981 and 2016: Effects of better patient supervision," Epilepsy & Behavior, vol. 92, 2019, pp. 1–4.

[4] G. Rabending and U. Runge, "The electronic treatment diary for epileptic patients (epivista®) - a new instrument for therapeutical treatment," vol. 7, 2007, pp. 273–280.

[5] D. Blum, J. Eskola, J. Bortz, and R. Fisher, "Patient awareness of seizures," Neurology, vol. 47, no. 1, 1996, pp. 260–264.

[6] C. Hoope, A. Poepel, and C. E. Elger, "Epilepsy: accuracy of patient seizure counts," Achives of neurology, vol. 64, no. 11, 2007, pp. 1595–1599.

[7] A. C. Inan, M. A. Montenegro, S. Jacob, K. Eck, C. Chiriboga, and F. Gilliam, "Seizure frequency in children with epilepsy: factors influencing accuracy and parental awareness," Seizure, vol. 18, no. 7, 2009, pp. 524–529.

[8] J. Gotman, "Automatic seizure detection: improvements and evaluation," Electroencephalography and Clinical Neurophysiology, vol. 76, no. 4, 1990, pp. 317–324.

[9] Y. Park, L. Luo, K. K. Parhi, and T. Netoff, "Seizure prediction with spectral power of eeg using cost-sensitive support vector machines," Epilepsia, vol. 52, no. 10, 2011, pp. 1761–1770.

[10] A. Ulate-Campos, F. Coughlin, M. Gainza-Lein, I. S. Fernández, P. Pearl, and T. Loddenkemper, "Automated seizure detection systems and their effectiveness for each type of seizure," Seizure, vol. 40, 2016, pp. 88–101.

[11] J. Lockman, R. S. Fisher, and D. M. Olson, "Detection of seizure-like movements using a wrist accelerometer," Epilepsy & Behavior, vol. 20, no. 4, 2011, pp. 638–641.

[12] I. Conradsen, P. Wolf, T. Sams, H. B. Sorensen, and S. Beniczky, "Patterns of muscle activation during generalized tonic and tonic–clonic epileptic seizures," Epilepsia, vol. 52, no. 11, 2011, pp. 2125–2132.

[13] A. Van de Vel et al., "Long-term home monitoring of hypermotor seizures by patient-worn accelerometers," Epilepsy & Behavior, vol. 26, no. 1, 2013, pp. 118–125.

[14] A. Dalton, et al., "Development of a body sensor network to detect motor patterns of epileptic seizures," IEEE Transactions on Biomedical Engineering, vol. 59, no. 11, 2012, pp. 3204–3211.

[15] S. Beniczky, T. Polster, T. W. Kjaer, and H. Hjalgrim, "Detection of generalized tonic–clonic seizures by a wireless wrist accelerometer: a prospective, multicenter study," Epilepsia, vol. 54, no. 4, 2013, pp. e58–e61.

[16] S.-K. Lin, Y.-S. Lin, C.-Y. Lin, and H. Chiueh, "A smart headband for epileptic seizure detection," in Healthcare Innovations and Point of Care Technologies (HI-POCT), 2017 IEEE, 2017, pp. 221–224.

[17] J. Arends et al., "Multimodal nocturnal seizure detection in a residential care setting: A long-term prospective trial," Neurology, vol. 91, no. 21, 2018, pp. e2010–e2019.

[18] F. Fürbass et al., "Automatic multimodal detection for long-term seizure documentation in epilepsy," Clinical Neurophysiology, vol. 128, no. 8, 2017, pp. 1466–1472.

[19] T. M. Nijsen, R. M. Aarts, J. B. Arends, and P. J. Cluitmans, "Automated detection of tonic seizures using 3-d accelerometry," in 4th European Conference of the International Federation for Medical and Biological Engineering, pp. 188–191.

[20] C. Shearer, "The crisp-dm model: the new blueprint for data mining," Journal of data warehousing, vol. 5, no. 4, 2000, pp. 13–22.

[21] S. Chandra, R. Bhat, H. Singh, and D. Chauhan, "Detection of brain tumors from mri using gaussian rbf kernel based support vector machine," IJACT, vol. 1, no. 1, 2009, p. 46.

[22] J. Friedman, T. Hastie, and R. Tibshirani, The elements of statistical learning. Springer series in statistics New York, USA, 2001, vol. 1, no. 10.

[23] A. Mannini, D. Trojaniello, A. Cereatti, and A. M. Sabatini, "A machine learning framework for gait classification using inertial sensors: Application to elderly, post-stroke and huntingtons disease patients," Sensors, vol. 16, no. 1, 2016, p. 134.

[24] A. Subasi, "EEG signal classification using wavelet feature extraction and a mixture of expert model," Expert Systems with Applications, vol. 32, no. 4, 2007, pp. 1084–1093.

[25] Y. Huang, C. Yi, W. Peng, H. Lin, and C. Huang, "A study on multiple wearable sensors for activity recognition," in Dependable and Secure Computung, 2017 IEEE Conference on, 2017, pp. 449–452.

[26] R. Agrawal and B. Ram, "A modified k-nearest neighbor algorithm to handle uncertain data," in IT Convergence and Security (ICITCS), 2015 5th Conference on, 2014, pp. 1–4.

[27] J. Kirchner, S. Faghih-Naini, P. Agdemir, and G. Fischer, "Sensor selection for classification of physical activity in long-term wearable devices," in IEEE SENSORS, 2018, pp. 1–4.

[28] A. Mannini and A. M. Sabatini, "Machine learning methods for classifying human physical activity from on-body accelerometers," Sensors, vol. 10, no. 2, 2010, pp. 1154–1175.

[29] Y.-Z. Yoon et al., "Cuff-less blood pressure estimation using pulse waveform analysis and pulse arrival time," IEEE Journal of biomedical and health informatics, vol. 22, no. 4, 2018, pp. 1068–1074.

First Experiences Implementing Predictive Analytics Tools in a Clinical Routine Setting

Examples from the Heidelberg University Hospital

Oliver Klar
Department of Medical Information Systems
University Hospital Heidelberg
Heidelberg, Germany
oliver.klar@med.uni-heidelberg.de

Raluca Dees
Department of Medical Information Systems
University Hospital Heidelberg
Heidelberg, Germany
raluca.dees@med.uni-heidelberg.de

Rasim Atakan Poyraz
Department of Medical Information Systems
University Hospital Heidelberg
Heidelberg, Germany
atakan.poyraz@med.uni-heidelberg.de

Gerd Schneider
Department of Medical Information Systems
University Hospital Heidelberg
Heidelberg, Germany
gerd.schneider@med.uni-heidelberg.de

Oliver Heinze
Department of Medical Information Systems
University Hospital Heidelberg
Heidelberg, Germany
oliver.heinze@med.uni-heidelberg.de

***Abstract*—The rise of Artificial Intelligence(AI) is ubiquitous. In healthcare it is seen as a key technology supporting clinicians in their daily routine. The PART research project (Predictive Analytics of Robustness Testing) aims to develop an AI driven system which has the focus on networked device monitoring, profitability analysis and predictive maintenance in a clinical environemt. However before thinking about sophisticated AI algorithms at Heidelberg University Hospital we experienced a variety of difficulties according to medical device data acquisition from various manufactures and data protection which have to be solved first. This paper focuses on those difficilitues in a very early stage of the project and makes suggestions for suitable solutions.**

Keywords-clinical artificial intelligence; artificial intelligence in healthcare; monitoring networked medical devices.

I. Background

The University Hospital Heidelberg is a maximum care center running most of its patient record systems electronically. An important part of the information systems architecture is the integration of medical devices especially in the operating room [1]. Having those devices integrated, it is important to know which device is up and running, which one has got a problem and needs maintenance. The Predictive Analytics for Robustness Testing (PART) project aims to develop a monitoring system for networked medical devices of various manufactures in everyday clinical practice. The focus is on profitability analysis, predictive maintenance and monitoring of networked medical devices using data mining strategies. The monitoring system should manage a wide range of devices from various manufacturers within the clinic. This work presents first experiences in designing and implementing such a vendor-independent monitoring system facing the real world setting of a university hospital. Subsequently, the complexity of the development will be briefly outlined from the perspective of a hospital.

II. Methods & Results

There are several obstacles according to medical devices, device data and data protection that have to be taken in advance of realizing a powerful monitoring system. First, one has to address problems caused by heterogeneity. Medical devices are mostly, due to reasons of independence, from different manufacturers. This ranges from infusion pumps to the latest CT or MRI scanners. Even though there are standards for networked medical devices in operation rooms (e.g., IEEE 11073) [2][3], the communication of these medical devices works mostly via proprietary interfaces and protocols and manufacturers are very reluctant disclosing those interfaces, or implementing given standards. Further, expensive medical devices like CT scanners usually have extensive maintenance contracts which include that maintenance, repair and service may only be performed by a service engineer of the manufacturer itself. Collecting relevant data from such medical devices, e.g., getting information about the condition and operating status is demanding. Gathering data by additional attached IoT sensors in a sterile environment like an operation room is under serve restrictions due to aspects like patient safety. This makes it very difficult to just gather data from each networked device and analyse it. In PART the current question is not which data mining algorithms fit the most for our needs, the question is where is the data coming from in the first place. Hence, we are looking in all directions and started working with simulated medical device data as well to get familiar with the data mining approaches. Another subject is data protection and privacy. By monitoring medical devices, collecting and analysing data it could be possible to draw conclusions about patients, treatment and the work of clinical personnel itself. This is sometimes seen very critically by the clinic staff and requires a close examination and further steps like anonymization of the medical device data. Although networked medical devices in clinical environments produce a high volume of data, it is quite challenging, as described above, to access, evaluate and generate added value from this data treasure. In order to develop a multi-vendor system that uses AI for monitoring, profitability analysis and predictive maintenance, one has to address those mentioned issues first.

III. Discussion and Outlook

At Heidelberg University Hospital our strategy is to work closely together with device manufactures to address the former mentioned issues like heterogeneity, proprietary interfaces and protocols and lack of medical device data itself. Further with the experience made, we have lowered our aspiration on our way to a AI driven system for predictive maintenance and start with simple descriptive analytics on simulated medical device data. When established, such a monitoring system has several benefits for the clinic. It should make the complete IT infrastructure more robust and stable. Detecting problems of networked medical devices in early stages or in a best case scenario, before they are going to happen, saves maintenance time and costs. Further, with workload statistics of the medical devices one gets a good tool for tracking the usage and can adapt the inventory accordingly.

Acknowledgment

This work is funded by the German Federal Ministry of Education and Research under reference number 16KIS0785 as part of the PART project. We thank all partners within the PART project.

References

[1] R. Dees, et al. „Implementing, Connecting, and Evaluating a Standard-Based Integrated Operating Room within a German University Hospital." *ACI Open, Vol. 02, Nr. 01, DOI: 10.1055/s-00034447Georg Thieme Verlag KG, Stuttgart, Deutschland*, Januar 2018.

[2] M. Kasparick, S. Schlichting, F. Golatowski and D. Timmermann. "Medical DPWS: New IEEE 11073 Standard for safe and interoperable Medical Device Communication." 2015: 223–228.

[3] M. Max, et al. "From SOMDA to application – integration strategies in the OR.NET demonstration sites." *Biomedizinische Technik. Biomedical Engineering*, Feb 01, 2018: 69-80.

Evaluation of Machine Learning Algorithms to Detect Irregular Health States in Wearable Sensor Generated Data

Reto Wettstein[1,2]
[1]Department of Medical Information Systems
University Hospital Heidelberg
Heidelberg, Germany
e-mail: reto.wettstein@med.uni-heidelberg.de

Christian Fegeler[2]
[2]Department of Medical Informatics
Heilbronn University of Applied Sciences
Heilbronn, Germany
e-mail: christian.fegeler@hs-heilbronn.de

***Abstract*—Wearable devices facilitate continuous monitoring of personal health data. However, automated health state analysis based on this data is challenging in various aspects. This work presents preliminary algorithm evaluation results for health state irregularity detection based on a continuous data sample collected by an in-ear heart rate and body temperature sensor. The results show that a One-Class Support Vector Machine could be suitable for the task.**

Keywords–Algorithm Evaluation; Anomaly Detection; Health State; Wearable Generated Data.

I. INTRODUCTION

Mobile devices like smart watches and fitness trackers are becoming an integral part of our lives. This facilitates continuous monitoring and analysis of personal health data outside of clinical environments [1]. There are already applications which use the ability of wearable devices for specific disease monitoring, like the heart arrhythmia detection functionality by the Apple Watch [2] or the Empatica Embrace 2 seizure detection bracelet [3]. But, none of them considers a person's overall health state. Based on this background, we have built a prototype of a real-time monitoring system for automated irregular health state detection [4]. The centerpiece of this system is a machine learning server component, deciding whether measurements are normal or indicate a change in a persons health state.

Interpretation of sensor health data is challenging. Physiological Response Patterns (PRP) depend on many factors like activities, the environmental context or demographic data and can change over time [5]–[7]. Therefore, PRPs can not be described in general terms. Algorithms have to be trained on a person related basis. Additionally, it is often difficult to collect and access irregular PRPs [8] and thus, this data is not available during training of algorithms. Furthermore, not only the accuracy of applied algorithms plays a key role, also sensitivity and specificity need to be taken into account to reduce, for example, alarm fatigue [9].

However, anomaly detection algorithms could be one type of algorithms used to classify individual PRPs as either normal or irregular. They have been successfully used in other domains (e.g., credit card fraud detection or measurement error detection) where irregular data is not available or can change over a period of time [10].

The objective of this work is to evaluate four anomaly detection algorithms in the context of wearable sensor generated health data. The aim is to verify whether anomaly detection algorithms, already successfully used in other fields than medicine, are suitable for the above mentioned system to detect irregularities in continuously measured health data like body temperature and heart rate.

The remainder of this work is structured as follows: Section II describes the approach for data collection, data preparation and evaluation of the selected algorithms. In Section III, the classification results of the algorithms are presented. Finally, a conclusion and an outlook about future work is given in Section IV.

II. METHODS

The anomaly detection algorithms have been selected so that they are based on different mathematical concepts. The selected ones are Local Outlier Factor, Isolation Forest, One-Class Support Vector Machine and Autoencoder. For assessment of these algorithms, a 72 hour-long data sample of a healthy 28 year old male subject (N = 1) was recorded using the prototype. Utilizing an in-ear sensor, the vital signs body temperature and heart rate were measured in 5 second intervals. For later division of the collected data into training and test set, the measurements were labeled according to the performed activity. After collection of the sample, the data was split into 2 minutes long time-series, having an overlap of 30 seconds (N = 6200). Since generation of irregular health data is not possible at the push of a button, the measurements during the activities sport, metro and eating were regarded as artificial irregularities. All the remaining measurements were considered as normal. Training of the algorithms was based on a data-driven approach supplement with two statistical features (i.e., mean and standard deviation of heart rate and body temperature). Only the normal data was used for training. Finally, the anomaly detection algorithms were evaluated on the artificial irregular data using confusion matrices to calculate the metrics accuracy, sensitivity and specificity.

III. RESULTS

The performance evaluation of the algorithms was done in an overall setting and individually for each type of irregular data. In the overall setting (see Table I), the algorithms performed with an accuracy higher than 80 %. With the exception of the Isolation Forest, specificity was higher than sensitivity. All algorithms showed a specificity higher than 88 %. For sensitivity, the algorithms reached results better than 76 %. The best overall results were achieved using the One-Class Support

TABLE I. OVERALL RESULTS SHOWING THE CONFUSION MATRIX (−1 IRREGULAR, +1 NORMAL), ACCURACY, SENSITIVITY AND SPECIFICITY OF EACH ALGORITHM USING ALL TYPES OF IRREGULAR DATA COMBINED.

		LOCAL OUTLIER FACTOR		ISOLATION FOREST		ONE-CLASS SVM		AUTO-ENCODER	
Confusion Matrix		−1	+1	−1	+1	−1	+1	−1	+1
	−1	142	22	148	16	144	20	125	39
	+1	14	150	24	140	13	151	19	145
Accuracy		89.02 %		87.80 %		89.94 %		82.32 %	
Sensitivity		86.59 %		90.24 %		87.80 %		76.22 %	
Specificity		91.46 %		85.37 %		92.07 %		88.41 %	

TABLE II. SPECIFIC RESULTS SHOWING THE CONFUSION MATRIX (−1 IRREGULAR, +1 NORMAL), ACCURACY, SENSITIVITY AND SPECIFICITY FOR EACH ALGORITHM USING EACH TYPE OF IRREGULAR DATA INDIVIDUALLY.

			SPORT		METRO		EATING	
LOCAL OUTLIER FACTOR	**Confusion Matrix**		−1	+1	−1	+1	−1	+1
		−1	63	1	37	15	42	6
		+1	6	58	5	47	3	45
	Accuracy		94.53 %		80.77 %		90.63 %	
	Sensitivity		98.44 %		71.15 %		87.50 %	
	Specificity		90.63 %		90.38 %		93.75 %	
ISOLATION FOREST	**Confusion Matrix**		−1	+1	−1	+1	−1	+1
		−1	64	0	51	1	33	15
		+1	9	55	8	44	7	41
	Accuracy		92.97 %		91.35 %		77.08 %	
	Sensitivity		100 %		98.08 %		68.75 %	
	Specificity		86.15 %		84.62 %		85.42 %	
ONE-CLASS SVM	**Confusion Matrix**		−1	+1	−1	+1	−1	+1
		−1	64	0	45	7	35	13
		+1	5	59	5	47	3	45
	Accuracy		96.09 %		88.46 %		83.33 %	
	Sensitivity		100 %		86.54 %		72.92 %	
	Specificity		92.19 %		90.38 %		93.75 %	
AUTO-ENCODER	**Confusion Matrix**		−1	+1	−1	+1	−1	+1
		−1	64	0	29	23	32	16
		+1	7	57	6	46	6	42
	Accuracy		94.53 %		72.12 %		77.08 %	
	Sensitivity		100 %		55.77 %		66.67 %	
	Specificity		89.06 %		88.46 %		87.50 %	

Vector Machine with 89.94 % accuracy, 87.80 % sensitivity and 92.07 % specificity.

Regarding each type of irregular data individually (see Table II), the activity sport was identified best in all four algorithms. The Local Outlier Factor and the Autoencoder performed better for the type eating than for the type metro. The One-Class Support Vector Machine and the Isolation Forest reached better results for the type metro than for the type eating.

IV. CONCLUSION AND FUTURE WORK

This work shows preliminary results regarding the ability of anomaly detection algorithms to classify PRPs of activities collected by wearable devices as either normal or irregular. The use of these algorithms in combination with the prototype could potentially enable individuals to identify aggravations of their health earlier and thus, seek medical attention earlier.

The advantage of the selected algorithms is that they consider the great inequality in the distribution of the two kinds of data, normal and irregular. By changing the deciding threshold between normal and irregular time-series measurements, it is possible to take influence on the sensitivity and specificity of an algorithm. This is of most importance in medical applications and allows to change the focus between not missing any true positive or not having to many false positive classifications. For specific applications, this tradeoff would have to be individually reviewed. The most promising results for prototype use were shown by the One-Class Support Vector Machine. However, the other algorithms have also shown positive results, so that a majority vote could be considered, if computationally reasonable.

Further research is needed to assess whether the same results can be achieved for more subjects and if irregularities caused by an imminent or already occurring disease could also be detected with high accuracy, sensitivity and specificity. Additionally, adding more monitoring resources for vital signs, such as blood pressure and respiratory rate, could improve the results.

REFERENCES

[1] M. J. Bietz et al., "Opportunities and challenges in the use of personal health data for health research," Journal of the American Medical Informatics Association, vol. 23, no. e1, 2016, pp. e42–e48.

[2] Apple Inc., "Using Apple Watch for Arrhythmia Detection," 2018, URL: https://www.apple.com/healthcare/site/docs/Apple_Watch_Arrhythmia_Detection.pdf [accessed: 2019-01-31].

[3] Empatica Inc., "Embrace 2, Medical quality technology for epilepsy management," 2018, URL: https://www.empatica.com/en-eu/embrace2/ [accessed: 2019-01-31].

[4] R. Wettstein and C. Fegeler, "Real-Time Body Temperature and Heart Rate Monitoring System for Classification of Physiological Response Patterns using Wearable Sensor and Machine Learning Technology," Department of Medical Informatics, Heilbronn University of Applied Sciences, unpublished.

[5] C. Bouchard and T. Rankinen, "Individual differences in response to regular physical activity," Medicine and science in sports and exercise, vol. 33, no. 6 Suppl, 2001, pp. s446–s451; discussion s452–s453.

[6] Z. Obermeyer, J. K. Samra, and S. Mullainathan, "Individual differences in normal body temperature: longitudinal big data analysis of patient records," BMJ (Clinical research ed.), vol. 359, 2017, p. j5468.

[7] F. Shamout, D. Clifton, and T. Zhu, "Age- and Sex-Based Early Warning Score," 2017, URL: http://www.robots.ox.ac.uk/~davidc/pubs/transfer_fs.pdf [accessed: 2019-01-31].

[8] C. H. Lee and H.-J. Yoon, "Medical big data: promise and challenges," Kidney research and clinical practice, vol. 36, no. 1, 2017, pp. 3–11.

[9] S. Sendelbach and M. Funk, "Alarm Fatigue, a patient safety concern," AACN Advanced Critical Care, vol. 24, no. 4, 2013, pp. 378–386.

[10] H. P. Kriegel, P. Krger, and A. Zimek, "Outlier Detection Techniques," Tutorial at the 16th ACM International Conference on Knowledge Discovery and Data Mining, 2010, URL: http://www.dbs.ifi.lmu.de/~zimek/publications/KDD2010/kdd10-outlier-tutorial.pdf [accessed: 2019-01-31].

mHealth Solution for Remote Intrapartum Monitoring: A Feasibility Study

Hiwale Sujitkumar, Krishnan Navaneetha, Addepalli Jeevan Ram, and Ulman Shrutin
Philips Research India,
Philips Innovation Campus, Manyata Tech-Park,
Bengaluru, India-560045
e-mail: sujit.hiwale@philips.com

Abstract—**The primary objective of this study was to evaluate feasibility of a mHealth-based Digital Labour and Delivery Solution (DLDS) for information exchange among healthcare professionals for remote intrapartum monitoring and decision making. The inclusion criteria for the study were a live-singleton pregnancy with cervical dilatation ≥ 4 cm but < 8 cm at the time of admission, and presenting without any complication necessitating an immediate intervention. Throughout labour, all the subjects were monitored using conventional workflows. After each assessment, the in-charge doctor took one management decision from four possible options- (1) "Wait and watch"; (2) "Accelerate the labour"; (3) go for "Assisted vaginal delivery"; and (4) go for "Caesarean section". For each subject, clinical history, examination, and decision details were entered in the DLDS. A doctor located remotely was asked to use the DLDS application to review two records per subject and take one of the four management decisions. The effectiveness of the DLDS for intrapartum information exchange was evaluated by comparing the decisions taken by a remote doctor using the DLDS to that of decisions taken by a doctor in a labour room. In total, 110 subjects were enrolled for the study. The overall agreement between the two doctors for 220 independent decision points was 0.764 using unweighted Cohen's kappa and 0.723 using weighted Cohen's kappa statistic. The substantial agreement between the two doctors for intrapartum decision making demonstrates the feasibility of the DLDS for remote intrapartum monitoring and decision making. However, further investigation is required to assess effectiveness and safety of DLDS for a general purpose remote intrapartum monitoring.**

Keywords-feasibility study; inter-observer variability; intrapartum; mHealth; obstetrics; partograph, telemedicine.

I. INTRODUCTION

Intense monitoring and prompt decision making is very important during the intrapartum phase due to a short time interval between onset of complications and time to intervene. This is further emphasized by the fact that complications during this phase are responsible for almost 42% of maternal mortality and 23% of neonatal mortality [1]. Effective monitoring during the intrapartum phase needs an effective collaboration between the healthcare professionals. Among various issues, which have an adverse impact on teamwork during delivery, poor communication patterns have been identified as one of the most important issues. This is evident by a fact that issues in communication have been identified as the root cause in 72% of total cases related to infant deaths and injuries during delivery [2].

Poor communication is usually a result of a poor transmission or exchange of information. Paper based methods and telephonic communication are two conventional methods of communication during intrapartum monitoring. However, these methods have limitations when it comes to clear and real-time information exchange during intrapartum and have been shown to be either inadequate or cumbersome for this purpose [3][4].

To improve communication during intrapartum care, many tools such as partograph, digital partograph have been introduced. These are shown to be effective but underutilized due to time constrain and a stiff learning curve [5]–[8]. To standardize telephonic communication, techniques such as Situation–Background–Assessment–Recommendation (SBAR) have been proposed. However, SBAR is a difficult technique to learn and practice, and requires extensive education and training for effective implementation [9]. Moreover, none of these techniques provide an integrated solution for intrapartum monitoring, making them of limited use for a conventional setup.

Structured and instant communication are areas where Information and Communications Technology (ICT) can play a major role. In recent times, mobile devices (smartphone and tablet) have emerged as one of the most important enablers of ICT in healthcare. Considering the need gaps in intrapartum communication and the potential of mobile devices for telehealth, we have designed a Digital Labour and Delivery Solution (DLDS). DLDS is a tablet-based solution designed for systematic information gathering and sharing during intrapartum monitoring.

The primary objective of this study was to evaluate feasibility of the DLDS in information exchange among health care professionals for remote intrapartum monitoring and decision making. We had a primary hypothesis that a remote doctor can be equally adept at decision making if he is provided with all the necessary information. The effectiveness of the DLDS was evaluated by comparing the decisions taken by a remote doctor using intrapartum information provided by the DLDS to that of decisions taken by a doctor in a labour room (in-charge doctor).

The rest of the paper is arranged as follows, Section II provides details of the study protocol and statistical methodology. The study results are summarized in Section III. Section IV provides commentary on overall results and their possible implications for clinical practice. Section V concludes with the most important findings and future work directions.

II. MATERIAL AND METHODS

This section provides details about the study protocol and statistical methodology.

A. *Study design*

This observational study was conducted in a medical college hospital in Mysuru (Mysore), India in 2016. Inclusion criteria for the study were a live-singleton pregnancy with cervical dilatation $\geq$ 4 cm but < 8 cm at the time of admission to a labour room. All the cases with planned caesarean section or cases with complication(s) or indication(s), which require immediate intervention or where a trial of labour is contraindicated were excluded. The study was conducted in accordance with local regulations after approval of an institutional review board. Subjects were enrolled only after obtaining informed consent in writing

B. *Study protocol*

All the enrolled subjects were managed as per the established clinical workflows and protocols of the hospital. The subjects were regularly assessed by an in-charge doctor (doctor involved in active management of a subject). After each assessment, the in-charge doctor took one management decision from four possible options- (1) "Wait and watch", i.e., to continue the expectant management without any active intervention; (2) "Accelerate the labour", i.e., accelerate the labour process either by means of artificial rupture of membranes or by medication; (3) go for "Assisted vaginal delivery", i.e., use of forceps or vacuum extraction method for delivery; and (4) go for "Caesarean section".

All the subjects and newborns were monitored up to 24 hours after delivery for any adverse outcomes. Outcomes monitored included obstructed labour, uterine rupture, post-partum haemorrhage, stillbirth, early neonatal mortality, Apgar score at five minutes, and newborn's admission to a Neonatal Intensive Care Unit (NICU).

For each subject, complete clinical history, examination, investigation details and management decision for each assessment were entered in the DLDS. To prevent any influence of the DLDS on clinical workflow and patient management an additional nurse (not actively involved with patient management) was appointed for data entry in the DLDS.

C. *DLDS application*

DLDS is developed as a monitoring and communication solution for labour, delivery and immediate post-partum care. DLDS is a tablet-based solution built on an Android platform and allows secured sharing of information over a Wi-Fi network. Its intuitive design and user interface allows systematic and easy entry of the past and present history, examination and investigation details of the patient with an option to customize entry fields. It also provides an advanced visualization for various clinical trends and partograph. DLDS can be used as a stand-alone delivery solution or could be integrated with maternal telehealth platforms such as Mobile Obstetrics Monitoring [10]. For the study, two DLDS tablets were used; the one in the labour room was designed to anonymize and securely transmit information to the other tablet over a wireless network connection.

D. *Workflow of the remote doctor*

A doctor who is not involved in the management of any of the study subjects was assigned as a 'remote doctor'. To ensure that there is no discrepancy in decision making due to skill and knowledge differences, a doctor with a similar profile as the in-charge doctor was selected as a remote doctor. The remote doctor was asked to use the second DLDS application to review case records (without management decision information) and enter one of the four management decisions in the DLDS. For each subject, the remote doctor reviewed the first record (at the time of admission) and the last record before any active intervention or delivery.

E. *Statistical analysis methodology*

The decisions taken by both the doctors for each case record were extracted from the two DLDS applications. The agreement between the in-charge doctor and the remote doctor on the four types of management decisions was assessed using the Cohen's kappa statistics. However, as different types of management decisions have different implications in clinical practice, it is important to study not only overall agreement between the two doctors but also an extent of disagreement for individual decisions. This is important as some decisions are closer to each other when compared to other decisions (e.g., a decision to go for "Caesarean section" is much closer to a decision to go for "Assisted vaginal delivery" in comparison to a decision of "Wait and watch").

As Kappa analysis does not account for the difference in decision types, weighted Kappa analysis was used for this purpose. The weights used to grade the differences in decisions are presented in Table I. The agreement scale proposed by Landis and Koch was used to grade the agreement between the doctors [11].

TABLE I. WEIGHT MATRIX FOR DECISION GRADING

Decision taken by in-charge doctor	***Decision taken by the remote doctor***			
	Wait and watch	***Accelera-te labour***	***Assisted vaginal delivery***	***Caesarea n section***
Wait and watch	0	1	2	3
Accelerate the labour	1	0	1	2
Assisted vaginal delivery	2	1	0	1
Caesarean section	3	2	1	0

All statistical analyses were done using Microsoft Office Excel-2016 and R (version 3.4.4).

III. Results

The study results are summarized in this section.

A. *Demographic characteristics of the study population*

In total, 110 subjects were enrolled for the study. The mean maternal age was 24.21 ± 2.69 year, with a mean body mass index of 24.48 ± 2.08 kg/m^2. The nulliparous women constituted 30.43% of the study population. Gestational age was in the range of 37 to 41.6 weeks (median = 39.55 weeks). The mean birth weight of the neonates was 3037.98 ± 345.25 g, with a range of 2320 g to 4040 g.

B. *Intrapartum monitoring and labour outcomes*

Throughout labour, all the subjects were monitored using the conventional workflows and protocols of the hospital. None of the cases had any significant antenatal complication. The average duration of labour was 7 hours 3 minutes (± 63 minutes). On an average, each subject was assessed 15.63 (± 0.518) times during labour, which comes out to be one assessment per 28 minutes. During each assessment, vital parameters, examination details and management decision for a subject were entered in the DLDS application.

Five cases were delivered by caesarean section. Two cases were delivered by forceps extraction method due to failure to progress. The rest of the cases were delivered vaginally. None of the cases had any adverse intrapartum or immediate postpartum outcome. All the neonates had Apgar score of eight or more at five minutes and none of them required admission to a NICU.

C. *Agreement between the two doctors for the management decisions*

The remote doctor was asked to review 220 records (two records per case) using the DLDS. The confusion matrix of the four management decisions taken by both the doctors is summarized in Table II. It was observed that for the "Wait and watch" decision the remote doctor was in a perfect agreement with the decisions of the in-charge doctor in 91.15% of total records; for "Accelerate the labour" this agreement was 88%. Agreements for "Assisted vaginal delivery" and "Caesarean section" were 50% and 20%, respectively. Nevertheless, small sample sizes in these two categorizes makes it difficult to draw a valid conclusion.

TABLE II. Decision agreement between the doctors

Decision taken by in-charge doctor	***Decision taken by the remote doctor (using DLDS)***				***Total***
	Wait and watch	***Accelera-te labour***	***Assisted vaginal delivery***	***Caesarea n section***	
Wait and watch	103	10	0	0	113
Accelerate the labour	11	88	0	1	100
Assisted vaginal delivery	1	0	1	0	2
Caesarean section	2	2	0	1	5
Total	117	100	1	2	220

The overall agreement between the two doctors for all the decisions combined was 0.764 using unweighted Cohen's kappa statistics. The weighted Cohen's kappa between the two doctors was 0.723.

IV. Discussion

One very important aspect of the modern intrapartum care is a teamwork approach towards pregnancy and delivery. Despite this, it has been observed that the intrapartum care is still troubled with poor team work. In a study conducted by Guise et al., they observed that less than 50% doctors and less than 37% nurses in labour rooms rated their teamwork as adequate [12], indicating the need to address this issue on a priority basis. The difference in knowledge, skill set and communication style among nurses and doctors has been shown to be responsible for this poor teamwork [13].

One very crucial part of communication is transmission or exchange of information in a structured way for effective decision making. Unfortunately, the existing modes of intrapartum communications are shown to be insufficient for this purpose. Telephonic communication is universally available and offers advantages of real time communication. However, it is highly skill and experience dependent [3] and can lead to misunderstanding due to miscommunication [14]. The paper-based methods are simple to follow but are non-standardized, static and prone to manual errors; making them a less reliable medium for information exchange [4][15].

Partograph, a paper-based tool has shown to be an effective method for monitoring labour progress but unfortunately, it is underutilized. Time constraints, staff shortage, lack of knowledge and negative attitude among healthcare providers are some of the obstacles, which are known to hinder an appropriate use of the partograph [5]–[7]. The digital alternatives to partograph are trying to solve these problems by having an easy way of partogram plotting and ability to share it remotely [8]. However, no single solution exists yet, which comprehensively addresses communication needs during intrapartum care. This is the area where ICT devices can play a major role. Emergence of affordable smartphones, increased computational capacity, wider coverage and faster data transfer speed have provided a further boost to the use of the mobile as a platform for delivery of healthcare services. DLDS is one such mHealth solution for intrapartum care.

The main objective of this study was to evaluate feasibility of the DLDS application for remote intrapartum monitoring and decision making. This was done by comparing decisions taken by a remote doctor using the DLDS to that of the in-charge doctor. In this regard, a substantial agreement was observed between the two doctors for intrapartum decision making. This demonstrates the feasibility of the DLDS for remote intrapartum monitoring and decision making. However, as only seven cases were delivered by a non-vaginal route, it is difficult to generalize findings of this study to mode of deliveries other than vaginal.

It was observed that the agreement between the doctors for non-operative mode of deliveries was significantly higher than for operative delivers. This finding is in line with the published literature, where complete agreement for caesarean section decision has been observed to be about 65% [16]. Nevertheless, the lower agreement for operative deliveries (in particular more decisions of "Assisted vaginal deliveries" and "Caesarean section" by the in-charge doctor) needs further investigation. This could be due to the remote doctor missing some crucial information or the doctor in-charge getting negatively influenced by real-life factors such as stress of other emergencies to attend, lack of sleep, or pressure from the healthcare workers or patients.

Small sample size from a single center and recruitment of just one doctor in the labour room and one for remote assessment are two important limitations of our study. However, as this was a feasibility study we first wanted to test and verify our concept before conducting a large study with multiple doctors. Despite having a small sample size, we compared 220 independent decisions points between the two doctors. Furthermore, as none of the cases in our study had any adverse outcome, it was not possible to assess adequacy and quality of information provided by the DLDS to the remote doctor in such situations. Nevertheless, it was observed that the remote doctor could use the DLDS application for decision making for all the study cases.

On the study design, the use of an additional nurse for data entry is likely to have contributed to better and more comprehensive data gathering, which may not have been possible in conventional workflows. However, having a complete and accurate data entry is prerequisite for any digital solution and it is bound to have some change in the existing workflow. It also brings the advantage of enhanced patient safety by improving the communication, comprehensiveness, and organization of patient notes [17]. Moreover, it has been also indicated that introduction of digital records are likely to reduce risk and liability for obstetric providers, especially in the intrapartum care [18].

V. CONCLUSION

The strength of this study lies in being one of the first studies where the feasibility of a telehealth solution for remote intrapartum monitoring and decision making has been studied systematically. The finding of this study could serve as an important input for further research in this area. In the future, we would like to extend this work on a larger sample size with recruitment of more remote doctors.

To conclude, our study has demonstrated a substantial agreement in the decisions taken by a remote doctor using the DLDS and intrapartum decisions taken by a doctor in a labour room. This supports the hypothesis that it is possible to remotely monitor intrapartum labour progress and take appropriate decisions if a remote doctor is provided with all necessary information. It further supports use of telehealth solutions such as DLDS for remote intrapartum monitoring. Considering limited resources and shortage of trained healthcare workers in the developing countries, we believe that there is a huge need for intrapartum telehealth solutions in such countries.

ACKNOWLEDGMENT

We duly acknowledge help of the staff of JSS, Medical college, Mysuru, India for this study.

REFERENCES

[1] J. E. Lawn *et al.*, "Two million intrapartum-related stillbirths and neonatal deaths: where, why, and what can be done?," *Int. J. Gynaecol. Obstet. Off. Organ Int. Fed. Gynaecol. Obstet.*, vol. 107 Suppl 1, pp. S5-18, S19, Oct. 2009.

[2] J. Commission, "Sentinel Event Alert, Issue# 30, Preventing Infant Death and Injury During Delivery," *Wash. DC Jt. Comm.*, 2004.

[3] P. Giesen *et al.*, "Safety of telephone triage in general practitioner cooperatives: do triage nurses correctly estimate urgency?," *Qual. Saf. Health Care*, vol. 16, no. 3, pp. 181–184, 2007.

[4] T. Schabetsberger *et al.*, "From a paper-based transmission of discharge summaries to electronic communication in health care regions," *Int. J. Med. Inf.*, vol. 75, no. 3–4, pp. 209–215, Mar. 2006.

[5] "World Health Organization partograph in management of labour. World Health Organization Maternal Health and Safe Motherhood Programme," *Lancet*, vol. 343, no. 8910, pp. 1399–1404, Jun. 1994.

[6] T. Lavender, A. Hart, and R. M. D. Smyth, "Effect of partogram use on outcomes for women in spontaneous labour at term," *Cochrane Database Syst. Rev.*, vol. 7, p. CD005461, 2013.

[7] S. Tayade and P. Jadhao, "The impact of use of modified who partograph on maternal and perinatal outcome," *Int. J. Biomed. Adv. Res.*, vol. 3, no. 4, pp. 256–262, 2012.

[8] H. Underwood, J. Ong'ech, G. Omoni, S. Wakasiaka, S. R. Sterling, and J. K. Bennett, "Improving partograph training and use in Kenya using the partopen digital pen system," in *Biomedical Engineering Systems and Technologies*, Springer, 2014, pp. 407–422.

[9] B. B. Pope, L. Rodzen, and G. Spross, "Raising the SBAR: how better communication improves patient outcomes," *Nursing2014*, vol. 38, no. 3, pp. 41–43, 2008.

[10] "Philips - Mobile Obstetrics Monitoring Maternal telehealth software," *Philips*. [Online]. Available: https://www.philips.co.id/healthcare/product/HC867055/mobile-obstetrics-monitoring-maternal-telehealth-software. [Accessed: 09-Jan-2019].

[11] J. R. Landis and G. G. Koch, "The measurement of observer agreement for categorical data," *Biometrics*, vol. 33, no. 1, pp. 159–174, Mar. 1977.

[12] J.-M. Guise and S. Segel, "Teamwork in obstetric critical care," *Best Pract. Res. Clin. Obstet. Gynaecol.*, vol. 22, no. 5, pp. 937–951, 2008.

[13] C. D. Beckett and G. Kipnis, "Collaborative communication: integrating SBAR to improve quality/patient safety outcomes," *J. Healthc. Qual.*, vol. 31, no. 5, pp. 19–28, 2009.

[14] E. Joffe, J. P. Turley, K. O. Hwang, T. R. Johnson, C. W. Johnson, and E. V. Bernstam, "Evaluation of a problem-specific SBAR tool to improve after-hours nurse-physician phone communication: a randomized trial," *Jt. Comm. J. Qual. Patient Saf.*, vol. 39, no. 11, pp. 495–501, 2013.

[15] K. Nagpal *et al.*, "Information transfer and communication in surgery: a systematic review," *Ann. Surg.*, vol. 252, no. 2, pp. 225–239, 2010.

[16] S. A. Pillai *et al.*, "Decisions to Perform Emergency Caesarean Sections at a University Hospital: Do obstetricians agree?," *Sultan Qaboos Univ. Med. J.*, vol. 16, no. 1, pp. e42-46, Feb. 2016.

[17] J. Cowan, "Clinical governance and clinical documentation: still a long way to go?," *Br. J. Clin. Gov.*, vol. 5, no. 3, pp. 179–182, 2000.

[18] J. Stausberg, D. Koch, J. Ingenerf, and M. Betzler, "Comparing paper-based with electronic patient records: lessons learned during a study on diagnosis and procedure codes," *J. Am. Med. Inform. Assoc.*, vol. 10, no. 5, pp. 470–477, 2003.

Towards a Unified Framework for Distributing Health Awareness Message Using Social Media Platforms

Elaf Alsisi
Manufacturing Department
Cranfield University
Milton Keynes, United Kingdom
email:e.alsisi@cranfield.ac.uk

Ahmed Al-Ashaab
Manufacturing Department
Cranfield University
Milton Keynes, United Kingdom
email:a.al-ashaab@cranfield.ac.uk

Hongmei He
Manufacturing Department
Cranfield University
Milton Keynes, United Kingdom
email:h.he@cranfield.ac.uk

Abstract— **The use of social media as a platform to increase people health awareness has increased considerable in the last recent years. Although a growing research has demonstrated the tremendous value of using social media for promoting health awareness among the public, there is lack of framework that guide health organisations to design and deliver effective health awareness. The aim of this paper is to firstly assess the current process of publishing health awareness messages health organisations and subsequently provide a suitable framework for disseminating such messages via social media platforms. To this end, an initial literature review of the current best practices of social media was carried out and then a survey was developed and administered to selected health organisations seeking their current practices of publishing health awareness message. Based on the analysis of their information, a proposed framework of publishing health awareness message was developed. Initial results show the adaptability and importance of a unified social media context-aware framework in disseminating health awareness message to public. This paper concludes by identifying directions for future research.**

Keywords— Social Media; Health Promotion; Health Awareness; E-health, healthcare

I. INTRODUCTION

The rapid increase number of people with lifestyle health problems and infectious diseases created considerable pressure and challenges on national health systems across the globe. The reactive strategy of providing more health services to treat the increased number of public appears to be less effective [1]. This has driven policymakers to search for more effective proactive strategies to face this enormous challenge. One strategy, inspired by Benjamin Franklin axiom that "an ounce of prevention is worth a pound of cure'', is public health communication – conveying messages to public increase health awareness. The importance of this reactive strategy is illustrated in the emergence of health communication as a scholarly field on it is own that separated itself from communication studies, health education and promotion and related disciplines [2].

One aspect of public health communication which has received increasing attention is the channels (media) through which health message can be effectively delivered to, and reach, a wide range of relevant audience. In this respect, health organisations have traditionally focused on the printed media (e.g., poster, leaflet, etc.) and traditional media (e.g., SMS, phone calls, etc.) as the media through which that transfer their health messages.

With increasing people usability of, and access, to the Internet, health organisations have relatively recently turned to social media to promote health awareness among the public [3]. Social media (e.g., Facebook, Twitter and Snapchat, etc.) have enabled greater space for collaboration and facilitate the creation and exchange of information [4]. However, although the use of social media to convey health message has been hailed as a valuable tool due to several advantages (e.g. cost-effective) over the conventional media, the use of this new form of public health communication pose challenges [5]. The increased number in types of social media (e.g., Facebook, Twitter) and content of health message in these media create unmanageable source of information can create confusion for individuals and affect their capacity to obtain, process, and understand the basic health information and services needed to make appropriate health decisions [5]. However, there is limited studies in the current literature. Therefore, there is an urgent need for transformative the design (in term of content) and media) to reach to a guiding framework to help health organisation to effectively deliver health message and fully obtain the advantages of social media.

The aim of this paper is firstly to assess the current process of publishing health awareness message in health organisations and subsequently to provide a suitable unified framework for disseminating such messages via social media platforms. To this end, an initial literature review of the current best practices of social media was carried out and then a survey was developed and administered to selected health organisations seeking their current practices of publishing health awareness message the survey has been conducted for three months. Based on the analysis of their information, a proposed unified framework of publishing health awareness message was developed.

The remainder of this paper is structured as follows. In the next section, a brief review of the literature on health communication and social media is provided. Section 3 analyses the current process of publishing health awareness message in health organisations. Section 4 presents a proposed framework for publishing health awareness message. The paper concludes and provides future research to the public directions in Section 5.

II. LITERATURE REVIEW

The public health communication has emerged as a modern strategy to change public health behaviour by raising the awareness of risk diseases. Public health communication refers to 'the scientific development, strategic dissemination, and critical evaluation of relevant, accurate, accessible, and understandable health information communicated to and from intended audiences to advance the health of the public' [2]. Thus, the process of communicating health awareness message starts by ensuring an accurate content of the message to effectively delivering the message. This involves collecting information on health diseases from multiple sources and then transform them into an informative message. Having designed the content of the message, the next step is to select an effective media to deliver it. Health organisations have traditionally focused on the printed media and traditional media (e.g., SMS, phone calls, etc.) as the media through which that transfer their health messages. In addition, mass media has been used to expose a high percentage of population to health promotion message. This media involves a variety of form to convey health messages, including television, newspaper, radio, booklets, billboards, posters, and leaflets.

The transformation from Web 1.0 to the second generation of the World Wide Web (Web 2.0) facilitates information sharing, interoperability, user-centred design and collaboration [4]. The development of Web 2.0 has led to the web applications that generate social interaction and exchange of information between individuals. Social media includes collaborative projects, blogs, content communities, social networking sites, virtual game worlds and virtual social worlds are examples of these web applications [6]. Social media applied to almost every aspect of people interaction and has recently been used in public health communication [5]. The development of e-health promotion via social media platforms plays a significant role in the contribution to reinforcement the public health at the level of individual and society [7]. Since the Internet has expanded accessibility to a wide range of data, its advancement to Web 2.0 has given a contributory place where everyone can create, participate and, interact with others [8]. Nowadays, social networking sites including Twitter, Facebook, and Instagram, are popular online platforms which deployed as a new means for delivering health awareness knowledge to the public [5][9]. It enables individuals or groups of people to communicate, interact with each other, share information, collaborate and exchange content [10][11]. It can be categorised into various types, including social networking sites (e.g. Facebook and LinkedIn), blogs (e.g., WordPress and Blogger), microblogging (e.g., Twitter), media sharing sites (e.g., YouTube, Instagram and Flickr) and Wiki (e.g., Wikipedia) [11][12]. Twitter allows users to communicate and interact with followers in messages called a tweet with a maximum length of 280 characters. A tweet can be included with a photo or a video and link to the primary source (URL) [13][14]. According to global Statics [15], Twitter has an estimated 330 million active users per month. There are many features that reinforce the level of engagement in Twitter, such as retweets, likes, mentions and replies [16].

III. THE CURRENT PROCESS OF PUBLISHING HEALTH AWARENESS MESSAGE

The current process of publishing health awareness message has been captured at research visits in 10 participating healthcare organizations located in Kingdom of Saudi Arabia, namely; Ministry of Health in Riyadh, Ohud Hospital in Madinah, King Saud Medical City in Riyadh, King Fahad Medical City in Riyadh, King Faisal Specialist Hospital & Research Centre in Riyadh, King Fahad Armed Forces Hospital in Jeddah, King Abdulaziz Medical City in Jeddah, King Fahad Hospital in Madinah, King Fahad Specialist Hospital in Dammam, and Saad Specialist Hospital in Khobar. Selection of these organizations is primarily due to these organizations having a robust organizational culture clearly identified by their employees. Also, these 10 organizations possess a highly tested and highly reliable medical information database that were collected by health educators and physicians. Finally, these organizations utilize a diverse means of spreading health awareness messages; hence, an added motivation for choosing these 10 organizations. A mixed approach was adopted in order to acquire the results of the process of publishing such messages. Firstly, structured interviews were conducted by the researcher in 10 organizations with employees working in different departments, namely; Health Education Department, Media Administration, and the administration of identity and production. Secondly, a questionnaire was designed based on the literature review and distributed by email where different types of question/item was included. Finally, using IBM SPSS Statistics 25, data analysis was performed that captured the data influence and impact. Typical questions in the questionnaire include the following: 1. what are the most common diseases that public need to be continuously aware of? 2. What are social media platforms that are in use in your organization? 3. Who does formulate the models of scientific material of the health awareness message? 4. Who is in charge of designing the health awareness message after receiving the scientific material? 5. Who is in charge of publishing the health awareness message after being designed by the responsible party? 6. Which of the following do the health education specialists use to evaluate the health awareness message on targeted people in short term? 7. How important and effective are the following activities when assessing the impact of the health awareness message in long term? 8. How frequent and effective using the following media representation in the content of health awareness message? 9. How frequent and effective using the following component in the content of health awareness message?

The process varied in terms of the health problem targeted by the awareness message (e.g., diabetes), the media of publishing health awareness message (e.g., traditional channels, social media, etc.), the writer of the scientific material health awareness message, the designer of the health awareness message, the publisher of the health awareness message, the evaluation of the impact of the health awareness message on society in the short and long term, the media content of the health awareness message (e.g., text, video, etc.) and the content elements of the health awareness message (e.g., organization's badge).

The health awareness messages designed by the participated health organisations targeted various health problems (see Figure 1). The most frequent health problems covered is obesity (98%) followed by diabetes (94%) and then high blood pressure (88%). It is clear that the most frequently covered problems represent the most common and life-threating problems. The least frequent health problems depression (64%), liver disease (64%) and eye disease (62%). The remaining health problems with their frequencies are illustrated in Figure 1.

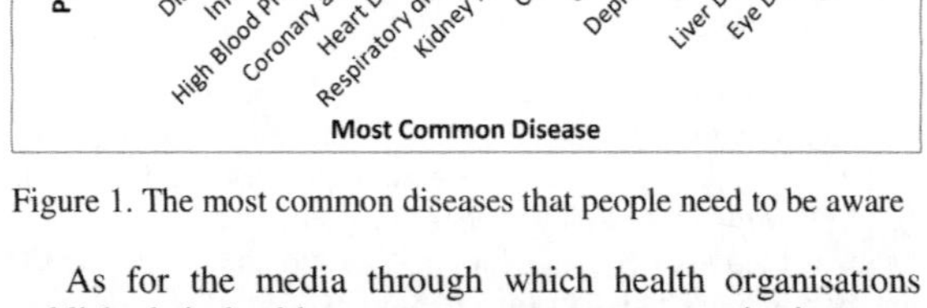

Figure 1. The most common diseases that people need to be aware

As for the media through which health organisations publish their health awareness message, organisations use three types of media namely, printed media (e.g., poster, leaflet, etc.), traditional media (e.g., SMS, phone calls, etc.) and social media (e.g., Facebook, Twitter and Snapchat, etc.). Our analysis revealed that the printed media techniques still dominated as media to publish health awareness message. However, social media platforms are gaining momentum as are being, on average, used more than traditional media. The participated health organisations reported that a wide range of social media platforms are being employed to deliver health awareness message to reach a large number of people (refer to Figure 2). The most frequently used social media platform is Twitter followed by Snapchat and then WhatsApp. On the other hand, the most effective platforms are Twitter, WhatsApp and Snapchat respectively based on the answers of the respondents. Noticeably, although Facebook is more effective than YouTube, it is being used less than YouTube. The remaining social media platforms and their associated frequency of use and effectiveness are presented in the diagram.

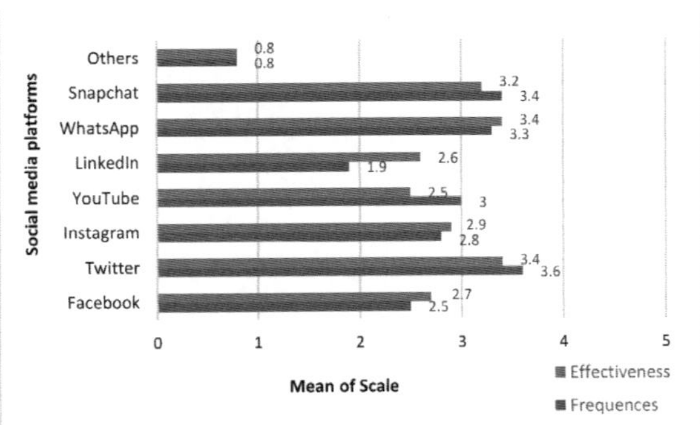

Figure 2. The frequency and effectiveness of social media techniques

The participated health organizations reported that the writer of the scientific material of the health awareness message can one from 10 different writers (see Figure 3). The most frequently used writer is certified health websites followed by health education specialist and then doctors/physicians (see blue bars). However, the most effective writer according to the sampled organizations is doctors/physicians, certified health websites and health education specialist, respectively. It can be observed that although is one of the most effective writers, the frequency of use is relatively low compared with certified health websites and health education specialist. The remaining writers with their frequency of use and effectiveness are shown in Figure 3.

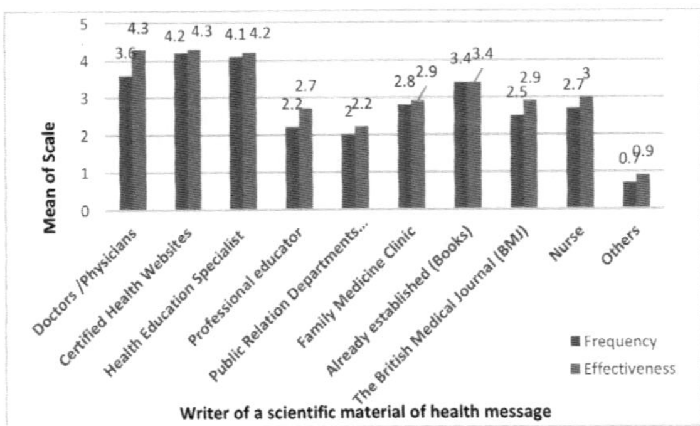

Figure 3. The writer of the scientific material of the health awareness message

The participated health organizations revealed that the design of the health awareness message is managed by different parties within and without these organisations (refer to Figure 4). The most frequent responsible party being adopted for designing the health awareness message is graphical designer followed by Health Education Department and then Advertising and Design Agency. Concerning their effectiveness, the more frequent used parties are the most effective parties as clearly depicted in Figure 4.

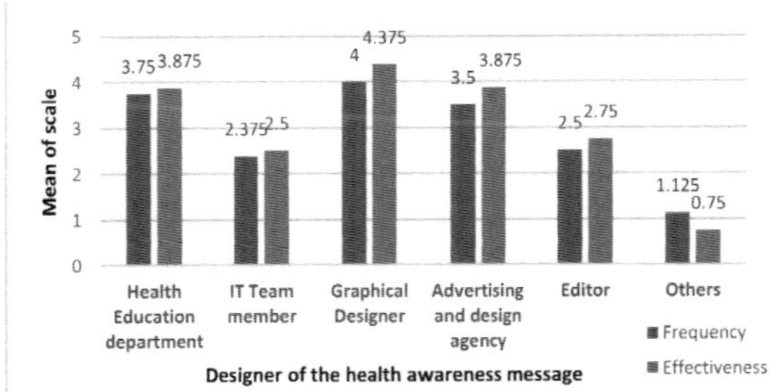

Figure 4. The designer of the health awareness message

The participated health organisations indicated that the publishing of the health awareness message can be accomplished by different parties (see Figure 5). The most frequent responsible party who in charge of publishing the health awareness message is Public Relation Departments followed by Health Education Department and then Doctors/Physicians and IT Team. As for the effectiveness of publishing through these parties, the results show the same pattern of distribution as the most frequently used. In other words, the most frequently used parties are the most effective parties.

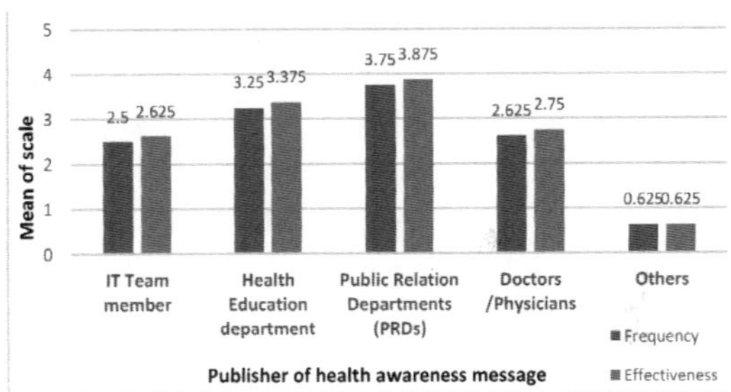

Figure 5. The publisher of the health awareness message

The participated health organisations use different evaluation methods of the impact of the health awareness message on society on both the short-term (see Figure 6) and long-term. The organisations give high importance to 'direct questions', 'immediate evaluation of teaching session', 'teach-back method' and 'seek feedback from the patient', respectively, as methods of evaluating the impact of the health awareness message on the short-term (see blue bars).

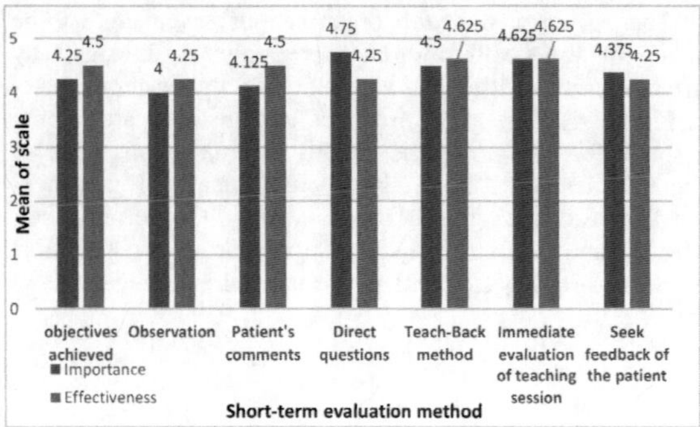

Figure 6. The evaluation methods of the impact of the health awareness message on society on the short-term

With respect to their effectiveness, it is clearly shown that 'teach-back method' and 'immediate evaluation of teaching session' followed by 'patient's comments' and then 'direct questions', 'observations' and 'seek feedback from the patient' (see red bars). In general, although 'objectives achieved' and 'patient's comments' evaluation methods are more effective compared to other methods, organizations attach less importance to them compared to less effective methods. As for the long-term evaluation of the impact of health awareness message, organizations use a variety of methods. Table 1 shows the importance of each method based on its influence on targeted people. The most emphasized means reported by the organizations was by checking the increased amount of number of people who are doing early detection of specific diseases. The remaining methods, such as using statistical methods, are relatively equally important for these organizations.

Table 1. Methods of evaluating the impact of the health awareness message

Number	Method	Importance
1	Use one of any statistical methods	3.5
2	The amount of drug consumed are reduced	2.62
3	Severity of disease's conditions is reduced	3.25
4	Diminishing incidence rates of the disease through addressing the risk factors causing the disease	3.62
5	Increase the number of people who are doing the early detection and of specific disease and its complications	4.12
6	Increase the number of people who are doing the comprehensive detection	3.62
7	Others	1.25

In term of the media content of the health awareness message, the participated health organisations use a variety of media contents (refer to Figure 7). Including a photo in the health awareness message was the most commonly used content by the organisations followed by text, cartoon and video, based on the mean of frequency (see blue bar). This was expected given that these media, in particular, fully capture individuals' attention. This is clearly evidenced when we asked about how effective these media contents. The respondents indicated that insert text, cartoon, and cartoon drawing are the most effective media content with the remaining are relatively equally effective (see red bar). Surprisingly, although adding a photo to the health message was the most frequently used, it was the least effective media content. The remaining media contents with their associated frequency of use and effectiveness are detailed in Figure 7.

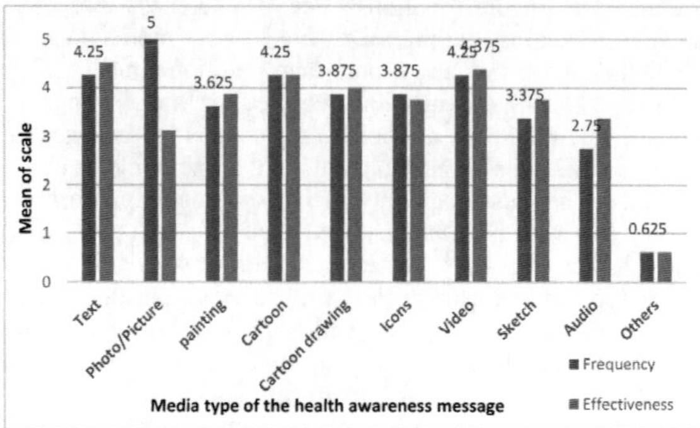

Figure 7. The media content of the health awareness message

Finally, the involved health organisations in this study reported that they use different content elements of the health awareness message (refer to Figure 8). Based on the mean of the respondents' answers, the most frequently used element is including the organization's badge and appropriate exercise.

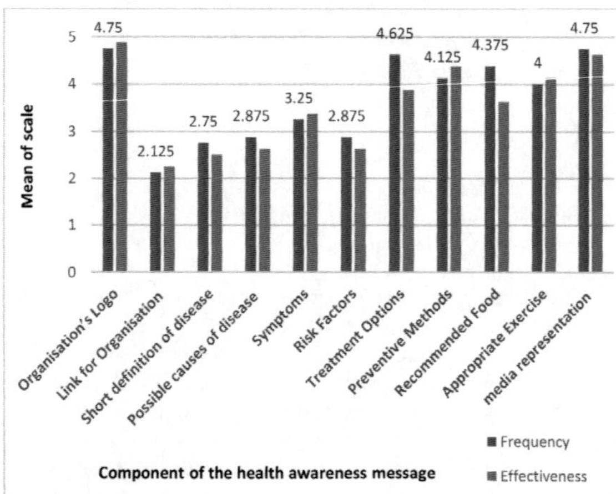

Figure 8. The content elements of the health awareness message

As for effectiveness, inserting an organization's badge and recommended food are the most effective elements according to the sampled organisations, while inserting link for the organisation's website is the least effective element. The remaining elements and their associated frequency of use and effectiveness.

IV. A PROPOSAL OF USING SOCIAL MEDIA TO SPREAD THE HEALTH MESSAGE

In the previous section, a detailed analysis of the current process of publishing health awareness message in 10 organizations was provided. This analysis is shown as an entity of the process that lead to develop a framework of distributing the health awareness message (see Figure 9). The analysis highlighted the need for solid and unified framework that guides health organizations in publishing health message using social media platforms. The analysis revealed that organizations often used multiple social platforms and parties

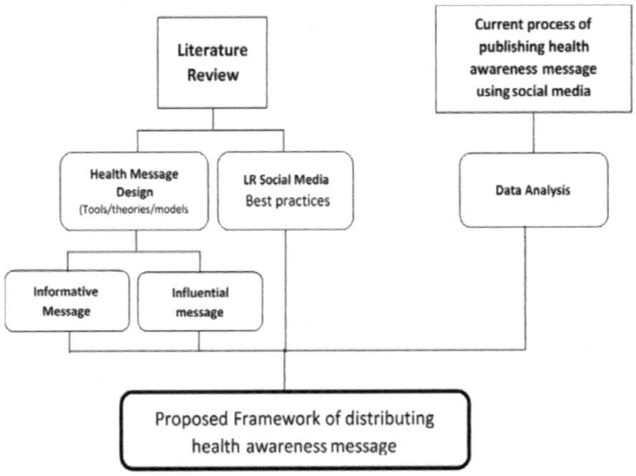

Figure 9. The process of developing the framework

to publish health organizations. Although this might provide more cover of a wide range of audience, it might generate information overload and thus affect the patients' abilities to seek information from the message. Although social media have considerable potential as effective platforms for health promotion and awareness, these media require careful application as may not always achieve their intended outcomes [17][18][19]. Therefore, publishing health awareness message should be a systematic and organized process [20][21].

In this study, an initial framework has been suggested in order to distribute the health awareness knowledge to public in an effective manner (refer to Figure 10). The framework suggests that publishing health awareness message should be organized and follow through seven stages with sub stages. In the first stage, the health disease or problem that needs health awareness message designed for should be determine. The determination process is based on health organizations awareness priorities set up and highlighted in their health communication strategy and plan. Once the health disease or issue is specified, the second stage is to collect authentic, accurate and relevant material on it. In this stage, information and material can be gathered from multiple sources such as certified health websites, health education specialist or doctors/physicians. In the third stage, the design stage, consists of two sub-stages: content design and technical design. In the content design, the relevant information on the health disease should be only selected and included in the message. This can also be accomplished by health education specialist or doctors/physicians. In the technical design, media content should be selected and included. Content media can be text, photo, video, audio, cartoon, and cartoon drawing. This should be based on the social media being used. However, the most effective media content are text, cartoon and cartoon drawings. In this stage, graphical designer or specialist can be used. In the fourth stage, the health organizations should decide on the responsible for publishing the message. IT department, health education department or public health department can be responsible for this stage. In the fifth stage, organizations can employ a wide range of social media platforms to deliver their health messages.

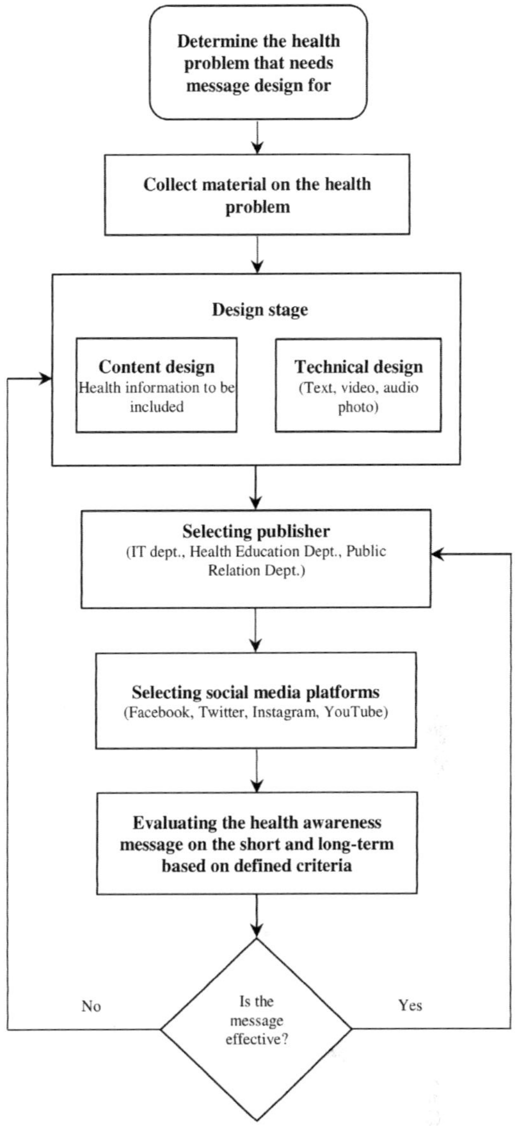

Figure 10. A proposed framework for publishing health awareness using social media

The most effective platforms are Facebook, Twitter, WhatsApp and Snapchat. Since Twitter has seen exponential growth in the use of health promotion, it is important to find out the factors that attribute tweets to be effective and impactful in delivering the health awareness message. Twitter allows users to communicate and interact with followers in messages called a tweet with a maximum length of 280 characters. A tweet can be included with a photo or a video and link to the primary source (URL) [13][14]. According to global Statics [15], Twitter has an estimated 330 million active users per month. There are many features that reinforce the level of engagement in Twitter such as retweets, likes, mentions and replies [16]. In the sixth stage, organizations evaluate the impact and effectiveness of the health awareness message on the short and long-term. On the short-term, organizations can use observations, patient's comment on social media or direct questions to assess the impact of health awareness message. On the long-term, organization can use one of any statistical methods, the amount of drug consumed is reduced, severity of disease's conditions is reduced or increased the number of people who are doing the early detection of specific disease and its complications to evaluate the effectiveness of the health awareness message. The results of the evaluation process determine the actions of health organizations regarding the health message. If the message was effective, the organizations can then contain publishing the message. However, if the message is ineffective, the organizations should go back and improve both the content and technical design of the message

V. CONCLUSION AND FUTURE WORK

This paper aims to enhance our understanding of the process of publishing health awareness message via social media platforms. To this end, a survey was administered to selected health organizations to understand the current state and practices of publishing health awareness message to the public. The analysis of current practices revealed that the process lacked a systematic and unified approach to publish health message. The health organizations varied considerable in terms of the social media platforms used, content writer, designer, publisher and content elements. Although the use of wide range of social media platforms may enable organizations to reach a wide number of audiences, it may affect their ability to receive and understand resulted from information overload. To address this, this paper developed a unified framework for publishing health awareness message.

The area of social media in health communication is still in its infancy [22][23][24]. Future research could explore factors influencing individual's intentions to use social media for health information. For example, one could draw on the Theory of Planned Behaviours to understand how attitudes, social pressure (family and friends) and behavioural control variables may influence their intentions to use and look for health information on social media platforms. Another research opportunity is to investigate, using experiments methods, attributes of effective health awareness message from the individual's perspective. This will inform the health policymakers and practitioners design of effective health message.

REFERENCES

[1] L. S. Suggs, C. McIntyre, W. Warburton, S. Henderson, and P. Howitt, "Communicating health messages: a framework to increase the effectiveness of health communication globally," Report of the WISH Communicating Complex, Institute of Global Health Innovation, Imperial College London, 2015.

[2] J. M. Bernhardt, "Communication at the core of effective public health," American Journal of Public Health, vol. 94, no. 12, pp. 2051-2053, 2004.

[3] W. Y. S. Chou, Y. M. Hunt, and B. W. Hesse, "S'ocial media use in the United States: implications for health communication," Journal of Medical Internet Research, vol. 11, no. 4, 2009.

[4] M. M. Al-Debei, E. Al-Lozi, and A. Papazafeiropoulou, "Why people keep coming back to Facebook: Explaining and predicting continuance participation from an extended theory of planned behaviour perspective," Decision Support Systems, vol. 55, no. 1, pp. 43-54, 2013.

[5] S. A. Moorhead et al., "A New Dimension of Health Care : Systematic Review of the Uses , Benefits , and Limitations of Social Media for Health Communication," Journal Of Medical Internet Research, vol. 15, no. 4, pp. 1–16, doi: 10.2196/jmir.1933, 2013.

[6] A. M. Kaplan, and M. Haenlein, "Users of the world, unite! The challenges and opportunities of Social Media," Business Horizons, vol. 53, no. 1, pp. 59-68, doi: 10.1016/j.bushor.2009.09.003, 2010.

[7] Y. Albalawi, and J. Sixsmith, "Identifying Twitter in fl uencer pro fi les for health promotion in Saudi Arabia," Health Promotion International, vol. 32, no. 3, pp. 456-463, doi: 10.1093/heapro/dav103, 2015.

[8] A. Benetoli et al., "Professional Use of Social Media by Pharmacists : A Qualitative Study," Journal Of Medical Internet Research, vol. 18, no. 9, doi: 10.2196/jmir.5702, 2016.

[9] C. Ferguson et al., "Social media : A tool to spread information : A case study analysis of Twitter conversation at the Cardiac Society of Australia & New Zealand 61st Annual Scientific Meeting 2013," Collegian Journal of the Royal College of Nursing Australia, vol. 21, no. 2, pp. 89–93, doi: 10.1016/j.colegn.2014.03.002, 2014.

[10] E. Brusilovskiy, G. Townley, G. Snethen, and M. S. Salzer, "Social media use, community participation and psychological well-being among individuals with serious mental illnesses," Computers in Human Behavior, vol. 65, pp. 232-240, doi: 10.1016/j.chb.2016.08.036, 2016.

[11] J. Salem, H. Borgmann, and D. G. Murphy, "Integrating Social Media into Urologic Health care: What Can We Learn from Other Disciplines?," Current Urology Reports, vol. 17, no. 2, pp. 1-7, doi: 10.1007/s11934-015-0570-2, 2016.

[12] F. J. Grajales, S. Sheps, K. Ho, H. Novak-Lauscher, and G. Eysenbach, "Social media: A review and tutorial of applications in medicine and health care," Journal of Medical Internet Research, vol. 16, no. 2, doi: 10.2196/jmir.2912, 2014.

[13] K. Wagner, *Twitter is testing a big change: Doubling the length of tweets from 140 to 280 characters.* [Online]. Available from: https://www.recode.net/2017/9/26/16364002/twitter-longer-tweets-character-limit-140-280 , 2017.

[14] G. B. Colbert, J. Topf, and M. A. Sparks, "The Social Media Revolution in Nephrology Education," Kidney International Reports, vol. 3, no. 3, pp. 519-529, doi: 10.1016/j.ekir.2018.02.003, 2018.

[15] Statista, *Most famous social network sites worldwide as of October 2018, ranked by number of active users (in millions).* [Online]. Available from: https://www.statista.com/statistics/272014/global-social-networks-ranked-by-number-of-users/, 2018.

[16] Y. A. Albalawi, and J. Sixsmith, "Exploring the diffusion of tweets designed to raise the road safety agenda in Saudi Arabia," Global Health Promotion, vol. 24, no. 4, pp. 5-13, doi: 10.1177/1757975915626111, 2016.

[17] H. Korda, and Z. Itani, "Harnessing social media for health promotion and behavior change," Health promotion practice, vol. 14, no. 1, pp. 15-23, doi: 10.1177/1524839911405850, 2013.

[18] C. Hawn, "Take two aspirin and tweet me in the morning: how Twitter, Facebook, and other social media are reshaping health care," Health affairs, vol. 28, no. 2, pp. 361-368, doi: 10.1377/hlthaff.28.2.361, 2009.

[19] M. L. Antheunis, K. Tates, and T. E. Nieboer, "Patients' and health professionals' use of social media in health care: motives, barriers and expectations," Patient Education and Counseling, vol. 92, no. 3, pp. 426-431, 2013.

[20] R. Thackeray, B. L. Neiger, A. K. Smith, and S. B. V. Wagenen, "Adoption and use of social media among public health departments," BMC Public Health, vol. 12, no. 242 , 2012.

[21] A. B. Heldman, J. Schindelar, J. B. Weaver, "Social media engagement and public health communication: implications for public health organizations being truly "social"," Public Health Reviews, vol. 35, no. 13, 2013.

[22] J. Fisher and M. Clayton, "Who gives a tweet: assessing patients' interest in the use of social media for health care," Worldviews on Evidence - Based Nursing, 9(2), 100-108, 2012.

[23] C. Hanson et al., "Use and acceptance of social media among health educators," American Journal of Health Education, vol. 42, no. 4, pp. 197-204, 2011.

[24] M. A. Wakefield, B. Loken, and R. C. Hornik, "Use of mass media campaigns to change health behaviour," The Lancet, vol. 376, no. 9748, pp. 1261-1271, 2010.

Remote Head CT Evaluation for Acute Stroke Diagnosis Using a Smartphone: Reliability and Diagnostic Equivalence with a Primary Medical Interpretation Workstation

Antonio Salazar
Laboratory of Telemedicine
University of Los Andes
Bogotá, Colombia
e-mail: ant-sala@uniandes.edu.co

Aníbal Morillo
Department of Diagnostic Imaging
University Hospital Fundación Santa Fe de Bogotá
Bogotá, Colombia
e-mail: ajmorillo@gmail.com

Manuel Granja
Lyerly Neurosurgery
Baptist Neurological Institute
Jacksonville, Florida, USA
e-mail: mfgranja@gmail.com

Oscar Torres
Department of Diagnostic Imaging
University Hospital Fundación Santa Fe de Bogotá
Bogotá, Colombia
e-mail: oscartoteles@yahoo.es

Nicolás Useche
Department of Diagnostic Imaging
University Hospital Fundación Santa Fe de Bogotá
Bogotá, Colombia
e-mail: jnuseche@gmail.com

Natalia Rueda
Department of Diagnostic Imaging
University Hospital Fundación Santa Fe de Bogotá
Bogotá, Colombia
e-mail: nrueda_ruiz@hotmail.com

Sonia Bermúdez
Department of Diagnostic Imaging
University Hospital Fundación Santa Fe de Bogotá
Bogotá, Colombia
e-mail: sonia.bermudez@radiologiafsfb.org

Brenda Ropero
Department of Diagnostic Imaging
University Hospital Fundación Santa Fe de Bogotá
Bogotá, Colombia
e-mail: bv.ropero47@uniandes.edu.co

***Abstract*— The aim of this study was to evaluate the equivalence of head computed tomography (CT) interpretations performed with either a diagnostic workstation or a smartphone in an emergency telestroke service. After institutional review board approval, a factorial design with 1504 interpretations was used (188 patients, 4 radiologists, and 2 reading systems). The variables evaluated included the following image findings: presence of hemorrhagic lesion, imaging contraindications for the administration of intravenous tissue plasminogen activator (tPA), ischemic lesion in the anterior cerebral artery (ACA), middle cerebral artery (MCA), and/or posterior circulation (PC) territory, hyperdense MCA, and a dichotomized score of the well-known Alberta Stroke Program Early CT Score (ASPECTS). The statistical equivalence between each variable was studied for all reading systems, and the reliability was analyzed using the Fleiss' kappa coefficient. The statistical equivalence ($P < 0.05$) was achieved at a 5% difference for all variables. To claim equivalence, the differences between the two reading systems were obtained and ranged from 0.5% to 4% - the minimum for hemorrhagic lesions and maximum for hyperdense MCA. Intraobserver agreements were classified as moderate, good or very good, with kappa values ranging from 0.52 to 0.94. In addition, we obtained a maximum agreement on the hemorrhagic lesions and a minimum agreement on the presence of ischemic lesions in ACA; this outcome deserves a particular analysis. Finally, we conclude that after providing radiologists with real clinical scenarios, the diagnostic performance for detecting acute stroke is likely equivalent regardless of the use of a smartphone or a diagnostic workstation. Mobile solutions are feasible alternatives for the interpretation of head CT images in patients with acute stroke and can be used as a handy tool in the development of more efficient telestroke services.**

Keywords-stroke; telestroke; teleradiology; smartphone; reliability; equivalence.

I. INTRODUCTION

Stroke is an acute neurologic dysfunction of vascular origin involving focal areas in the brain. It is considered a major cause of death and disability in developed and developing countries [1][2].

Patients with acute stroke symptoms who arrive at the emergency room require a comprehensive initial evaluation including the time of symptom onset, baseline risk factors, severity of their neurologic condition [3]-[5] and their potential for recuperation based on several clinical predictors, most importantly, an ischemic versus hemorrhagic etiology [6]-[9]. This task requires high levels of clinical and radiological expertise. For patients with ischemic stroke, determining whether to perform an endovascular thrombectomy and/or treat the patient with the administration of intravenous tissue plasminogen activator (tPA) is always a challenging task [10]. However, few eligible patients receive these treatment modalities due to their geographic distances from primary stroke centers or the limited availability of vascular neurologists and neuroradiologists to define the eligibility to receive this treatment.

In most hospitals in Colombia, there are no neuroradiologists at all. Similarly, in our location, which is a Joint Commission International (JCI)-certified primary stroke center with endovascular thrombectomy capabilities, there are not enough neuroradiologists *in situ* to support a telestroke network. To increase the availability of neuroradiologists, mobile solutions using smartphones should be evaluated.

Noncontrast head CT is the most widely used first imaging technique in patients with acute stroke symptoms [11]. This examination allows us to establish whether the stroke is hemorrhagic or ischemic and for ischemic strokes, determine whether it is acute or chronic; in addition, it allows professionals to rule out any contraindications to tPA administration from the imaging point of view. These contraindications may be the presence of the following: intra-axial neoplasm, intracranial neoplasm, arteriovenous malformation, aneurysm, hemorrhagic transformation of an ischemic infarct, and infarction > 1/3 of middle cerebral artery territory, which may be estimated using the ASPECTS [12].

The aim of this study was to evaluate the reliability and diagnostic equivalence of head CT interpretations when using a smartphone compared to a primary diagnostic interpretation workstation in an emergency telestroke service.

Next, we describe how the variables were determined, the characteristics of the remote equipment used, the methodology for data collection and the statistical tools for the analysis of the results; in Section 2, we describe the results according to the intraobserver agreement with respect to the diagnostic performance of stroke. Finally, we describe our experience with the reliability of the remote smartphone device and the projections of teleradiology in a society that seeks to efficient attention in the cases of stroke.

II. MATERIALS AND METHODS

The Institutional Review Board (IRB) of our institution approved this retrospective study and waived the requirement of informed consent. We employed a factorial design with repeated measures.

A. Sample

Patients with symptoms of acute stroke who presented to the emergency room for urgent evaluation between 2013 and 2018 were included in the study. The patients were randomly selected without repetition. Cases with image artifacts were excluded. The cases consisted of head CT examinations stored in our hospital Picture Archiving and Communication System (PACS), which were acquired using a General Electric LightSpeed 64 slice CT scanner (General Electric Healthcare, GE Medical Systems, Milwaukee, WI, USA), with 100 kV, 10 mAs, axial: 5 mm, sagittal: 3 mm, FOV: 26 cm, pixel spacing: 0.469, and matrix: 512 x 512.

B. Observers and interpretation variables

Four neuroradiologists were selected as observers (three with over ten years of experience and one with four years of experience in neuroradiology). They were asked to evaluate the presence of hemorrhagic lesions, the confidence in the presence of any ischemic lesion in the ACA territory, MCA territory, PCA territory, and the confidence in the presence of a hyperdense MCA. The confidence in the presence of these conditions was ranked using the following scores: 0, definitely absent; 1, most likely absent; 2, cannot decide; 3, most likely present; and 4, definitely present. For all cases in which an ischemic lesion was detected in the MCA territory (scores 3 or 4), time evolution was also inquired (i.e., acute, subacute, chronic). For acute lesions, the ASPECTS score was reported by selecting regions with infarcts in the MCA territory, obtaining a score that ranged from 0–10. Finally, the presence of one or more imaging contraindications for tPA administration, such as an intra-axial neoplasm, intracranial neoplasm, arteriovenous malformation, aneurysm or hemorrhagic transformation of an ischemic infarct, was also evaluated.

C. Display monitors and viewer software

The routine reading system for CT interpretations in our hospital is a PACS workstation with a Digital Imaging and Communication in Medicine (DICOM)-compliant viewer software Agfa IMPAX 6.5 (AGFA HealthCare, Mortsel, Belgium). Images were displayed using an E-2620 BARCO monitor (BARCO N.V, Kortrijk, Belgium), which is a 2-megapixel (MPx) LCD medical grayscale display, DICOM-compliant, dot pitch of 0.249 mm, with a spatial resolution of 1600 x 1200 pixels, a maximum luminance of 700 cd/m2, and an 8-bit grayscale. This reading system, hereafter referred to as Medical-IMPAX, was used as the reference reading system in this study.

As a mobile alternative, a Samsung Galaxy S8 Plus (Samsung Electronics, South Korea) smartphone, with a display of 146.5 mm (5.8"), 570 pixels per inch, a spatial resolution of 1440 x 2960 pixels, and a maximum luminance of 1000 cd/m2 was selected. The viewer software used on this smartphone was the Agfa XERO

Viewer 3.0 (Agfa HealthCare, Mortsel, Belgium) software. This reading system is hereafter referred to as Smartphone-XERO.

D. *Procedure*

Each radiologist read all cases using both the Medical-IMPAX and the Smartphone-XERO systems. At each reading, the radiologist determined the variables mentioned in the section "Observers and interpretation variables". The two reading software packages provided image manipulation tools to adjust the window/level, zoom and multiplanar reformation presentation. These tools were available for all images and could be used at the observer's discretion to improve the image interpretations.

The radiologists were blinded to the patient name, any individualizing items and the original image report. Data collection was performed using a web-based platform, and image readings were stored in a secured database. This software randomizes cases and guides the radiologist throughout the complete report, thus ensuring the integrity and completeness of the data. Relevant clinical data such as sex, age, main neurological symptoms and relevant past medical history (e.g., diabetes, hypertension, headache, Parkinson's disease, Alzheimer's disease, sleep apnea/hypopnea syndrome, or cardiac arrhythmia) were also available. Readings were performed over the course of one year in two or four-hour sessions per reader, with no time limitations for each case. This was a counterbalanced study for the reading systems used by each radiologist in each session.

E. *Data analysis*

The confidence scores were dichotomized to evaluate both reliability and equivalence. Scores from 0–2 were classified as negative, and scores from 3–4 were classified as positive. Patients with an ASPECTS score ≤ 5 are not eligible to receive tPA treatment. Thus, we dichotomized the ASPECTS score into two categories: 0 if the score ranged from 0–5 (a contraindication for tPA administration) and 1 if the score ranged from 6–10 (indicating eligibility for the administration of the tPA treatment). This variable was named "dichotomized-ASPECTS".

The variables presence of hemorrhagic lesions and presence of any imaging contraindications to tPA administration were not dichotomized as they were already binary (i.e., 0: negative, 1: positive).

Reliability was evaluated in terms of intraobserver agreements of interpretation (cases rated by the same observer using different reading systems), with the Fleiss' kappa coefficient [13]. The kappa coefficients were ranked as defined by Altman [14]: "very good", ($\kappa = 1$ to 0.81); "good", ($\kappa = 0.8$ to 0.61); "moderate", ($\kappa = 0.6$ to 0.41); "fair", ($\kappa = 0.4$ to 0.21); and "poor", ($\kappa < 0.2$). For these calculations, STATA 13.0 software (Stata Corp, College Station, TX, USA) was used.

The dichotomized variables were evaluated to determine their statistical equivalence by means of generalized estimated equations (GEE) [15] using IBM SPSS Statistics 19 software (IBM Corp., Armonk, NY, USA). To evaluate equivalence, mean differences and standard errors were obtained from the GEE analysis. The hypothesis test for equivalence was as follows: the null hypothesis H_0 was $|\text{Mean Difference (I-J)}| - \delta = 0$, and the alternative hypothesis H_a was $|\text{Mean Difference (I-J)}| - \delta < 0$, where I and J are the two reading systems compared and δ (delta) is the maximum allowable difference permitted to claim equivalence, as suggested by several authors in recent years [16]-[19]. We calculated a $(1-2\alpha)\%$ confidence interval for all comparisons, which is also a method to evaluate equivalence [18][19]. The significance level was set to 5% (i.e., $\alpha = 0.05$), and δ was set to 0.05 (5%). Finally, we calculated the required value of δ to claim equivalence for each variable (named δ_{eq} in our result tables).

The reading time was also recorded by the software to evaluate the equivalence between the two reading systems. As this variable is continuous, the mean differences and their standard errors were obtained from an ANOVA with the IBM SPSS Statistics 19 software.

III. RESULTS

There were 90 (47.87%) males and 98 (52.13%) females in the sample. The ages of the patients ranged from 30–97 years, with a mean age of 71.3 years (standard deviation of 15) overall, with a mean age of 69.1 years for males and 73.3 years for females.

To carry out the reliability and equivalence evaluation, each variable must be set by all four observers. The detection of any hemorrhagic lesion was set by all observers; hence, there were 188 patients for this variable and 1504 readings (188 patients by 4 observers by 2 systems). Of these 188 patients, 28 were classified as hemorrhagic lesions by all observers, and 160 were evaluated for other findings (e.g., contraindications to the tPA administration, infarct in the anterior, middle or posterior cerebral artery territories, and hyperdense middle cerebral artery). For these variables, 1280 readings were included in the reliability and equivalence evaluations. Finally, for patients with acute infarct in the middle cerebral artery territory, the ASPECTS score was calculated, ending up with 118 patients with ASPECTS score set by all radiologists; therefore, 944 readings were included in the evaluations.

A. *Intraobserver agreement*

The intraobserver agreements between the Medical-IMPAX and Smartphone-XERO reading systems, i.e., when each case was interpreted the same way using both the Medical-IMPAX and the Smartphone-XERO reading systems, by the same observer, are presented in Table I. There was very good intraobserver agreement on the hemorrhagic lesion detection, ($\kappa = 0.94$, $P < 0.001$).

There was a moderate intraobserver agreement on the detection of contraindications to tPA administration (presence of intra-axial neoplasm, intracranial neoplasm, arteriovenous malformation, aneurysm, or hemorrhagic transformation of an ischemic infarct), ($\kappa = 0.55$,

$P < 0.001$). There was a good intraobserver agreement on the dichotomized-ASPECTS (0-5; 6-10) in the MCA territory ($\kappa = 0.66$, $P < 0.001$).

TABLE I. INTRAOBSERVER AGREEMENT BETWEEN THE MEDICAL-IMPAX AND SMARTPHONE-XERO READING SYSTEMS

Imaging findings	Fleiss' Kappa[a]	Standard Error	Agreement[b]
Hemorrhagic lesion	0.94	0.037	Very Good
Contraindications to tPA administration	0.55	0.039	Moderate
Anterior cerebral artery territory infarction	0.52	0.040	Moderate
Middle cerebral artery territory infarction	0.68	0.040	Good
Posterior cerebral artery territory infarction	0.63	0.040	Good
Hyperdense middle cerebral artery	0.62	0.040	Good
Dichotomized-ASPECTS	0.66	0.046	Good

[a] Readings were performed on the two systems by four observers. All values were significant ($P < 0.001$) [b] as defined by Altman [14]. Intraobserver agreement between both reading systems showed a very good agreement for the diagnosis of hemorrhagic lesions and an appropriate agreement for those variables that are the most important for stroke in which administration of tPA determines the survival of the patients.

TABLE II. EQUIVALENCE TESTS FOR THE IMAGING FINDINGS BETWEEN THE MEDICAL-IMPAX AND SMARTPHONE -XERO READING SYSTEMS

Imaging findings	Mean difference[a]	SE	(1-2α)% Confidence Interval for equivalence testing		δ_{eq} (%)
			Lower	Upper	
Hemorrhagic lesion	-0.001	0.002	-0.005	0.002	0.51
Contraindications to tPA administration	-0.003	0.007	-0.015	0.009	1.52
Anterior cerebral artery territory infarction	0.005	0.011	-0.013	0.022	2.23
Middle cerebral artery territory infarction	0.003	0.016	-0.023	0.029	2.91
Posterior cerebral artery territory infarction	0.005	0.014	-0.018	0.028	2.78
Hyperdense Middle Cerebral Artery	0.011	0.018	-0.018	0.040	4.01
Dichotomized-ASPECTS	0.017	0.009	0.002	0.032	3.23

[a] Equivalence tests for all variables were statistically significant, all $P < 0.05$, at a difference of 5% ($\delta = 0.05$). SE = Standard error of the mean difference, α = significance of the test (0.05), δ = difference of the means set to test equivalence, δ_{eq} = minimum δ required to achieve equivalence. The null and alternative hypotheses were $|difference| - \delta = 0$ and $|difference| - \delta < 0$, respectively.

A moderate intraobserver agreement on the presence of an ACA territory infarction was also observed ($\kappa = 0.52$, $P < 0.001$). In contrast, good intraobserver agreements on the presence of an MCA territory infarction and the presence of a PCA territory infarction were observed, with $\kappa = 0.68$ and 0.63, respectively (both $P < 0.001$). In the presence of a hyperdense middle cerebral artery, a good intraobserver agreement was observed ($\kappa = 0.62$, $P < 0.001$).

B. *Equivalence tests for the imaging findings*

The mean differences between the variables using both reading systems, the standard error of the differences, and the (1-2α)% confidence interval for equivalence testing, at a delta (δ) difference of 5%, are presented in Table II. In addition, the minimum δ required to claim equivalence (δ_{eq}) was calculated and presented in the last column.

Equivalence tests for all variables were statistically significant, all $P < 0.05$, showing equivalence at a difference of 5%. The largest (1-2α)% confidence interval was for the presence of a hyperdense middle cerebral artery and the smallest was for the detection of a hemorrhagic lesion. The minimum δ required to claim equivalence (δ_{eq}) for each variable ranged from 0.0051 (0.51%) for the detection of hemorrhagic lesions to 0.0401 (4.01%) for the detection of hyperdense middle cerebral artery.

Both a low mean difference and a low minimum δ required to claim equivalence were observed for the presence of contraindications to tPA administration and for dichotomized-ASPECTS ($\delta_{eq} = 1.52\%$ and 3.23%, respectively).

C. *Reading time*

The mean reading time was 114.31 s for the Medical-IMPAX system and 143.39 s for the Smartphone-XERO. The mean difference between the two reading systems was -29.08 s (standard error = 5.4). The equivalence tests for this variable produced a significant ($P < 0.001$) result at a reading time delta of 60 s. The minimum δ required to claim equivalence (δ_{eq}) for reading time was 38 s.

IV. CONCLUSION

High intraobserver agreements between the Medical-IMPAX and Smartphone-XERO reading systems were observed, with kappa values ranging from 0.52 to 0.94. This suggests that every time that a radiologist reads the same patient image using any of the two reading systems, either the Medical-IMPAX or the Smartphone-XERO, there are no differences in the initial patient outcome. A very good intraobserver agreement was observed for hemorrhagic lesions ($\kappa = 0.94$). These findings are consistent with other studies that used a tablet computer as a mobile solution [20]-[24].

Previous studies have reported nonsignificant differences in the detection of ischemic lesions between tablet computers and primary workstations [20][22][25][26]. However, the variables and evaluation methods used in these studies were different from those used in our analysis, and it is important to note that other similar studies did not include a smartphone as a diagnostic device.

To the best of our knowledge, this is the first study that used reliability and equivalence statistical methods to identify potential diagnostic differences between a smartphone and a primary diagnostic workstation in reading head CT examinations in a telestroke context prior to intravenous thrombolysis. Furthermore, when the gold standard is available, a diagnostic accuracy evaluation, using Receiver Operating Characteristics (ROC) curves, will be conducted.

The results of our study indicate that the patients who were not eligible for tPA administration based on imaging criteria, e.g., those having ASPECTS ≤ 5, or the presence of an intra-axial neoplasm, intracranial neoplasm, arteriovenous malformation, aneurysm, or hemorrhagic transformation of an ischemic infarct, were well detected when performed by experienced radiologists using both reading systems.

One limitation of this study centers on the illumination conditions. Readings using the Medical-IMPAX system were performed in diagnostic rooms with controlled ambient light levels. In contrast, readings using the Smartphone-XERO system were performed without controlling ambient light levels. Nevertheless, this situation is more realistic for a telestroke system in which a radiologist is asked to interpret a head CT as soon as possible wherever he or she is located.

The smartphone used in this study was an average-sized mobile phone compared to the ones used today, in which the XERO viewer software allows adjustments of window/level and changes over reconstruction planes (sagittal, coronal or axial), similar to a PACS workstation. Radiologists indicated the absence of ergonomic problems because the use of the smartphone was comparable with the daily use of personal devices. Finally, they reported better performance when making adjustments using a mobile pen than when they did it tactilely. Interpretation using the smartphone spent an average of 29 seconds more than a PACS workstation, which is a reading time that does not affect the goals stipulated in our protocol for stroke code and is negligible compared with the transport time of a neuroradiologist to the hospital in our city.

In conclusion, there was no superiority of any specific reading system on the evaluated clinical variables. In addition, a high intraobserver agreement was documented for the same variables, providing evidence that the Medical-IMPAX and Smartphone-XERO reading systems may be interchangeable without any reliability loss.

This study provides evidence that the Smartphone-XERO reading system can be used for acute stroke diagnosis based on head CT examinations, ruling out possible imaging contraindications to tPA administration and ASPECTS quantification.

In the statistical design of this study, the radiologist and the reading systems were fixed factors because they were not selected at random; therefore, our results only apply to them. Nevertheless, as neuroradiologists are highly specialized readers of neurological images, we expect that our results may be extrapolated to other groups of neuroradiologists. Similarly, the reading systems in radiology must be DICOM-compliant, which allows us to generalize our results to other reading software or medical displays. In contrast, the smartphone display used in this study may be significantly different when compared to other smartphone displays; hence, our results only apply to smartphones with "retina" display or similar displays.

As future work, the web platform for reading the Head CT of patients with suspicion of acute stroke will be used to train radiologists for these cases, for example, to train them in evaluating the infarct size using the ASPECTS and finding contraindications for tPA administration. In addition, the data stored will be used with machine learning and data mining techniques to understand and improve the reading process of neuroradiologists and the subsequent search of diagnostic aids using image processing.

ACKNOWLEDGMENT

We thank our institutions and the National Department of Science, Technology and Innovation of Colombia for funding this study (Grant 1204-744-55680).

REFERENCES

[1] E. J. Benjamin, S. S. Virani, C. W. Callaway, A. M. Chamberlain, A. R. Chang, S. Cheng, et all. Heart Disease and Stroke Statistics—2018 Update: A Report From the American Heart Association. *Circulation*. 2018;137:e67–e492.

[2] G. Saposnik, O. H. Del Brutto, and f. t. I. S. o. C. Diseases, "Stroke in South America: A Systematic Review of Incidence, Prevalence, and Stroke Subtypes," *Stroke*, vol. 34, pp. 2103-2107, September 1, 2003 2003.

[3] G. Teasdale, A. Maas, F. Lecky, G. Manley, N. Stocchetti, and G. Murray, "The Glasgow Coma Scale at 40 years: standing the test of time," *Lancet Neurol*, vol. 13, pp. 844-54, Aug 2014.

[4] L. K. Kwah and J. Diong, "National Institutes of Health Stroke Scale (NIHSS)," *J Physiother*, vol. 60, p. 61, Mar 2014.

[5] A. K. Boehme , C. Esenwa , M. S.V. Elkind. Stroke Risk Factors, Genetics, and Prevention. *Circulation Research*. 2017;120:472–495

[6] K. Ghandehari, F. Ahmadi, S. Ebrahimzadeh, K. Shariatinezhad, and K. Ghandehari, "Assessment of ABCD(2) scale in patients with transient ischaemic attack or stroke," *Neurol Neurochir Pol*, vol. 46, pp. 421-7, Sep-Oct 2012.

[7] F. O. Lima, G. S. Silva, K. L. Furie, M. R. Frankel, M. H. Lev, E. C. Camargo, *et al.*, "Field Assessment Stroke Triage for Emergency Destination: A Simple and Accurate Prehospital Scale to Detect Large Vessel Occlusion Strokes," *Stroke*, vol. 47, pp. 1997-2002, Aug 2016.

[8] G. Y. Lip, R. Nieuwlaat, R. Pisters, D. A. Lane, and H. J. Crijns, "Refining clinical risk stratification for predicting stroke and thromboembolism in atrial fibrillation using a novel risk factor-based approach: the euro heart survey on atrial fibrillation," *Chest*, vol. 137, pp. 263-72, Feb 2010.

[9] S. Runchey and S. McGee, "Does this patient have a hemorrhagic stroke?: clinical findings distinguishing hemorrhagic stroke from ischemic stroke," *Jama*, vol. 303, pp. 2280-6, Jun 9 2010.

[10] M. Kulcsar, S. Gilchrist, and M. G. George, "Improving Stroke Outcomes in Rural Areas Through Telestroke Programs: An Examination of Barriers, Facilitators, and State Policies," *Telemedicine and e-Health,* vol. 20, pp. 3-10, 2014/01/01 2013.

[11] E. C. Jauch, J. L. Saver, H. P. Adams, Jr., A. Bruno, J. J. Connors, B. M. Demaerschalk, *et al.*, "Guidelines for the early management of patients with acute ischemic stroke: a guideline for healthcare professionals from the American Heart Association/American Stroke Association," *Stroke,* vol. 44, pp. 870-947, Mar 2013.

[12] J. H. W. Pexman, P. A. Barber, M. D. Hill, R. J. Sevick, A. M. Demchuk, M. E. Hudon, *et al.*, "Use of the Alberta Stroke Program Early CT Score (ASPECTS) for Assessing CT Scans in Patients with Acute Stroke," *American Journal of Neuroradiology,* vol. 22, pp. 1534-1542, September 1, 2001 2001.

[13] J. Fleiss, B. Levin, and M. Paik, *Statistical Methods for Rates and Proportions*, 3rd Edition ed.: Wiley, 2003.

[14] D. G. Altman, *Practical Statistics for Medical Research*: Chapman and Hall/CRC, 1991.

[15] J. C. Nelson and M. S. Pepe, "Statistical description of interrater variability in ordinal ratings," *Statistical Methods in Medical Research,* vol. 9, pp. 475-496, October 1, 2000 2000.

[16] W. Chen, N. A. Petrick, and B. Sahiner, "Hypothesis Testing in Noninferiority and Equivalence MRMC ROC Studies," *Academic Radiology,* vol. 19, pp. 1158-1165, 2012.

[17] H. Jin and Y. Lu, "A non-inferiority test of areas under two parametric ROC curves," *Contemporary Clinical Trials,* vol. 30, pp. 375-379, 2009.

[18] J.-P. Liu, M.-C. Ma, C.-y. Wu, and J.-Y. Tai, "Tests of equivalence and non-inferiority for diagnostic accuracy based on the paired areas under ROC curves," *Statistics in Medicine,* vol. 25, pp. 1219-1238, 2006.

[19] N. A. Obuchowski, "Testing for equivalence of diagnostic tests," *AJR. American journal of roentgenology,* vol. 168, pp. 13-17, 1997.

[20] S. Tewes, T. Rodt, S. Marquardt, E. Evangelidou, F. K. Wacker, and C. von Falck, "Evaluation of the use of a tablet computer with a high-resolution display for interpreting emergency CT scans," *Rofo,* vol. 185, pp. 1063-9, Nov 2013.

[21] J. B. Park, H. J. Choi, J. H. Lee, and B. S. Kang, "An Assessment of the iPad 2 as a CT Teleradiology Tool Using Brain CT with Subtle Intracranial Hemorrhage Under Conventional Illumination," *Journal of Digital Imaging,* vol. 26, pp. 683-690, 2013.

[22] P. D. McLaughlin, F. Moloney, S. B. O'Neill, K. James, L. Crush, O. Flanagan, *et al.*, "CT of the head for acute stroke: Diagnostic performance of a tablet computer prior to intravenous thrombolysis," *J Med Imaging Radiat Oncol,* Feb 03 2017.

[23] A. J. Salazar, N. Useche, M. Granja, A. J. Morillo, and S. Bermúdez, "Ruling Out Brain CT Contraindications prior to Intravenous Thrombolysis: Diagnostic Equivalence between a Primary Interpretation Workstation and a Mobile Tablet Computer," *International Journal of Telemedicine and Applications,* vol. 2017, p. 7, 2017.

[24] S. Panughpath, S. Kumar, and A. Kalyanpur, "Utility of mobile devices in the computerized tomography evaluation of intracranial hemorrhage," *Indian Journal of Radiology and Imaging,* vol. 23, pp. 4-7, January 1, 2013 2013.

[25] A. Bhatia, S. Patel, G. Pantol, Y.-Y. Wu, M. Plitnikas, and C. Hancock, "Intra and Inter-Observer Reliability of Mobile Tablet PACS Viewer System vs. Standard PACS Viewing Station-Diagnosis of Acute Central Nervous System Events," *Open Journal of Radiology,* vol. Vol.03No.02, p. 8, 2013.

[26] K. Yoshimura, T. Nihashi, M. Ikeda, Y. Ando, H. Kawai, K. Kawakami, *et al.*, "Comparison of Liquid Crystal Display Monitors Calibrated With Gray-Scale Standard Display Function and With γ 2.2 and iPad: Observer Performance in Detection of Cerebral Infarction on Brain CT," *American Journal of Roentgenology,* vol. 200, pp. 1304-1309, 2013/06/01 2013.

Ambient Monitoring System for Urination

Toshifumi Tsukiyama

School of Information Environment
Tokyo Denki University
Adachi-ku, Tokyo, Japan
Email: t-tsuki@mail.dendai.ac.jp

***Abstract*—Urination is one of vital quotidian activities that is closely involved in maintaining the healthy life of the elderly. The number of times and the time of day that an elderly person urinates are important data to detect a latent, dangerous situation that may require medical treatment. In this paper, we propose a sensor system for monitoring urination at home. Since the bathroom of a house is shared by elderly persons and their family members, the monitoring system has to detect urination but also identify a person who uses the bathroom. Urination is usually accompanied by the use of tap water from a toilet's flush tank. Urination can be made recognizable by monitoring the flow of water in the toilet bowl. The monitoring system uses the water flow sensor that is attached to the toilet. Active Radio Frequency IDentification (RFID) tags are used for identification of persons. The subject of the monitoring wears an active RFID tag when he/she uses the bathroom. The ID code of the tag is read by the tag reader contactlessly without any actions, when a person carrying a tag gets closer to the tag reader. Our system reports his/her records of urination to a local nursing center without explicit user interaction. The proposed sensor system is available at a reasonable cost and can be installed easily in any type of housing. No interaction by the user is required. The monitoring system has been installed in a volunteer's housing and some results from this experimental trial will be presented.**

Keywords–Urination; The elderly; Water flow sensor; Active RFID tag; Vibration sensor.

I. INTRODUCTION

We are confronted with an increasing population of the elderly, many of whom are apt to suffer from latent dangerous situations that may require medical attention. However, the number of home-care nurses available for regular home visits is limited to check their health status. Thus, new care services, such as those that use monitoring systems [1][2], are needed to cut costs in health care while still providing adequate medical treatment for the elderly.

Since urination is one of vital quotidian activities that is closely involved in maintaining a healthy lifestyle, we focus on urination as a sign of latent dangerous situations. As you know, it is not uncommon to pass urine more frequently in the senior years. This is partly due to age-related changes in the bladder muscle. However, sometimes, frequent urination is a sign of some underlying disease. Humans urinate an average of four to eight times a day; thus, if an individual makes much less or much more frequent use of the toilet, then some illness or problem can be suspected [3]. If one wakes two or more times before dawn to urinate, this can also be a sign of a poor health state [4]. Common causes of such symptoms are a urinary tract infection, diabetes, stroke or other neurological diseases, chronic kidney disease, dehydration, and so on. Therefore, some of these causes of frequent urination or nocturia can be serious and an early diagnosis can play a significant role in treating the condition.

The number of times and the time of day that the elderly urinate are important data to investigate and treat them accordingly, and could help the doctor to make a correct diagnosis for them. However, it is very boring and troublesome to record the frequency of urination with time stamps. We can not expect the elderly themselves or their family members to record them correctly. Some systems for monitoring urination have been proposed [5] but they are supposed to be used in the process of taking care of patients with bladder dysfunction or in critical care. The assistance of medical professions is needed to operate such a system. In this paper, we propose a simple, easy monitoring system for urination at home. Our system records the number of times and the time of day that an individual urinates, and reports his/her records to a local nursing center without explicit user interaction. The proposed sensor system is available at a reasonable cost and can be installed easily in any type of housing. No interaction by the user is required. No personal data, such as photographs or video recording are saved in the system or transmitted.

This paper is organized as follows. In Section 2, we give a brief outline of the proposed monitoring system. Section 3 describes the technical aspect of the monitoring system. Section 4 describes the implementation of the monitoring system and experimental results. Concluding remarks are given in Section 5.

II. OVERVIEW OF THE PROPOSED MONITORING SYSTEM

Our monitoring system is supposed to be set up in the bathroom of a house, which is shared by an elderly couple or an elderly person and his / her family members. The monitoring system has to detect urination but also identify a person who uses the bathroom. An Active Radio Frequency IDentification (RFID) tag is used for identification of persons. The subject of the monitoring wears an active RFID tag when he/she uses the bathroom. RFID tags emit Radio Frequency (RF) signals periodically with an unique ID code. The ID code can be read by the tag reader contactlessly without any actions, when a person carrying a tag gets closer to the tag reader. Urination is usually accompanied by the use of tap water from a toilet's flush tank. The proposed monitoring system uses a sensor which can detect the flow of water in the toilet. The sensor was originally developed in the previous work[2] but some parts of the sensor are redesigned to connect the tag reader.

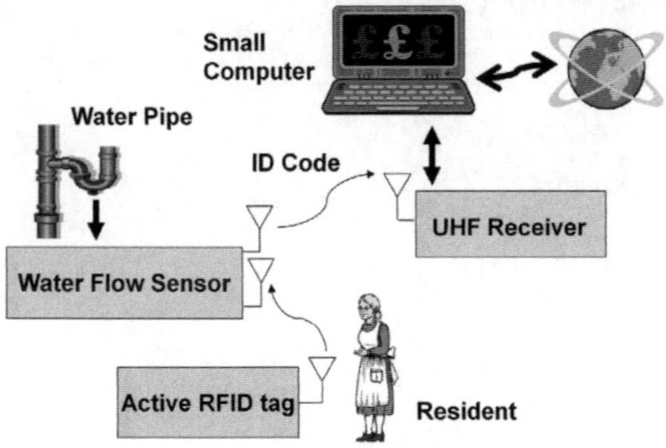

Figure 1. Functional deployment of the monitoring system.

Figure 1 provides the functional deployment of the monitoring system. The monitoring system consists of three main components: a water flow sensor equipped with an UHF (2.4GHz) transmitter and a tag reader, an active RFID tag, and a small computer with an UHF receiver and WiFi. When a resident uses the bathroom, the water flow sensor transmits UHF signals to the receiver connected to the computer at one-second intervals while water is flowing into the flush tank. If the resident is carrying a tag, the water flow sensor transmits the ID code emitted by the tag. If not, the default ID code is sent.

The computer records the receipt time of the ID code sent from the water flow sensor. The program installed in the computer derives the time of tap water use from the time stamps of the received ID codes. The data is collected and compiled, day by day or week by week. The computer sends a report to a local nursing center via the Internet. If there is something wrong with the urination record, then, a caregiver will visit the resident to verify his/her condition. If a reasoning program is installed in the computer, the program could report signs of declining health to a local nursing center, when there are major variations between the patterns of the normal and actual urination [6].

III. Equipments for the Monitoring System

This section describes the technical aspect of the functional blocks shown in Figure 1. Our system adopts a RFID system to see if a person in the bathroom is the subject being monitored or not. There are two types of RFID systems: passive RFID system and active RFID system. Passive RFID systems use tags with no internal power source and instead are powered by the electromagnetic energy transmitted from a RFID tag reader. The tags are very light and small but the reader becomes relatively large because a big antenna is necessary for transmitting the electricity to tags. Active RFID systems use battery-powered tags that continuously broadcast their own signal like radio beacons. The tag readers become the small size, such as a USB memory stick. Although active RFID systems are compact and fairly uncomplicated, many of the ready-made tag readers are supposed to be connected to computers through standard I/O ports, such as Bluetooth. It is very difficult to embed such tag readers into sensor

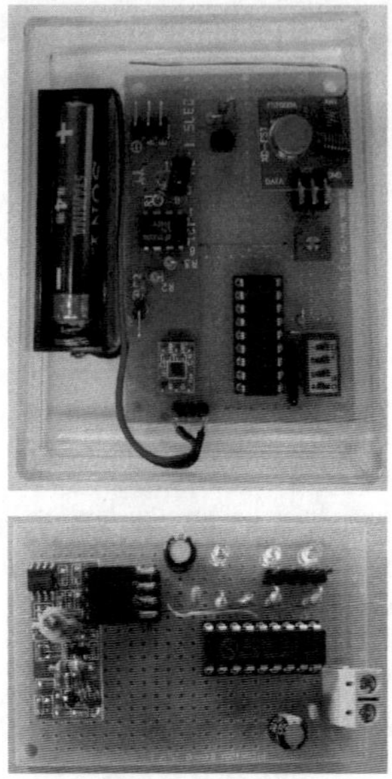

Figure 2. Active RFID tag (the upper image) and the tag reader (the lower image).

systems without a microprocessor from a technical view point. Therefore, we designed an active tag system which suits the monitoring system. The tag reader is easy to integrate other electric circuits in the water flow sensor. In our system, we simply need to confirm the presence of a tagged person in a small area like a bathroom. Since the tag system plays the role of a non-contact proximity sensor, the read range of the tag must be shorter than a few meters. Therefore, the active tag is set to a lower transmit power not to exceed the read range. The number of a family member is usually two or three people. The ID code of 2-bit length is sufficient to recognize one of the family members.

Figure 2 illustrates a prototype of the active RFID tag and tag reader. The tag consists of an UHF (433MHz) transmitter module, an IC chip for encoder, and a timer which switch on the transmitter at one or two second interval. The tag reader module is comprised of an UHF receiver module and an IC chip for decoder. The tag emits a weak RF signal periodically with a 2-bit code. If the tag is adjacent to the tag reader within a few meters, the reader outputs the parallel data of 2 bits.

Figure 3 illustrates a prototype of the water flow sensor. A water flow sensor consists of a vibration sensor, a sound-activated switch, a tag reader, and an UHF (2.4GHz) transmitter. Mechanical vibration in the range of 0.5 to 10KHz occurs at the water pipe to the flash tank while tap water is running. The vibration sensor is attached to the water pipe to a flush tank to pick up the mechanical vibrations. A ready-made contact microphone based on piezoelectric effect can be applied to the vibration sensor. The contact microphone are originally used for tuning musical instruments, such as guitars, and available for less than 5 euros in the market. The microphone is clip-on and the point of the clip is machined to fit the shape of a water pipe. Since similar vibrations are observed on the surface of a toilet bowl while tap water is running, a metal disk with

Figure 3. Prototype of the water flow sensor which consists of a vibration microphone, a sound-activated switch, the tag reader (the upper image), a UHF transmitter (the lower image), and a battery.

piezo material used for buzzers is stuck on the toilet bowl with adhesive tape, in the case of wall toilets where the tank and water pipe are conceal with a wall. Figure 4 shows examples of the vibration microphone clipped onto the water pipe to the flush tank of a toilet, and the metal disk stuck with adhesive tape on the surface of a toilet bowl.

When the sensor detects mechanical vibrations from the toilet, the sound-activated switch turns on the transmitter. Consequently, the transmitter sends a radio frequency signal (2.4GHz) with a 2-bit ID code from the tag. The code "00" is reserved as the default code. If the resident is not carrying a tag the default ID code is sent. The signal is transmitted at one-second intervals while the water continues to flow through the pipe. The transmitter can send signals indoors to a range of up to about 15 m, which is sufficient in a normal house. The sensors have a rechargeable lithium polymer (Li-Po) battery (6600 mAh) built-in and can keep functioning for about 12 months.

Figure 5 illustrates the small computer (Raspberry Pi B+) with a UHF receiver. The receiver receives UHF signals with the ID code from the water flow sensor, and the ID code is transferred to the computer through a USB port. Programs installed in the computer are written in C and run on the Pidora (a kind of the Linux for the Raspberry Pi). The number of times that the ID codes were transmitted from the sensor indicates the amount of time that there is running tap water, and this is proportional to the amount of water used because ID codes are transmitted steadily at one-second intervals while the water is running. The received ID codes are accumulated at half-hour intervals to obtain a distribution of the duration of tap water use.

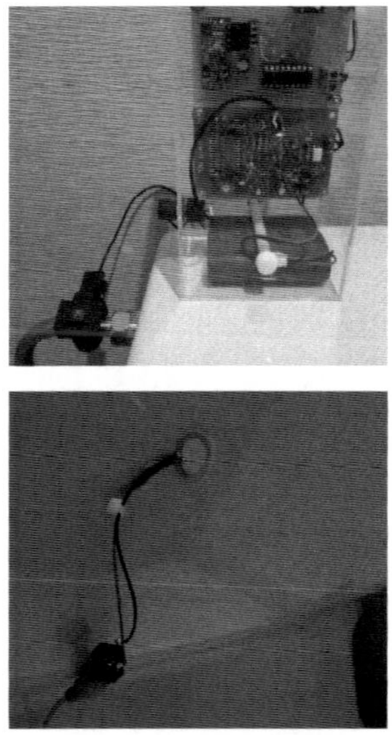

Figure 4. Contact microphone clipped onto the water pipe to the flush tank of a toilet (the upper image), and the metal stuck on the surface of a toilet bowl (the lower image).

Figure 5. Small computer with a UHF receiver inserted in its USB port.

IV. Implementation and Experimental Results

A monitoring system was installed in a real housing environment and an experiment was conducted to see whether the sensor system can acquire the time and frequency of water use from the resident's quotidian activity. The house had a living area of 108 square meters (roughly 15 m by 8 m), and the water flow sensor was set on the water pipe connected to the toilet's flush tank in the bathroom. The distance between the sensor and the computer was 5 m. The residents were a 70-year-old man and 67-year-old woman. He filled the role of the subject being monitored and wore the RFID tag with an ID code of "01". When she or guests used the bathroom the default ID code "00" was assigned because they did not have any RFID tags. First, the sensitivity and reliability of the monitoring system were checked. The system could detect gently running water, such as would be poured into a glass.

Figure 6(a) illustrates an example of the distribution of the duration of water flow during the day at the residence's

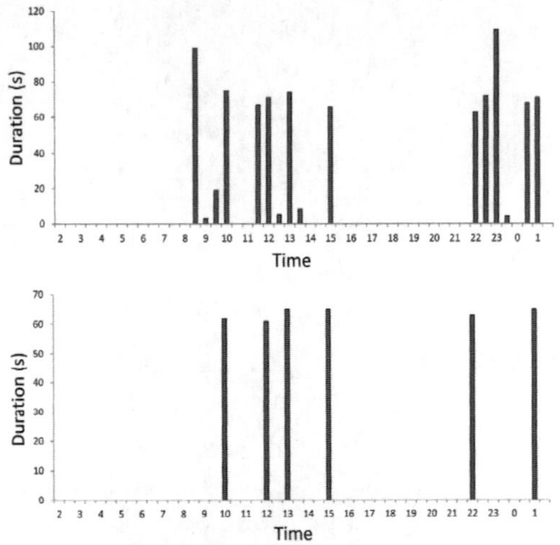

Figure 6. (a)Distribution of the duration of water flow into the toilet's flush tank (the upper image), and (b) the use record of the subject being monitored (the lower image).

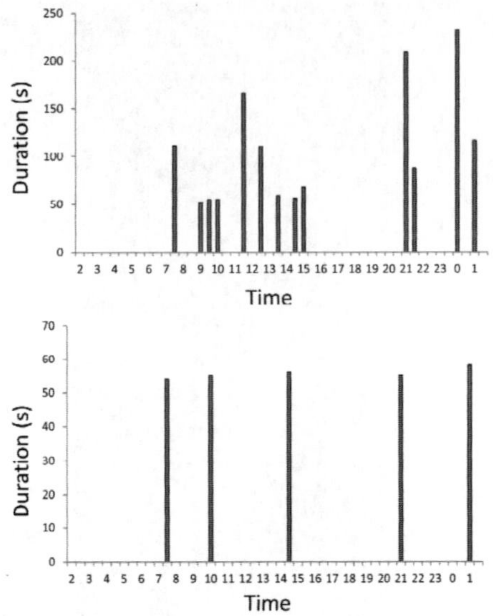

Figure 7. (a)Distribution of the duration of water flow into the toilet's flush tank (the upper image), and (b) the use record of the subject being monitored (the lower image).

toilet. In this case, the residents stayed in the house all day. The horizontal axis of the graph denotes the time at half-hour intervals starting at 2 a.m. The vertical axis indicates the duration of water use in seconds. The water flow sensor transmitted ID codes at one-second intervals while water was flowing into the flush tank. It depends on the water pressure, but it takes 60 seconds from 40 to fill the tank with water. Figure 6(a) expresses that the bathroom was used 13 times by the family members during the day. Figure 6(b) illustrates the distribution of the duration of water flow which was extracted from the distribution shown in figure 6(a), based on the ID code "01". The graph shows the use record of the bathroom by the subject being monitored, and expresses that the subject being monitored urinated 7 times during the day, at an average interval of 2.5 hours.

Another example of the distribution of the duration of water flow is illustrated in figure 7(a). In the case, their son's family with kids visited the home and stayed overnight, and the family member, including the subject being monitored went out from 3 to 9 p.m. for shopping or something. The figure expresses that the bathroom was used 22 times during the day. Figure 7(b) shows the use record of the bathroom by the subject being monitored. The graph expresses that the subject being monitored urinated 5 times or more during this day, at an average interval of 3.5 hours.

Each night at 2 a.m., the program sent out a daily report to the appointed addresses by e-mail, which denotes a time log of urination for the day. At the same time, all the data gathered to the monitoring system was saved to the ftp server on the Internet, and was deleted from the monitoring system.

V. CONCLUSION

Urination is one of vital quotidian activities that is closely involved in maintaining the healthy life of the elderly. The number of times and the time of day that an elderly person urinates are important data to detect a latent, dangerous situation that may require medical treatment. There is a demand for a reasonably priced, noncontact monitoring system that can directly recognize such quotidian activity. In addition, the monitoring system should be affordably installed into any type of housing. To meet these demands, we proposed the water flow sensor with a vibration microphone that can be easily clipped on to a water pipe leading to a flush tank of a toilet. Since the bathroom of a house is shared by elderly persons and their family members, the monitoring system has to detect urination but also identify a person who uses the bathroom. Active RFID tags were adopted for identification of the subject of the monitoring. We made a prototype of the monitoring system from electronic parts that are all available in the market. The prototype has been installed in a volunteer's housing and useful results were derived from the experimental trial.

REFERENCES

[1] S. C. Mukhopadhyay, and G. S. Gupa, "Sensor and Robotics Environment for Care of the Elderly," in Proceedings of IEEE International Workshop on Robotics and Sensors Environment, 2007, pp. 1-6.

[2] T. Tsukiyama, "In-home health monitoring system for solitary elderly," Procedia Computer Science 63, 2015, pp. 229-235.

[3] "Frequent Urination: Causes and Treatments," 2007, URL: http://www.webmd.com/urinary-incontinence-oab/frequent-urination-causes-and-treatments [accessed: 2018-10-31].

[4] V. Varilla, R. V. Samal, D. Galindo, and J. Ciocon, "Nocturia in the elderly: A wake-up call," Cleveland Clinic Journal of Medicine, Vol. 78, No. 11, 2011, pp. 757-764.

[5] A. Otero, F. Palacios, T. Akinfiev, and A. Apaikov, "A Low Cost Device for Monitoring the Urine Output of Critical Care Patients," 2010, pp. 10714-10732.

[6] M. Skubic, R. D. Guevara, and Marilyn Rants, "Testing Classifiers for Embedded Health Assessment," Springer-Verlag LNCS 7251, 2012, pp. 198-205.

Investigations on the Impact of Anthropomorphism and Gamification on Breast Cancer Survivors' Expressed Preferences in a Physical Activity Promotion Intervention

João P. Monteiro*
Carolina T. Lopes
Nuno C. Duarte

INESC TEC / Faculdade de Engenharia
Universidade do Porto
Rua Dr. Roberto Frias s/n
4200-465 Porto, Portugal
*Email: jpsm@ieee.org

André T. Magalhães

Serviço de Cirurgia Geral
Centro Hospitalar de São João
Alameda Prof. Hernani Monteiro
4200-319 Porto, Portugal

Hélder P. Oliveira

INESC TEC / Faculdade de Ciências
Universidade do Porto
Rua Campo Alegre 1021/1055
4169-007 Porto, Portugal

***Abstract*—Among the most common breast cancer treatment-associated effects, it is possible to recognize a high prevalence of arm/shoulder restricted mobility and arm swelling that deteriorate the upper-body function and may lead to chronic lymphedema. Within such a perspective, the need has been identified for breast cancer survivors to sustain a constant and specific physical activity. For that, homebased programs are being presented as a desirable path to be followed. Concurrently, an emergence of new technologies has led to major changes in society. For instance, technology from the video games industry has been used with emphasis in the recovery, and follow-up stages, to evaluate and motivate the patient after treatment. The present work aims to evaluate a set of contextual interfaces that use data acquired with a colour and depth sensor to monitor, and provide real-time feedback to, a given user. Furthermore, fundamental design guidelines from serious games are explored within the context of developing a system aid for physical follow-up care in the form of a set of exercises selected by the medical community. The proposed interfaces were evaluated in a clinical setting with a group of breast cancer survivors.**

Keywords–Patient-empowerment services; Preventive Systems; Self-management systems; Monitoring systems.

I. INTRODUCTION

While contributing for improved overall survivorship, contemporary breast cancer treatment techniques may result in several impairments in women's upper-body function and, consequently, contribute to a decreased quality of life [1]. As Breast Cancer Survivors (BCS) are living longer, the adverse effects resulting from the cancer treatment are more frequent. Upper body morbidity (e.g., decreased range of motion, muscle strength, pain and lymphedema) are among the most prevalent side effects. Regarding lymphedema alone, a swelling condition resulting from lymphatic ablation commonly associated with breast cancer treatment, it has been estimated that over 1 million BCS in the US and 10 million women worldwide may meet the criteria for breast cancer-related lymphedema [1].

While the assessment of the oncological outcome of the cancer treatment can be easily objectively quantified by disease-free and overall survival rates, the same does not hold for functional aspects closely related to quality of life. Assessment of BCS symptoms and health-related quality of life outcomes are usually made using Patient Reported Outcome (PRO) questionnaires that quantify significant outcome variables from the patient's perspective [1].

A prospective surveillance model for BCS has been proposed, highlighting the importance of monitoring for functional and physical impairment commonly associated with treatment [2]. Notwithstanding, and though some methods for monitoring and assessing do exist, an integrated approach able to achieve early detection, promote risk-reduction and self-management, while engaging the user in an appropriate follow-up strategy, is still being reported as missing [3].

In Section II, an outline of topics related to the application of typical elements of game playing is presented in order to contextualize the proposed methodology, that is presented in Section III. The paper concludes with a discussion of results in Section IV, regarding key questions relating to the application of strategies of anthropomorphization and gamification as means to promote engagement to particular physical activities within the context of patient empowerment systems.

II. RELATED WORK

Engaging patients in their healthcare can be recognized as a paramount topic that has evolved through time also as a reflection of specific technological and societal contexts [4]. In this sense, growing trends of the quantified-self movement, personal health records tools dissemination and interactive video games that combine physical exercise with game-play and have a primary purpose other than entertainment present themselves as currently active research lines.

A. Gamification

Physical activity promotion programmes tested in patients with disabilities and impairment problems demonstrate that patients' functionality can improve with an intensive training split that is contextualised and oriented as a pursuit in the achievement of a well defined goal. However, this task division is prone to present a major set-back, which is the lack of interest of the patient in performing repetitive tasks [5].

On the other hand, it is possible to note that a game, overall, aims to offer the player a challenge of a physical or/and mental nature that can be completed using a set of rules, being able to install feelings of amusement or entertainment in the participant while returning feedback in a form of grades or scores, while possibly unlocking new challenges based on the feedback received. Video games have the same goals, only a computer is used as an intermediary [6].

The concept of serious games is one that is hard to define, but it usually refers to games used for training, advertising, simulation or education. A particular example of such a gamified approach, commonly referred to as exergaming or exergames, can be described as a type of video game, or multimedia interaction that requires the player to physically move in order to play [7]. With the evolution of video game acceptance by the general public, serious games have begun to surge, spreading into healthcare where they can eventually provide a more personalized experience to users, improving not just physical, but also mental aspects of care. This surge, and the evolution of visual computing, seems to enable the development of personalized home systems, which could objectively evaluate the patient's state, while motivating for continued physical activity. Specifically for rehabilitation, research has been done, where small game prototypes were tested for specific circumstances, such as upper limb rehabilitation. [8].

B. *Anthropomorphism*

Different elements can be considered to be included in serious games as strategies to promote improved adherence [9]. Of those, it is possible to highlight virtual representations of the self, through which players are presented to the possibility of assuming the role of a character in the game [10]. On the topic of player controlled game characters, the Illusion of Virtual Body Ownership (IVBO) considers the effect of game players experiencing a sense of artificial body parts to be their own, within the context of an Virtual Reality (VR) setting [11]. Previous research [11] tends to suggest that the IVBO may result from an interaction of both synchronous visual, motor and tactile sensory inputs, as well as pre-existing visual and proprioceptive body representation factors. Included in the group of the latter factors, is the virtual body realism in terms of visual human resemblance, or anthropomorphism [12]. On a related note, the Uncanny Valley appertains to a theorized relationship between humans and robots [13] (e.g., Fig. 1).

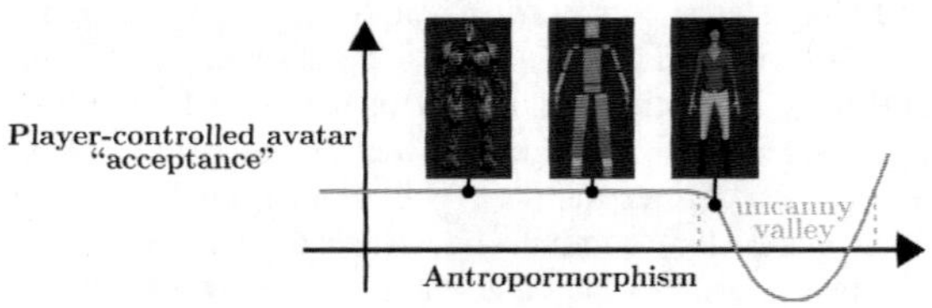

Figure 1. Example assessment of the strength of the IVBO in relation to the degree of anthropomorphism of an user controlled avatar, and the hypothesized uncanny valley effect [13], adapted from [11].

The hypothesis is that there should exist a positive relationship between how human a robot looks, and how comfortable people are with its appearance, up to the moment a robot would get too close to being human in appearance, without being fully human, at which point human reaction would became negative [13]. Its impact in game design has been evaluated although there seems to not exist absolute evidence to support, or disprove it [12].

III. Proposed Research Approach

While there are several games that include serious topics, the inclusion of serious game elements is not yet enough to induce learning or real-world action [14]. Overall, despite engagement being considered a valuable resource, research on patient engagement technologies regarding impact on health outcomes has been limited [15]. Given this, this work's main goal is to develop and assess a game to promote an adequate exercise routine for BCS, to be used independently, as a self-management system to support breast cancer survivorship while monitoring one's physical status. The overall architecture of the proposed system is outlined in Fig. 2.

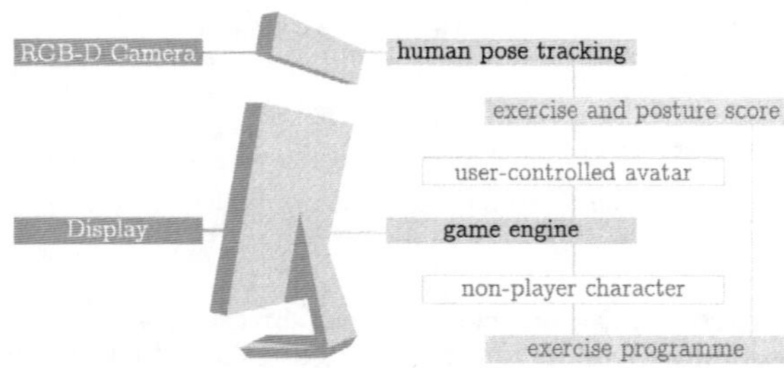

Figure 2. Overall architecture of the proposed system, comprising the use of colour and depth data to monitor, and provide real-time feedback to, a user in the context of a physical activity promotion for breast cancer survivors.

We consider the Microsoft Kinect as an easily accessible, Color and Depth (RGB-D) sensor-device that enables to monitor a user's movement and provide feedback through the usage of an avatar, so that the user is aware of the performed movement, aiming at promoting adherence to exercise [16]. Both versions of the Kinect range sensor, i.e., the Kinect^{SL}, which is based on the Structured Light principle, and the Time-of-Flight variant Kinect^{ToF}, were considered [17]. To create the game environment, Unity was selected as the game engine, given its accessibility and widespread use.

In this paper, we pursue the following main topics:

1) anthropomorphism as a strategy to engage, and
2) gamification as a mean to promote physical activity,

about which we present a body of exploratory work.

A. *Exercise programme selection*

A standardized exercise programme consisting of shoulder flexion, abduction, and horizontal adduction was selected in accordance to the National Institute for Health and Clinical Excellence (NICE) guidelines [18]. The individual exercises comprised in the programme are illustrated in Fig. 3.

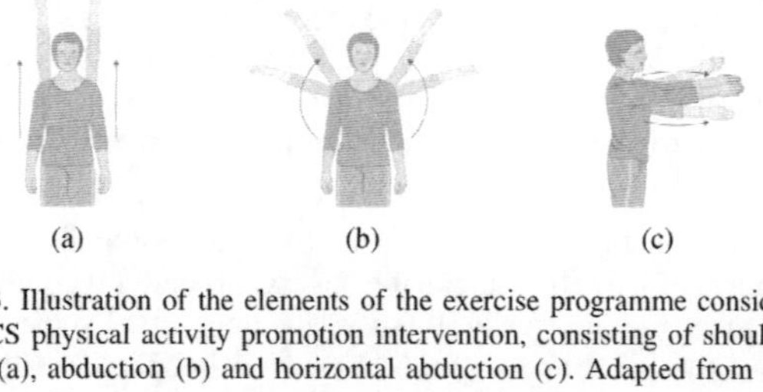

Figure 3. Illustration of the elements of the exercise programme considered for BCS physical activity promotion intervention, consisting of shoulder flexion (a), abduction (b) and horizontal abduction (c). Adapted from [18].

The exercise routine is composed of three sets, each comprising ten repetitions of one of the three exercises included in the programme, and small breaks between sets.

B. Expressed acceptance assessment

Analysis of engagement can be considered valuable in providing insights into game mechanisms that can then be applied to games for learning, or physical activity promotion [19], although not trivial to measure. In order to assess the acceptance of particular contexts of a given physical activity promotion intervention, a criteria set, based on [20], was used as basis for user expressed acceptance assessment. The criteria comprised the following aspects of testing:

- c_1) suitability for the task,
- c_2) information accessibility,
- c_3) continuity correction,
- c_4) visual pleasingness,
- c_5) self-descriptiveness,
- c_6) adequacy of user workload.

Based on that criteria, a questionnaire composed of six questions was formulated in Portuguese, and a five point scale, ranging from strong disagreement (1) to strong agreement (5), considered for range of response options.

C. Study on anthropomorphism

1) Participants and design: Seventy-two adults (mean age of the cohort was 57.79 ± 11.16 years, all female) participated. They were recruited via personal invitation from surgeon-led follow-up consultations of BCS. Written informed consent was obtained from all participants. All participants were fluent in Portuguese and did not get paid for their participation.

2) Procedure and materials: Participants were invited to participate in this study via personal invitation at the end of a follow up consultation at the Breast Center of São João Hospital during the period from the end of October until the beginning of December, 2016. Participants were informed that the study was part of the development of an aid designed to promote physical activity recommend for BCS. The recruited participants were prompted to use the system, in an adjacent room to the consultation room (as illustrated in Fig. 4).

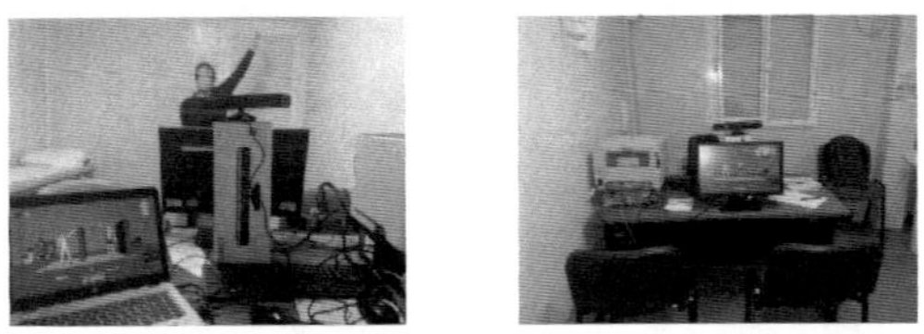

(a) Experimental set-up (b) Perspective of the user

Figure 4. Acquisition environment for the study on anthropomorphism and the tested system comprising a Kinect^{SL}, laptop and additional screen.

The architecture illustrated in Fig. 2 was adapted so that it would entail a Non-Player Character (NPC) in the form of a virtual assistant that exemplified the movements to be performed according to the established exercise programme while the user was exercising. The same programme would be repeated four times, considering additional breaks between routines, one for each of the considered levels of the user controlled avatar anthropomorphism (illustrated in Fig. 5).

Levels of user controlled avatar anthropomorphism (while exercising)	
no visual feedback	
skeleton	a_1)
humanoid avatar	a_2)
mirror	

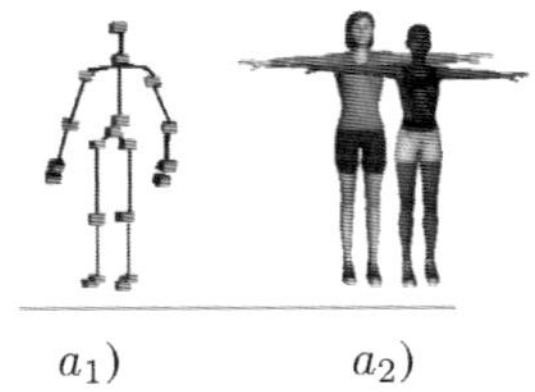

Figure 5. Illustration of the user controlled character: avatar based characters animated with the user's tracked movement with different levels of anthropomorphism a_1) and a_2).

After using the system, each patient was inquired of its satisfaction level of the usage of the system through a questionnaire that required the user to rate each of the tested interfaces according to a five point scale ranging from least preferred (1) to most preferred (5). Each session took approximately 30 minutes, comprising the usage of the system for the proposed exercise programme and the filling of the questionnaire.

3) Results: Each of the four interfaces were evaluated using the aforementioned score in a five point scale after the user completed the exercise programme using all of the proposed interfaces. Table I presents the mean expressed preferences for the user controlled character variations.

TABLE I. AVERAGE AND STANDARD DEVIATION (SD) OF EXPRESSED PREFERENCES OVER THE INTERFACES COMPRISING VARIATIONS FOR THE USER CONTROLLED CHARACTER TESTED BY SEVENTY-TWO BCS IN A CLINICAL SETTING.

		avatar		
	no visual feedback	skeleton	humanoid	mirror
Average	4.10	4.22	4.22	4.19
SD	0.77	0.88	0.77	0.82

Although it seems to not exist an abrupt drop on the collected expressed preference between evaluated interfaces with different levels of user controlled avatar anthropomorphism, both skeleton and humanoid examples seem to be preferred over the alternatives with either no visual feedback, or mirror-based feedback.

D. Study on gamification

1) Participants and design: Sixty-eight adults (mean age of the cohort was 59.09 ± 10.92 years, all female) participated. The same recruitment method mentioned in Subsection III-C (Study on anthropomorphism) was used. A sub group of 22% of participants (15 out of 68) were randomly assigned to receive printed information resources, in form of a pamphlet produced at the Breast Center of São João Hospital.

2) Procedure and materials: As in the study on anthropomorphism, participants were invited to participate after a surgeon-led follow-up consultation at the Breast Center of São João Hospital. The recruitment took place from the beginning of November until the end of December, 2017. Participants were informed about the study being part of the development of an aid to promote physical activity recommend for BCS, and prompted to use the system, in an adjacent to the consultation room (as illustrated in Fig. 6).

(a) Experimental set-up

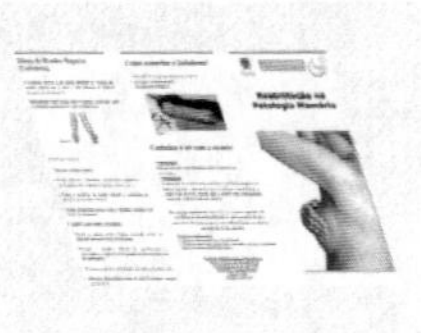

(b) Printed pamphlet

Figure 6. Side-by-side illustration of: a) Acquisition environment for the study on gamification for the tested system comprising a Kinect^{ToF}, laptop and additional screen; b) Printed pamphlet produced at the Breast Center of São João Hospital and distributed to BCS.

The architecture used for the study on anthropomorphism, was considered, and kept the NPC virtual assistant exemplifying the exercise programme. To provide real-time feedback of the user's own movement only a human avatar was used. Differently from the previous study, the user controlled avatar was animated with the human pose provided by a Kinect^{ToF}. Another novelty introduced by the Kinect version (and corresponding SDK and respective tools) is the gesture builder tool, which allows the creation of a database containing movements, which allows to perceive to which degree of completion is a given movement being performed. After building a library of the selected exercises, this was used to score the performance of the user. The normal scoring of the game attributed 1 point for every 1% of progress in each repetition, and a final score was presented as a percentage of the routine completed (the complete routine corresponds to 3000 points).

After the usage of the system, each patient was inquired to express level of acceptance that required the user to rate each of the previously identified criteria according to a five point scale ranging from strong disagreement (1) to strong agreement (5). Each session took approximately 10 minutes, which comprised the usage of the interface for the proposed exercise programme by the user and the filling of the questionnaire.

3) Results: Table II presents the mean expressed acceptance for the proposed Gamified Aid for Monitoring Exercise (GAME) with a humanoid player controlled character and an NPC virtual assistant, against an informative printed pamphlet. Of the total cohort of sixty-eight BCS, fifty-three were randomly assigned to use the GAME and fifteen assigned for being shown the printed pamphlet.

TABLE II. AVERAGE AND STANDARD DEVIATION (SD) OF EXPRESSED ACCEPTANCE FOR BOTH THE PROPOSED GAME, AND A PRINTED PAMPHLET CONTAINING INFORMATION ABOUT THE SELECTED EXERCISE PROGRAMME.

	GAME		pamphlet	
Criteria	Average	SD	Average	SD
c_1)	4.60	0.74	4.00	1.11
c_2)	4.72	0.50	4.08	1.27
c_3)	4.92	0.32	5.00	0.00
c_4)	4.88	0.29	4.60	0.84
c_5)	4.96	0.19	4.60	0.84
c_6)	4.88	0.39	4.00	1.11

In the context of the evaluation, it seems to exist a stronger agreement, across considered criterion, for the proposed GAME being a preferred medium over printed materials.

IV. Conclusion

The present work investigates the impact of providing real-time feedback to BCS within the context of a physical activity promotion intervention. A system comprised of a RGB-D sensor with a processing pipeline to monitor the user, and in that way animate a user controlled avatar, was considered.

In the first exploratory study, the effect of different levels of anthropomorphism of the user controlled avatar was investigated. Seventy-two BCS participated in the cohort. The results seem to agree with the hypothesised Uncanny Valley effect, in the sense that a more anthropomorphised representation of the self (a mirror), seems not to be the preferred interface. Although not possible to assess from the presented results, but also supported considering previous research, i.e., [11], subjectively constructed proprioceptive body representations of the self, seems to be an apparently worth considering factor in the context of BCS, with potential impact to adherence to systems using anthropomorphised avatars.

In the second study, a gamified approach considering an humanoid avatar and an NPC assistant was evaluated against a printed pamphlet. From a total of sixty-eight participants, a subgroup of 15 was randomly assigned to be shown the pamphlet containing information about appropriate care following breast cancer treatment, including the recommendation to perform simple exercises to be repeated throughout survivorship. The remaining participants played a game where an NPC assistant would exemplify the recommend exercise programme, while a humanoid avatar would replicate the user's movements, and in real-time provide feedback of the exercise being executed. Overall, the collected expressed acceptance suggests that the proposed gamified aid for monitoring exercise seems suitable for the task, informative, visual pleasing, self-descriptive, and providing an adequate workload to the user.

Acknowledgments

The authors would like to thank the direction, members, and users of the Breast Center of São João Hospital, that, valuably, supported and participated in the research.

This work is financed by the ERDF - European Regional Development Fund through the Operational Programme for Competitiveness and Internationalisation - COMPETE 2020 Programme within project POCI-01-0145-FEDER-006961, and by National Funds through the Portuguese funding agency, FCT - Fundação para a Ciência e a Tecnologia as part of project UID/EEA/50014/2013; and also by FCT within the Ph.D. grant number SFRH/BD/138823/2018.

References

[1] V. M. Boquiren, T. F. Hack, R. L. Thomas, A. Towers, W. B. Kwan, A. Tilley, E. Quinlan, and B. Miedema, "A longitudinal analysis of chronic arm morbidity following breast cancer surgery," Breast Cancer Research and Treatment, vol. 157, no. 3, 2016, pp. 413–425.

[2] N. L. Stout, J. M. Binkley, K. H. Schmitz, K. Andrews, S. C. Hayes, K. L. Campbell, M. L. McNeely, P. W. Soballe, A. M. Berger, A. L. Cheville, C. Fabian, L. H. Gerber, S. R. Harris, K. Johansson, A. L. Pusic, R. G. Prosnitz, and R. A. Smith, "A prospective surveillance model for rehabilitation for women with breast cancer," Cancer, vol. 118, no. S8, 2012, pp. 2191–2200.

[3] L. Lai, J. Binkley, V. Jones, S. Kirkpatrick, C. Furbish, P. Stratford, W. Thompson, A. Sidhu, C. Farley, J. Okoli, D. Beech, and S. Gabram, "Implementing the prospective surveillance model (PSM) of rehabilitation for breast cancer patients with 1-year postoperative follow-up, a prospective, observational study," Annals of Surgical Oncology, vol. 23, no. 10, 2016, pp. 3379–3384.

[4] W. Tauxe, "A tumour through time," Nature, vol. 527, no. 7578, 2015, pp. S102–S103.

[5] M. Simon, "Gamification and serious games for personalized health," Studies in Health Technology and Informatics, vol. 177, 2012, p. 8596.

[6] F. Laamarti, M. Eid, and A. E. Saddik, "An overview of serious games," Int. Journal of Computer Games Technology, vol. 2014, 2014, pp. 1–15.

[7] S. West, N. A. Borghese, M. Pirovano, R. Mainetti, R. van de Langenberg, and E. D. de Bruin, "Usability and effects of an exergame-based balance training program," Games for Health Journal, vol. 3, no. 2, 2014, pp. 106–114.

[8] B. Lange, S. Koenig, E. McConnell, C.-Y. Chang, R. Juang, E. Suma, M. Bolas, and A. Rizzo, "Interactive game-based rehabilitation using the microsoft kinect," in 2012 IEEE Virtual Reality (VR). IEEE, 2012.

[9] D. Thompson, "Designing serious video games for health behavior change: Current status and future directions," Journal of Diabetes Science and Technology, vol. 6, no. 4, 2012, pp. 807–811.

[10] M. Rice, R. Koh, Q. Lui, Q. He, M. Wan, V. Yeo, J. Ng, and W. P. Tan, "Comparing avatar game representation preferences across three age groups," in CHI 13 Extended Abstracts on Human Factors in Computing Systems on - CHI EA 13. ACM Press, 2013.

[11] J. Lugrin, J. Latt, and M. E. Latoschik, "Avatar anthropomorphism and illusion of body ownership in vr," in 2015 IEEE Virtual Reality (VR), 2015, pp. 229–230.

[12] T. Waltemate, D. Gall, D. Roth, M. Botsch, and M. E. Latoschik, "The impact of avatar personalization and immersion on virtual body ownership, presence, and emotional response," IEEE Trans. on Visualization and Computer Graphics, vol. 24, no. 4, 2018, pp. 1643–1652.

[13] M. Mori, K. MacDorman, and N. Kageki, "The uncanny valley [from the field]," IEEE Robotics & Automation Magazine, vol. 19, no. 2, 2012, pp. 98–100.

[14] K. Starks, "Cognitive behavioral game design: a unified model for designing serious games," Frontiers in Psychology, vol. 5, 2014, p. 28.

[15] J. E. Prey, J. Woollen, L. Wilcox, A. D. Sackeim, G. Hripcsak, S. Bakken, S. Restaino, S. Feiner, and D. K. Vawdrey, "Patient engagement in the inpatient setting: a systematic review," J Am Med Inform Assn, vol. 21, no. 4, 2014, pp. 742–750.

[16] J. Han, L. Shao, D. Xu, and J. Shotton, "Enhanced computer vision with microsoft kinect sensor: A review," IEEE Transactions on Cybernetics, vol. 43, no. 5, oct 2013, pp. 1318–1334.

[17] H. Sarbolandi, D. Lefloch, and A. Kolb, "Kinect range sensing: Structured-light versus time-of-flight kinect," Computer Vision and Image Understanding, vol. 139, oct 2015, pp. 1–20.

[18] J. Yarnold, "Early and locally advanced breast cancer: Diagnosis and treatment national institute for health and clinical excellence guideline 2009," Clinical Oncology, vol. 21, no. 3, 2009, pp. 159–160.

[19] E. A. Boyle, T. Hainey, T. M. Connolly, G. Gray, J. Earp, M. Ott, T. Lim, M. Ninaus, C. Ribeiro, and J. Pereira, "An update to the systematic literature review of empirical evidence of the impacts and outcomes of computer games and serious games," Computers & Education, vol. 94, 2016, pp. 178–192.

[20] K. S. Park and C. H. Lim, "A structured methodology for comparative evaluation of user interface designs using usability criteria and measures," International Journal of Industrial Ergonomics, vol. 23, no. 5-6, 1999, pp. 379–389.

An Educational Platform for Direct Communication between the National Competent Authority and Healthcare Professionals in Croatia

Katarina Gvozdanovic, Adam Galkovski, Darko Krnic, Arian Rajh, Dubravka Sudic
Agency for Medicinal Products and Medical Devices
Zagreb, Croatia
e-mail: {kgvozdanovic, dkrnic, arajh, dsudic}@halmed.hr, adam.galkovski@gmail.com

Abstract—**This article describes the OPeN system, a pharmacovigilance system developed by the Croatian Agency for Medicinal Products and Medical Devices, its current update, and gives insight into relevant information and the educational function of the application.**

Keywords - Education for Healthcare Professionals; Drug Safety; OPeN System; Pharmacovigilance

I. INTRODUCTION

The paper "OPeN: Linking the National Adverse Reactions Database with Clinical IT Systems in Croatia," presented at the Tenth International Conference on eHealth, Telemedicine, and Social Medicine (eTELEMED) in March 2018 in Rome, described the OPeN system, its functioning and role in the Croatian public healthcare system [1]. This paper describes the further investment of resources for the existing application and ongoing development of its informational and educational functions. Accordingly, this paper presents an update of the work presented at the eTELEMED conference. The educational module was planned as a result of Phase 3 of developing the OPeN system.

The OPeN system is a pharmacovigilance system developed by the Agency for Medicinal Products and Medical Devices of Croatia (HALMED). OPeN is an abbreviation of "online platform for electronic reporting of adverse drug reactions (ADRs)" (or in the Croatian language, *Online Platforma za elektroničku prijavu Nuspojava*). It is based on the Croatian National Adverse Reactions Database, maintained by HALMED. OPeN aims to achieve integration with the Croatian Central Health Information System (CEZIH), and which includes the exchange platform named Healthcare Networking Information System (HNIS), Electronic Healthcare Record (EHR), patient register, Register of Healthcare Professionals (HCPs), a messaging subsystem, online patient healthcare website, and specialized services related to e-prescriptions, e-laboratory referrals, e-booking for hospital admittance, and other interconnected functions [2].

Pharmacovigilance (PhV) as a discipline depends on spontaneous reporting, data mining and similar techniques, as well as direct reporting from clinical systems. Although Croatia is one of the top ten countries in the world for the number of Adverse Drug Reactions (ADRs) per million inhabitants, spontaneous reporting by HCPs and patients is generally recognized as an insufficient resource for pharmacovigilance protective mechanisms in public health [3]. The OPeN system is used for transferring PhV data to the national adverse reactions database, both Information technology (IT) solutions are complementary and are maintained by HALMED. Currently, OPeN is used by HALMED's internal experts and external HCPs. HCPs use the module to transfer PhV data whereas HALMED's staff use it for processing data.

This paper serves as a follow-up of our previously presented research. In Section II, we describe methods and functionalities of the new OPeN module. In Section III, we discuss the trustworthiness of the possible data sources on drug safety information. The goal of the new OPeN module and its key indicators are set and defined in Section IV. Finally, in Section V, we offer our conclusions related to the education of HCPs, reliable data sources and the role of national competent authorities (NCAs), which have drug marketing and post-marketing related mandate.

II. METHODS

In 2017 and 2018, HALMED developed a web-based application called OPeN with the aim of automating the reporting of ADRs by HCPs directly using proprietary IT systems at their workplaces. This simplifies the process of sending ADR data to HALMED. It also accelerates the process of assessing and processing ADR data given that ADRs are directly reported to the PhV department and do not have to pass through administrative channels prior to reaching the PhV department. As previously planned, Phase 3 of developing the OPeN system includes development of the educational module to serve as a platform for ongoing education of HCPs. The underlying idea is to offer HCPs the opportunity to learn about medicines at their workplace or home without the need to attending courses.

The module will feature a web interface with a login function requiring a user's first name, last name or medical license number and a password. Once a HCP is logged in, the OPeN system has been designed to identify them based on their license designation. A list of all active HCPs in Croatia has been integrated into OPeN and is updated every two weeks by the Croatian Institute for Public Health, which in cooperation with HALMED shares the information. The list of HCPs contains information on the HCP's first and last name, type of HCP (pharmacist, physician), specialization, and institution (including geographic data). Additional data, such as contact information is obtained from professional associations of pharmacists (Croatian Chamber of Pharmacists, CCPH) and physicians (Croatian Medical Chamber, CMC), which then enables HALMED to directly send links to target HCPs and referencing educational materials.

The basic workflow of the OPeN system is shown in Figure 1. After logging in, HCPs choose whether to access the ADR reporting or educational module. The educational module, which is the focus of this paper, contains webinars, surveys and questionnaires designed by HALMED's employees. Every test covers a different topic and is customized by staff from HALMED's pharmacovigilance department. Authors can choose from a number of questions in the particular test, type of questions (e.g., multiple choice, single choice, free text) with the ability to select the correct answer. In some cases, for specific safety issues, input may be sought from members of academia at the relevant faculties.

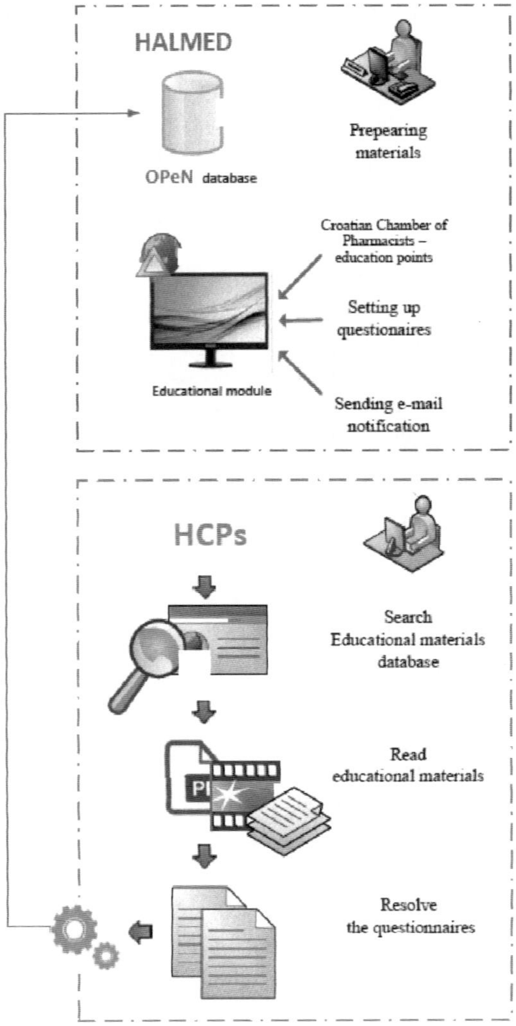

Figure 1. OPeN's educational module workflow

The involvement of academia, besides the opportunity to share their knowledge on relevant topics, additionally raises an awareness of the OPeN educational platform in an academic setting, and thus guaranteeing that future HCPs become familiar with the platform from the start of their careers. Once the test is created, a link referring to the test location is sent via e-mail to a prepared list of recipients or HCPs. The recipients are chosen based on their specialty, location or type of HCP (physicians or pharmacists). HCPs receive an e-mail with a link to the test and basic information on the topic, medicines to be covered, specific safety issues, and the like. By clicking on the link, they are directed to the OPeN educational module, more precisely, to a specific test and educational materials necessary for successful learning. Alternatively, they can log into the system and browse the educational module, without receiving a prior e-mail notification.

Every test, along with the available educational materials, is stored in the OPeN database and can be searchable by key words. Individual HCPs can view information on all available tests created by HALMED, regardless of whether they have taken the test or not. However, they can view only results from their own tests. If needed, they can retake the test any time. In certain situations, the only limited time frame given to HCPs to take the test is for particularly important safety issues requiring immediate communication and actions.

HALMED's employees have at their disposal additional functionalities, such as the ability to post various types of educational material (videos, films) to the module, and placing links to documents published on the HALMED website. The authors of the test set a threshold for passing the test, where each person that passes the test is awarded ongoing education points as a HCP. The grade system is proposed to and then agreed with the boards of HCP chambers.

HCPs will also be able to track the number of continuous education points they received on the OPeN platform, including points awarded in the ADR reporting module.

One of the main features of the OPeN education platform is the ability to analyze test results and assess the outcome of educational activities. All results can be analyzed for an individual HCP and also based on aggregated data (e.g., test results for physicians or pharmacists, specialist or non-specialists, hospitals or primary care, and other comparisons). This leads to precisely targeting knowledge gaps, better knowledge on medicines for HCPs and hopefully better patient safety.

III. DISCUSSION

Up to date, accurate and practical information is necessary in ensuring that drugs are prescribed safely and effectively. It also promotes improvements in prescribing practices, minimises harm to patients and facilitates decision making based on valid information [4]. Generally speaking, HCPs should have immediate access to high quality evidence-based information on medicines. This may include prescribing information or critical safety information on medicines.

HCPs have at their disposal various sources of information on medicines; however, not all sources are equally regarded as useful or relevant. There is usually the issue of credibility of the source, or updates and redundancy. Information on medicines can be communicated by different stakeholders including the pharmaceutical industry, scientists, regulators, professional chambers, learned societies, and the like. According to Molimard et al. [5], communicating information on medication often utilizes tools and messages that are not adapted to the target audience. This is further hindered due to a lack of knowledge in communication techniques and the fact that communication on safety information is subject to regulatory or legal requirements [5]. Various factors related to communication strategies, such as trust in the sender of information, may influence the acceptance of information [6].

According to a study published by Piening et al. [7], HCPs indicated that they preferred safety information to be issued by the Dutch regulatory agency (MEB), the Dutch Pharmacovigilance Center or their own professional associations via e-mails, in medical journals and using electronic prescribing systems - as the preferred channels for the distribution of information on drug safety.

The results of project SCOPE which aims to describe and improve pharmacovigilance practices in the EU indicate that current safety communication practices seemed to be broadly similar among European medicinal products regulators, i.e., NCAs [8][9]. NCA safety communication is mainly restricted to communicating safety issues directly through bulletins or newsletters and indirectly through approval and coordinating the dissemination of dear healthcare professional communication (DHPC) and educational materials [8]. DHCP and educational materials cover significant safety issues and are usually distributed by the pharmaceutical company following NCA content approval. Both communications are distributed to individual HCPs and may also be made available on the NCA website [8]. Awareness of the existence of these materials varies among HCPs in different EU member States (MSs) and even among different types of HCPs (e.g., general practitioners or cardiologists).

In Croatia, there are 3,800 active pharmacists and almost 14,500 physicians. HALMED has been actively involved in providing important safety information to HCPs at seminars, workshops and congresses, safety committee meetings as well on its website. The drawback of seminars, workshops and congresses has been the fact that they are resource intensive, whereas the difficulty in presenting information on the website is limited impact and reach based on our website analytics and metrics. In general, actions undertaken by HALMED have been somewhat restricted in terms of one-way safety communication to HCPs without the ability of measuring the impact of such activities. Upon consideration of all these factors, HALMED decided to develop an on-line platform for HCPs in order to enhance direct communication with HCPs and provide them with information on medicine-related fundamental safety issues. We are not aware that a similar platform exists in other EU member states. With regards to Croatia, there are a few educational platforms for HCPs, with "Pliva Zdravlje" being one of the most popular (https://www.plivazdravlje.hr/). This particular platform is operated by a local pharmaceutical company and covers topics relevant to physicians and pharmacist, while not necessarily related to medicines. However, some physicians view this platform as lacking credibility, something that is not be the issue with the HALMED education platform [7].

The HALMED platform is planned to be an extension of the already existing ADR reporting system for OPeN. The OPeN education module acts as an IT infrastructure providing two-way communication with HCPs. It also enables hosting of webinars, questionnaires and surveys as ongoing education on medicines for HCPs. Webinars are developed by employees at the HALMED pharmacovigilance department along with members of academia from relevant faculties. Topics covered include information on important safety issues regarding medicines such as restrictions on indications, severe adverse reactions, change to the benefit-risk ratio, and the like.

Expected benefits for HALMED include:

- Establishing closer cooperation with HCPs by providing a two-way communication platform
- Increasing the reach of safety communication on medicines
- Rationalization of internal resources
- Establishing HALMED as a trusted source of safety information
- Promoting the importance of medicine safety monitoring, including ADR reporting
- Measuring information uptake by measuring both process and knowledge indicators
- A convenient and modern education format appealing to younger HCPs
- The ability to measure demographic characteristics of respondents, establish a correlation with their knowledge in order to adapt and focus the safety message
- The ability to additionally focus on educational activities based on analysis of the results of taken tests
- Reduced administrative burdens – consecutive education points are automatically assigned to individual HCPs after completion of assignments, webinars and tests
- Increasing cooperation between HALMED and all areas of the Croatian health system, starting with academia where future HCPs receive their education and right through to professional chambers and the Institute of Public Health, thus ensuring that stakeholders cooperate on important safety issues

Expected benefits for HCPs include:

- Drug safety information provided by trusted source
- Up to date information on medicines safety

- Convenient communication and education format – electronic communication, the ability to set MS Outlook reminders
- Convenience of receiving education at one's own pace, without physically attending courses. This is especially convenient for HCPs in remote locations, like islands or rural areas
- Customized overview of educational history
- Continuous education points

IV. Results

Our aim is to establish a drug risk management system, able to face and respond to public health challenges in terms of safe drug use, with particular emphasis on issues such as antimicrobial resistance and vaccination hesitancy. With the help of HALMED's online tools, HCPs are informed about the latest safety information on safe drug use.

We have set ourselves a rather ambitious goal. In 2020 we plan to reach a minimum of 30% of all HCPs in Croatia. In 2021, our plan is to reach a minimum of 50% of all HCPs in Croatia through webinars, questionnaires and surveys using our platform. Further development of this tool will provide us with the opportunity to:

- Participate in national and international initiatives dealing with safe drug use, especially vaccine hesitancy and antimicrobial resistance
- For every additional risk minimization measure, an explanation will be provided via video or audio by a member of HALMED's PhV team
- Every emerging safety issue will incorporate a video or audio recording
- Based on our platform and authority, we will endeavor to increase vaccine coverage
- We will facilitate development of the vaccination information center on our platform
- Our platform enables us to support policies to combat antimicrobial resistance

Key performance indicators that indicate success include:

- Increased public trust in HALMED's competence
- An increase in the number of queries about safe drug use, especially vaccines and antimicrobial resistance
- Successful cooperation with national institutions and bodies, healthcare workers and patient associations
- An increased number of public campaigns with participation from HALMED, especially on issues regarding antimicrobial resistance and vaccination hesitancy

V. Conclusion

Development of an additional software educational module is an update and expansion of the existing OPeN system. As we had reasonably presumed based on previous cooperation with HCPs, and based on the study of Piening et al. shows, HCPs consider NCA to be a more reliable source of various material on drugs, which also includes educational material, when compared to other sources. Moreover, HCPs have begun using HALMED's online website, and are becoming acquainted with its interface and usability. HALMED has had a lot of experience in educating HCPs, and this function is derived from its mandate. Therefore, its modernization through the use of a modern portal tool, such as OPeN, was an obvious option. The expansion of the educational function of OPeN will help HALMED achieve the stated benefits and key performance indicators, take advantage of the stated opportunities and strengthen its role in the national health domain. Finally, the benefits for HCPs, in receiving a quality education in an efficient manner, will result in benefits to patients, improved patient care, and consequently, patient health.

References

[1] A. Rajh, D. Sudic, and K. Gvozdanovic, "OPeN: Linking the National Adverse Reactions Database with Clinical IT Systems in Croatia." In *eTELEMED 2018: The Tenth International Conference on eHealth, Telemedicine, and Social Medicine,* IARIA, March 2018, pp. 54-59.

[2] 2016 Annual Report Ericsson Nikola Tesla Group General Report. (2016). From: https://www.ericsson.hr/documents/20181/21890/ar2016_en_general.pdf/495d6bff-e782-4878-ba8a-6cfc9603f082 [Retrieved: December, 2018]

[3] HALMED. *Adverse Drug Reaction Report for 2016* [Online], From: http://www.halmed.hr/fdsak3jnFsk1Kfa/ostale_stranice/Izvjesce-o-nuspojavama-za-2016.pdf [Retrieved: December, 2018]

[4] European Medicines Agency. Guideline on good pharmacovigilance practices (GVP) Module XV—safety communication. From: http://www.ema.europa.eu/docs/en_GB/document_library/Scientific_ guideline/2013/01/WC500137666.pdf [Retrieved: December, 2018]

[5] M. Molimard et al. "Information and communication on risks related to medications and proper use of medications for healthcare professionals and the general public: precautionary principle, risk management, communication during and in the absence of crisis situations". Therapie. 2014 Jul-Aug; 69(4):355-66. doi: 10.2515/therapie/2014045. Epub 2014 Aug 8.

[6] M. K. Lindell, and R.W. Perry. "The protective action decision model: theoretical modifications and additional evidence". Risk Anal.2012;32(4):616–32.

[7] S. Piening et al. *Healthcare Professionals' Self-Reported Experiences and Preferences Related to Direct Healthcare Professional Communications.* Drug Saf. 2012 35: 1061. https://doi.org/10.1007/BF03261992

[8] S. T. de Vries et al. "Communication on Safety of Medicines in Europe: Current Practices and General Practitioners' Awareness and Preferences". Drug. Saf. 2017 40:729–742

[9] S. T. de Vries et al. "Safety Communication Tools and Healthcare Professionals'Awareness of Specific Drug Safety Issues in Europe: A Survey Study". Drug. Saf. 2018 41: 713–724

International Academy, Research,
and Industry Association (IARIA)
412 Derby Way
Wilmington, DE 19810

ISBN 978-1-5108-8363-5

CURRAN ASSOCIATES INC.
proceedings.com

Annual Session of Scientific Papers IMT ORADEA 2018

MATEC Web of Conferences Volume 184 (2018)

Baile Felix, Romania
31 May - 1 June 2018

edp sciences